AF352770

Advances in Hydrologic Forecasts and Water Resources Management

Advances in Hydrologic Forecasts and Water Resources Management

Editors

Fi-John Chang
Shenglian Guo

MDPI • Basel • Beijing • Wuhan • Barcelona • Belgrade • Manchester • Tokyo • Cluj • Tianjin

Editors
Fi-John Chang
National Taiwan University
Taiwan

Shenglian Guo
Wuhan University
China

Editorial Office
MDPI
St. Alban-Anlage 66
4052 Basel, Switzerland

This is a reprint of articles from the Special Issue published online in the open access journal *Water* (ISSN 2073-4441) (available at: https://www.mdpi.com/journal/water/special_issues/hydrologic_forecast).

For citation purposes, cite each article independently as indicated on the article page online and as indicated below:

LastName, A.A.; LastName, B.B.; LastName, C.C. Article Title. *Journal Name* **Year**, *Article Number*, Page Range.

ISBN 978-3-03936-804-4 (Hbk)
ISBN 978-3-03936-805-1 (PDF)

Contents

About the Editors

Fi-John Chang serves as a Distinguished Professor at National Taiwan University. He is also the Founding President of the Taiwan Hydro-Informatics Society. His academic expertise focuses on transdisciplinary fields across hydrological science, engineering, and environments. As an expert in technology revolution and the modernization of artificial intelligence, big data and data mining on hydro-informatics, he has successfully applied advanced technologies to key issues in integrated water resource management, eco-hydro-systems, water quantity/quality, etc., with research results acknowledged by international societies, including 210 papers. He was the recipient of the Outstanding Research Award from the Ministry of Science and Technology in Taiwan (2010 & 2018) and the International Award from the PAWEES (2014). He has drawn on his expertise in this cutting-edge domain in current roles as the principal investigator in governmental/NGO research projects, including the interdisciplinary studies on groundwater recharge, eco-hydrology, and flood inundation in Taiwan, as well as international collaborative projects on Food-Water-Energy (FEW) Nexus between Taiwan, USA, Japan, and Brazil and the inter-comparison appraisal of sustainable river management between Taiwan and France. More information about his research can be found at http://scholar.google.com/citations?hl=en&user=XZoDI_EAAAAJ.

Shenglian Guo graduated in Wuhan Institute of Hydraulic & Electric Engineering (1982) and was awarded his M.Sc (1986) and Ph.D. (1990) at the National University of Ireland. He is a professor of hydrology and water resources at Wuhan University, since 1993, a member of the Norwegian Academy of Technological Sciences, honorary chairman of IAHS Chinese National Commission, and the chief editor of the Journal of Water Resources Research. Prof. Guo is devoted to investigating the complex underlying mechanisms in hydrological and water resource systems, covering the topics about flood frequency analysis, hydrologic forecasting, cascade reservoir operation, adaptive water resources management, and the water–society system nexus. He is a highly respected and well-known scientist in China, and has a close collaborating network with Chinese water-related sectors. He has led over one hundred projects in hydrological and water resources fields, numerous works have been widely acknowledged. He has supervised about one hundred graduated students and has published 12 books and more than 500 academic papers, including 160 peer-reviewed papers in international journals such as Nature Communications, Water Resources Research, Journal of Hydrology, etc. He has 18 patents and 16 research projects and has received Science and Technology Advanced Awards or honors in China. More information about his research can be found at https://www.researchgate.net/profile/Shenglian_Guo.

Preface to "Advances in Hydrologic Forecasts and Water Resources Management"

The impacts of climate change on water resource management, as well as increasingly severe natural disasters over the last decades, have caught global attention. Reliable and accurate hydrological forecasts are essential for efficient water resource management and the mitigation of natural disasters. While the notorious nonlinear hydrological processes make accurate forecasts a very challenging task, it requires advanced techniques to build accurate forecast models and reliable management systems. One of the newest techniques for modeling complex systems is artificial intelligence (AI). AI can replicate the way humans learn and has great capability to efficiently extract crucial information from large amounts of data to solve complex problems. The fourteen research papers published in this Special Issue contribute significantly to the uncertainty assessment of operational hydrologic forecasting under changing environmental conditions and the promotion of water resources management by using the latest advanced techniques, such as AI techniques. The fourteen contributions across four major research areas: (1) machine learning approaches to hydrologic forecasting; (2) uncertainty analysis and assessment on hydrological modeling under changing environments; (3) AI techniques for optimizing multi-objective reservoir operation; (4) adaption strategies of extreme hydrological events for hazard mitigation. The papers published in this issue will not only advance water sciences but also help policymakers to achieve more sustainable and effective water resource management.

Fi-John Chang, Shenglian Guo
Editors

Editorial

Advances in Hydrologic Forecasts and Water Resources Management

Fi-John Chang [1,*] and Shenglian Guo [2,*]

[1] Department of Bioenvironmental Systems Engineering, National Taiwan University, Taipei 10617, Taiwan
[2] State Key Laboratory of Water Resources and Hydropower Engineering Science, Wuhan University, Wuhan 430072, China
[*] Correspondence: changfj@ntu.edu.tw (F.-J.C.); slguo@whu.edu.cn (S.G.)

Received: 18 June 2020; Accepted: 22 June 2020; Published: 24 June 2020

Abstract: The impacts of climate change on water resources management as well as the increasing severe natural disasters over the last decades have caught global attention. Reliable and accurate hydrological forecasts are essential for efficient water resources management and the mitigation of natural disasters. While the notorious nonlinear hydrological processes make accurate forecasts a very challenging task, it requires advanced techniques to build accurate forecast models and reliable management systems. One of the newest techniques for modelling complex systems is artificial intelligence (AI). AI can replicate the way humans learn and has the great capability to efficiently extract crucial information from large amounts of data to solve complex problems. The fourteen research papers published in this Special Issue contribute significantly to the uncertainty assessment of operational hydrologic forecasting under changing environmental conditions and the promotion of water resources management by using the latest advanced techniques, such as AI techniques. The fourteen contributions across four major research areas: (1) machine learning approaches to hydrologic forecasting; (2) uncertainty analysis and assessment on hydrological modelling under changing environments; (3) AI techniques for optimizing multi-objective reservoir operation; and (4) adaption strategies of extreme hydrological events for hazard mitigation. The papers published in this issue can not only advance water sciences but can also support policy makers toward more sustainable and effective water resources management.

Keywords: artificial intelligence; machine learning; water resources management; multi-objective reservoir operation; hydrologic forecasting; uncertainty; risk

1. Introduction

Natural disasters have been inclined to increase and become more severe over the last decades due to climate change. A preparation measure to cope with future floods is flood forecasting in each river basin for warning persons involved and for mitigating damages and the loss of life. Hydrological forecasting is essential for efficient water resources management and the mitigation of natural disasters such as floods and droughts. Establishing a viable hydrological forecasting model for communities at risk requires the combination of meteorological and hydrological data, forecast tools and trained forecasters. Forecasts must be sufficiently accurate to promote confidence so that communities and users will take effective actions when being warned. Multidisciplinary research and advanced methodologies in hydrological forecasts, especially in extreme floods and droughts, are widely implemented for water planning and management, which ultimately lead to improved optimum water resources management and effective control under a changing environment. Among them, artificial intelligence (AI) techniques are efficient tools for extracting the key information from complex highly dimensional input–output patterns and are widely used to tackle various hydrological problems such as flood forecasts discussed in this Special Issue [1–14]. Over the last decades, many studies have demonstrated that artificial intelligence (AI) techniques, such as machine learning (ML) methods, can produce flood forecasts in a few hours [15–19] while extending to seasonal forecasts many months in advance for larger river basins [20–24]. AI can also be an ideal tool for managing water resources in an ever-changing environment and for allowing water utility managers to effectively optimize multi-objective water resources revenues [25–30].

Reliable and accurate streamflow forecasts with lead-times from hours to days are critical to managing floods and to improving the efficiency of streamflow forecasts utilized for real-time reservoir operation. All forecasts, however, involve various degrees of uncertainty, which could be associated with meteorological data, hydrological model mechanics and parameters, or the model's errors in forecasts. To implement streamflow forecasts in real-time reservoir operation, we must deal with the uncertainty involved in streamflow forecasts. Although the forecast uncertainty plays an important role in reservoir operation and has been extensively studied in hydrology, there are comparatively less studies discussing the effect of forecast uncertainty on real-time reservoir operations [31–36]. This Special Issue aims at overcoming these challenges, addressing continuing efforts undertaken to gain insights on hydrological processes, dealing with the effect of forecast uncertainty, and engaging in more efficient water management strategies in a changing environment.

2. Summary of the Papers in the Special Issue

The papers in this Special Issue are well-balanced in terms of their focuses, encompassing hydrological forecasts, uncertainty analyses and water resources management. Three papers [1–3] address operational hydrological forecasts by using various machine learning (ML) methods. In [1], the authors propose an Internet of Things (IoT) machine learning-based flood forecast model to predict average regional flood inundation depth in a river basin in Taiwan, and they demonstrate how to on-line adjust the machine learning models so that the models' accuracy and applicability in multi-step-ahead flood inundation forecasts are promoted. They also highlight the combination of IoT and machine learning techniques could be beneficial to flood prediction. In [2], the authors introduce a general framework that fuses an unscented Kalman filter (UKF) post-processing technique with a recurrent neural network for probabilistic flood forecasting conditional on point forecasts. They declare that the proposed approach could overcome the under-prediction phenomena and alleviate the uncertainty encountered in data-driven flood forecasting so that model reliability as well as forecast accuracy for future horizons could be significantly improved. In [3], the authors propose a random forest (RF) model to predict the Normalized Difference Vegetation Index (NDVI) and explore its relationship with climatic factors. The results demonstrate that RF can be integrated into water resources management and can elucidate ecological processes in the Yarlung Zangbo river basin. These studies clearly indicate that machine learning techniques have a great capability to model the nonlinear dynamic features in hydrological processes, such as flood forecasts and NDVI, and IoT sensors are useful instruments for carrying out the monitoring of natural environments and enhancing hydrological forecasts.

Papers [4–6] report research on uncertainty analysis and assessment in hydrological modelling and forecasting. In [4], Hong's method is implemented to execute the point estimate method (PEM) in a case study that simulates water runoff using the ANUGA hydrodynamic model for an area in Glasgow, UK. The authors demonstrate that the Hong's method could more efficiently produce very similar probabilistic flood-inundation maps in the same areas as those of Monte Carlo (MC) simulation, where the Hong's method requires just three 11-minute simulation runs, rather than the 500 required for the MC simulation. In [5], the authors propose a multiple-criteria decision analysis method, namely the Generalized Likelihood Uncertainty Estimation-Technique for Order Preference by Similarity to Ideal Solution (GLUE-TOPSIS). The proposed method was implemented in the Storm Water Management Model (SWMM) and applied to the Dahongmen catchment in Beijing, China. They conclude that the proposed GLUE-TOPSIS is a valid approach to assessing the uncertainty of the urban hydrological model from a multiple objective perspective, which improves the reliability of model results in the urban catchment. In [6], the authors evaluate the parameter uncertainty for the Snowmelt Runoff

Model (SRM) based on different calibration strategies and its impact on a data-scarce deglaciating Yurungkash watershed in China. The results show that the future runoff projection contains a large amount of uncertainty and the onset of snowmelt runoff is likely to shift earlier in the year and the discharge over the snowmelt season is projected to increase.

Hydrological nonstationary has brought great challenges to the reliable applications of hydrological models with time-invariant parameters. Two papers [7,8] investigate the predictive ability and robustness of a hydrological model under changing environments. In [7], the authors propose a new method based on empirical mode decomposition (EMD) to synthesize and generate data which be interfered with the non-stationary problems. The new synthetic and historical flow data were used to simulate the water supply system of the Hushan reservoir in Taiwan, and the compared results show that the synthetic data are like the historical flow distribution. In [8], the authors investigate the predictive ability and robustness of a conceptual hydrological model (GR4J) with time-varying parameter under changing environments. The results show that the performance of streamflow simulation was improved when applying the time-varying parameters. Furthermore, the GR4J model with time-varying parameters outperformed the original GR4J model by improving the model robustness. Overall, these studies emphasize the importance of considering the parameter uncertainty of time-varying hydrological processes in hydrological modelling and climate change impact assessment.

Due to climate change, the importance of reservoirs is likely to increase, not only for water storage purpose but also for maximizing water use benefits and mitigating climate extremes. Four papers [9–12] employ advanced optimization methods to derive reservoir operating rules for multi-reservoir systems and/or optimize multi-objective reservoir operation. In [9], the authors conduct a multi-target single dispatching study on ecology and power generation in the lower Yellow River to solve the single-objective and the multi-objective optimal schema using the genetic algorithm (GA) and an improved non-dominated genetic algorithm (NSGA-II). The results provide a decision-making basis for the multi-objective dispatching of the Xiaolangdi reservoir and have important practical significance for further improvement on the ecological health of the lower Yellow River. In [10], the authors fuse the grey entropy method (GEM) with the Mahalanobis–Taguchi System (MTS) for selecting the optimal water level scheme at the Pankou reservoir in flood season. The results show that the optimal scheme selected by the proposed model can achieve greater benefits within an acceptable risk range and thus better coordinate the balance between risk and benefit, which verifies the feasibility and validity of the model. In [11], the authors show the advancement of the seasonal flow forecasts could provide the opportunity for reservoir operators to identify the early impoundment operation rules (EIOR) in the upper Yangtze river basin. Their results indicate the proposed GloFAS-Seasonal forecasts are skillful for predicting the streamflow condition according to the selected 20th and 30th percentile thresholds and the obtained seasonal forecasts and the early reservoir impoundment could enhance hydropower generation and water utilization. In [12], a novel enhanced gravitational search algorithm (EGSA) is proposed to resolve the multi-objective optimization model by considering the power generation of a hydropower enterprise and the peak operation requirement of a power system located on the Wujiang river of China. The results show that the EGSA method could obtain satisfying scheduling schemes in different cases for the multi-objective operation of hydropower system.

The early warning and post-assessment of extreme hydrological events are crucial for hazard mitigation. In [13], the authors explore the most effective flood control strategy for small and medium-scale rivers in highly urbanized areas. The probable cost-effective flood control scheme is to construct two new tributaries for transferring floodwater in the midstream and downstream of the Shegong river into the downstream of the Tieshan river. Their results indicate that flood control for small- and medium-scale rivers in highly urbanized areas should not simply consider tributary flood regimes but, rather, involve both tributary and mainstream flood characters from a whole region perspective. In [14], the authors report emergency disposal solutions for properly handling the landslide and dammed lake within a few hours up to days for mitigating flood risk. They present a general strategy to effectively tackle the dangerous situation created by a giant dammed lake with 770 million m^3 of water volume and formulate an emergency disposal solution for the 25 million m^3 of debris, composed of engineering measures of floodgate excavation and non-engineering measures of reservoirs/hydropower stations operation. The disposal solution not only reduces a large-scale flood (10,000-year return period, 0.01%) into a small-scale flood (10-year return period, 10%) but minimizes the flood risk with no death raised by the giant landslide.

3. Conclusions

Over the last several decades, substantial climate changes have occurred due to global warming. We also notice that artificial intelligence has been satisfactorily used to enhance our knowledge, to learn hydrological processes, and to engage in more efficient water management strategies under changing environmental conditions. The research papers published in this Special Issue contribute significantly to our understanding of the hydrological modelling approaches as well as water resources management. They can be categorized into four main subject areas: (1) machine learning methods for hydrologic forecasting; (2) uncertainty analysis and assessment on hydrological forecasts; (3) AI techniques for optimizing multi-objective reservoir operation; and (4) adaption strategies of extreme hydrological events for hazard mitigation. These papers presented novel methods to learn the complex hydrological processes and model hydrological forecasts, reduce models' uncertainty, and optimize water resources management. The selected manuscripts presented in this Special Issue make original contributions to addressing the state-of-the-art of artificial intelligence techniques, which provide a high level of research and practical information of implementing AI methods and strategies for accurate flood forecasts and reservoir operation, along with case studies from different regions of the world.

Author Contributions: Resources, F.-J.C., S.G.; writing—original draft preparation, F.-J.C., S.G.; writing—review and editing, F.-J.C., S.G.; supervision, F.-J.C., S.G. Both authors have read and agreed to the published version of the manuscript.

Funding: This research received no external funding.

Conflicts of Interest: The authors declare no conflict of interest.

References

1. Yang, S.N.; Chang, L.C. Regional inundation forecasting using machine learning techniques with the internet of things. *Water* **2020**, *12*, 1578. [CrossRef]
2. Zhou, Y.; Guo, S.; Xu, C.Y.; Chang, F.J.; Yin, J. Improving the reliability of probabilistic multi-step-ahead flood forecasting by fusing unscented Kalman filter with recurrent neural network. *Water* **2020**, *12*, 578. [CrossRef]
3. Chi, K.; Pang, B.; Cui, L.; Peng, D.; Zhu, Z.; Zhao, G.; Shi, S. Modelling the vegetation response to climate changes in the Yarlung Zangbo River basin using random forest. *Water* **2020**, *12*, 1433. [CrossRef]
4. Issermann, M.; Chang, F.J. Uncertainty analysis of spatiotemporal models with point estimate methods (PEMs)—The case of the ANUGA Hydrodynamic Model. *Water* **2020**, *12*, 229. [CrossRef]
5. Pang, B.; Shi, S.; Zhao, G.; Shi, R.; Peng, D.; Zhu, Z. Uncertainty Assessment of urban hydrological modelling from a multiple objective perspective. *Water* **2020**, *12*, 1393. [CrossRef]
6. Xiang, Y.; Li, L.; Chen, J.; Xu, C.Y.; Xia, J.; Chen, H.; Liu, J. Parameter uncertainty of a snowmelt runoff model and its impact on future projections of snowmelt runoff in a data-scarce deglaciating river basin. *Water* **2019**, *11*, 2417. [CrossRef]
7. Chu, T.Y.; Huang, W.C. Application of empirical mode decomposition method to synthesize flow data: A case study of Hushan Reservoir in Taiwan. *Water* **2020**, *12*, 927. [CrossRef]
8. Zeng, L.; Xiong, L.; Liu, D.; Chen, J.; Kim, J.S. Improving parameter transferability of GR4J model under changing environments considering nonstationary. *Water* **2019**, *11*, 2029. [CrossRef]
9. Bai, T.; Liu, X.; Ha, Y.P.; Chang, J.X.; Wu, L.Z.; Wei, J.; Liu, J. Study on the single-multi-objective optimal dispatch in the middle and lower reaches of Yellow River for river ecological health. *Water* **2020**, *12*, 915. [CrossRef]

10. Ji, C.; Liang, X.; Peng, Y.; Zhang, Y.; Yan, X.; Wu, J. Multi-dimensional interval number decision model based on Mahalanobis-Taguchi System with grey entropy method and its application in reservoir operation scheme selection. *Water* **2020**, *2*, 685. [CrossRef]

11. Chen, K.; Guo, S.; Wang, J.; Qin, P.; He, S.; Sun, S.; Naeini, M.R. Evaluation of GloFAS-Seasonal Forecasts for cascade reservoir impoundment operation in the upper Yangtze River. *Water* **2019**, *11*, 2539. [CrossRef]

12. Feng, Z.K.; Liu, S.; Niu, W.J.; Jiang, Z.Q.; Luo, B.; Miao, S.M. Multi-objective operation of cascade hydropower reservoirs using TOPSIS and gravitational search algorithm with opposition learning and mutation. *Water* **2019**, *11*, 2040. [CrossRef]

13. Liu, Z.; Cai, Y.; Wang, S.; Lan, F.; Wu, X. Small and medium-scale river flood controls in highly urbanized areas: A whole region perspective. *Water* **2020**, *12*, 182. [CrossRef]

14. Chen, G.; Zhao, X.; Zhou, Y.; Guo, S.; Xu, C.Y.; Chang, F.J. Emergency disposal solution for control of a giant landslide and dammed lake in Yangtze River, China. *Water* **2019**, *11*, 1939. [CrossRef]

15. Chang, F.J.; Hsu, K.; Chang, L.C. (Eds.) *Flood Forecasting Using Machine Learning Methods*; MDPI: Basel, Switzerland, 2019.

16. Chang, L.C.; Chang, F.J.; Yang, S.N.; Tsai, F.H.; Chang, T.H.; Herricks, E.E. Self-organizing maps of typhoon tracks allow for flood forecasts up to two days in advance. *Nat. Commun.* **2020**, *11*, 1–13. [CrossRef]

17. Kao, I.F.; Zhou, Y.; Chang, L.C.; Chang, F.J. Exploring a long short-term memory-based encoder-decoder framework for multi-step-ahead flood forecasting. *J. Hydrol.* **2020**, *583*, 124631. [CrossRef]

18. Zhou, Y.; Guo, S.; Chang, F.J. Explore an evolutionary recurrent ANFIS for modelling multi-step-ahead flood forecasts. *J. Hydrol.* **2019**, *570*, 343–355. [CrossRef]

19. Chang, L.C.; Chang, F.J.; Yang, S.N.; Kao, I.; Ku, Y.Y.; Kuo, C.L.; Amin, I.M.Z.M. Building an intelligent hydro-informatics integration platform for regional flood inundation warning systems. *Water* **2019**, *11*, 9. [CrossRef]

20. Jeong, D.I.; Kim, Y.O. Rainfall-runoff models using artificial neural networks for ensemble streamflow prediction. *Hydrol. Process.* **2005**, *19*, 3819–3835. [CrossRef]

21. Sun, A.Y.; Wang, D.; Xu, X. Monthly streamflow forecasting using Gaussian process regression. *J. Hydrol.* **2014**, *511*, 72–81. [CrossRef]

22. Kalra, A.; Ahmad, S.; Nayak, A. Increasing streamflow forecast lead time for snowmelt-driven catchment based on large-scale climate patterns. *Adv. Water Resour.* **2013**, *53*, 150–162. [CrossRef]

23. Kalra, A.; Miller, W.P.; Lamb, K.W.; Ahmad, S.; Piechota, T. Using large-scale climatic patterns for improving long lead time streamflow forecasts for Gunnison and San Juan River basins. *Hydrol. Process.* **2013**, *27*, 1543–1559. [CrossRef]

24. Turan, M.E.; Yurdusev, M.A. River flow estimation from upstream flow records by artificial intelligence methods. *J. Hydrol.* **2009**, *369*, 71–77. [CrossRef]

25. Chang, L.C.; Chang, F.J. Intelligent control for modelling of real-time reservoir operation. *Hydrol. Process.* **2001**, *15*, 1621–1634. [CrossRef]

26. Hossain, M.S.; El-Shafie, A. Intelligent systems in optimizing reservoir operation policy: A review. *Water Resour. Manag.* **2013**, *27*, 3387–3407. [CrossRef]

27. Ahmad, A.; El-Shafie, A.; Razali, S.F.M.; Mohamad, Z.S. Reservoir optimization in water resources: A review. *Water Resour. Manag.* **2014**, *28*, 3391–3405. [CrossRef]

28. Tsai, W.P.; Chang, F.J.; Chang, L.C.; Herricks, E.E. AI techniques for optimizing multi-objective reservoir operation upon human and riverine ecosystem demands. *J. Hydrol.* **2015**, *530*, 634–644. [CrossRef]

29. Chang, F.J.; Wang, Y.C.; Tsai, W.P. Modelling intelligent water resources allocation for multi-users. *Water Resour. Manag.* **2016**, *30*, 1395–1413. [CrossRef]

30. Uen, T.S.; Chang, F.J.; Zhou, Y.; Tsai, W.P. Exploring synergistic benefits of water-food-energy nexus through multi-objective reservoir optimization schemes. *Sci. Total Environ.* **2018**, *633*, 341–351. [CrossRef]

31. Li, X.; Guo, S.; Liu, P.; Chen, G. Dynamic control of flood limited water level for reservoir operation by considering inflow uncertainty. *J. Hydrol.* **2010**, *391*, 124–132. [CrossRef]

32. Zhao, T.; Cai, X.; Yang, D. Effect of streamflow forecast uncertainty on real-time reservoir operation. *Adv. Water Resour.* **2011**, *34*, 495–504. [CrossRef]

33. Yan, B.; Guo, S.; Chen, L. Estimation of reservoir flood control operation risks with considering inflow forecasting errors. *Stoch. Environ. Res. Risk Assess.* **2014**, *28*, 359–368. [CrossRef]

34. Zhao, T.; Zhao, J.; Lund, J.R.; Yang, D. Optimal hedging rules for reservoir flood operation from forecast uncertainties. *J. Water Resour. Plan. Manag.* **2014**, *140*, 04014041. [CrossRef]

35. Chen, L.; Singh, V.P.; Lu, W.; Zhang, J.; Zhou, J.; Guo, S. Streamflow forecast uncertainty evolution and its effect on real-time reservoir operation. *J. Hydrol.* **2016**, *540*, 712–726. [CrossRef]

36. Huang, K.; Ye, L.; Chen, L.; Wang, Q.; Dai, L.; Zhou, J.; Singh, V.P.; Huang, M.; Zhang, J. Risk analysis of flood control reservoir operation considering multiple uncertainties. *J. Hydrol.* **2018**, *565*, 672–684. [CrossRef]

Article

Regional Inundation Forecasting Using Machine Learning Techniques with the Internet of Things

Shun-Nien Yang and Li-Chiu Chang *

Department of Water Resources and Environmental Engineering, Tamkang University, New Taipei City 25137, Taiwan; aa22814946@yahoo.com.tw
* Correspondence: changlc@mail.tku.edu.tw; Tel.: +886-2-26258523

Received: 23 April 2020; Accepted: 29 May 2020; Published: 31 May 2020

Abstract: Natural disasters have tended to increase and become more severe over the last decades. A preparation measure to cope with future floods is flood forecasting in each particular area for warning involved persons and resulting in the reduction of damage. Machine learning (ML) techniques have a great capability to model the nonlinear dynamic feature in hydrological processes, such as flood forecasts. Internet of Things (IoT) sensors are useful for carrying out the monitoring of natural environments. This study proposes a machine learning-based flood forecast model to predict average regional flood inundation depth in the Erren River basin in south Taiwan and to input the IoT sensor data into the ML model as input factors so that the model can be continuously revised and the forecasts can be closer to the current situation. The results show that adding IoT sensor data as input factors can reduce the model error, especially for those of high-flood-depth conditions, where their underestimations are significantly mitigated. Thus, the ML model can be on-line adjusted, and its forecasts can be visually assessed by using the IoT sensors' inundation levels, so that the model's accuracy and applicability in multi-step-ahead flood inundation forecasts are promoted.

Keywords: machine learning model; Internet of Things (IoT); regional flood inundation depth; recurrent nonlinear autoregressive with exogenous inputs (RNARX)

---

1. Introduction

Flood is one of the most disruptive natural hazards, which causes significant damage to life, agriculture, and economy, and has a great impact on city development. Nowadays, flood tends to increase and become more severe as climate changes together with the rapid urbanization and aging infrastructure in cities. Early notification of flood incidents could benefit the authorities and public for devising preventive measures, preparing evacuation missions, and alleviating flood victims. A precautionary measure to cope with the upcoming flood is flood forecasting and warning involved persons, which would result in the reduction of damage and life lost. Flood forecast models have been developed over the last decades. Among them, physically based models have been commonly used and showed great capabilities for flood estimation, while they often require hydro-geomorphological monitoring datasets and intensive computation, which prohibits short-term prediction [1–3]. Statistical models, such as the multiple linear regression (MLR) [4–6] and autoregressive integrated moving average (ARIMA) [7–10] are also frequently used for flood modeling. Nevertheless, their capability for short-term forecasting has been restricted because of the nonlinear dynamic feature of storm events resulting in a lack of accuracy and robustness of the statistical methods [11].

Machine learning (ML) models have a great capability to model the nonlinear dynamic feature and have widely been used in hydrological issues, such as predicting the level of sewage in sewers [12]; arsenic concentration in groundwater [13]; or flood level prediction [14–19]. Among the ML models, nonlinear autoregressive models with exogenous inputs (NARX) [20] can adaptively learn complex flood systems and have been reported as valid for flood forecasting [21–26]. There is also relevant literature that applied it to regional flood forecasting. For instance, Shen and Chang [27] established a NARX model for flood forecasting in flood-prone areas in Yilan County, Taiwan and demonstrated that it has an error tolerance rate and can effectively suppress error accumulation in predicting the next 1–6 h; and Chang et al. [28,29] proposed the use of self-organizing map combing with recurrent nonlinear autoregressive with exogenous inputs (SOM-RNARX) for multi-time regional flood forecasting model and indicated the method could effectively model regional flood forecasting. It can provide the accuracy and reliability of the flood management system. The majority of these ML models have used rainfall as an input to make regional flood inundation forecasts. A drawback of these models was that they mainly relied on measurements from rain stations, which hinders the models from being sequentially adjusted due to lack of real observed inundation depth. Consequently, most existing models could not properly respond to a sudden flood and could not verify the resultant flood forecasts. Furthermore, the forecast was made based on present data that restrict it from determining flood inundation depths much further ahead. Consequently, there is a research gap from the perspective of ML modeling and the data monitoring system. In light of this, an analysis was conducted on the use of monitoring inundation depth data gathered from urban areas to forecast flooding with a view of on-line updating the model and mitigating the residuals between model outputs and real observed inundation depths.

Internet of Things (IoT) sensors are a useful means of carrying out the monitoring of rivers and other natural environments. They have attractive features: simple to install, low energy consumption, and high-precision sensors. The integration of a large number of IoT sensors can provide on-line comprehensive and broader information to effectively perform environmental monitoring and forecasting [30–34]. Recently, several studies have implemented IoT and big data [35–38] for flood forecasting. For instance, Chang et al. [39] proposed building an intelligent hydro-informatics integration platform to integrate ML models with sensors data for flood prediction. Sood et al. [40] proposed the Internet of Things (IoT) smart flood monitoring to predict floods and flood levels. Mishra et al. [41] combined IoT and deep learning to identify ditch or drainage channel blockage images. IoT has become one of the vital development projects in Taiwan lately. The authorities' IoT centers, (e.g., Water Resources Agency) can analyze the data and provide suitable countermeasures and rapidly transmit the data to regional control centers against flood. The integration and application of various sensors can offer a large amount of real-time monitoring data (e.g., inundation depths) for ML models' updating and testing.

This study intends to use the IoT sensor data to construct a machine learning-based embedded flood forecast model for predicting average flood depth in a river basin. The IoT sensor data will be sequentially fused into the ML model as extra input factors, so that the forecast model can be continuously updated and assessed, and the results can be closer to the current situation. This paper is organized as follows. The next section describes the proposed methodology. Subsequently, Sections 3 and 4 present the study area and practices relating to flood forecasting systems, respectively. Then, the results of visual assessments and numerical evaluations on studied areas are reported and discussed. Concluding remarks are given in the last section.

2. Methodology

We propose a methodology that couples machine learning models, i.e., the recurrent nonlinear autoregressive with exogenous inputs (RNARX) model [20], with the IoT sensor data for providing multi-step-ahead average regional inundated depth (ARID) during storm events. Various IoT sensor data were explored and implemented into the ML model as input factors to continuously update the model's parameters for better forecasting the ARID.

The architecture of the RNARX shown in Figure 1 is a three-layer dynamic neural network. The network was configured by using the model's outputs as parts of inputs, and its weights can be adjusted by using the conjugate gradient back-propagation learning algorithm. The major difference between the dynamic neural networks (e.g., RNARX) and the static neural networks (e.g., back propagation neural network, BPNN) is that dynamic neural networks use the network output of the current moment as one of the input factors for the next moment, so the model can effectively track the time-series features. The dynamic neural networks have been widely used in modeling time series, especially when there is no actual (observed) value of upcoming time, such as inundation depth. For instance, in this study, the average regional flood inundation depth in the urban area would be continuously predicted along the storm event, while there is no observed value in the coming hours. In model configuring stage, we used simulation results to train and validate the constructed model, while in actual applications, the average flooding depth of all grids, however, cannot be obtained. As known, the longer the forecast horizon, the greater the forecast error would be due to a lack of real monitoring of the inundation depth. In this study, a number of flood sensors implemented in the study area were explored for their effectiveness in modeling the multi-step flood inundation forecasts.

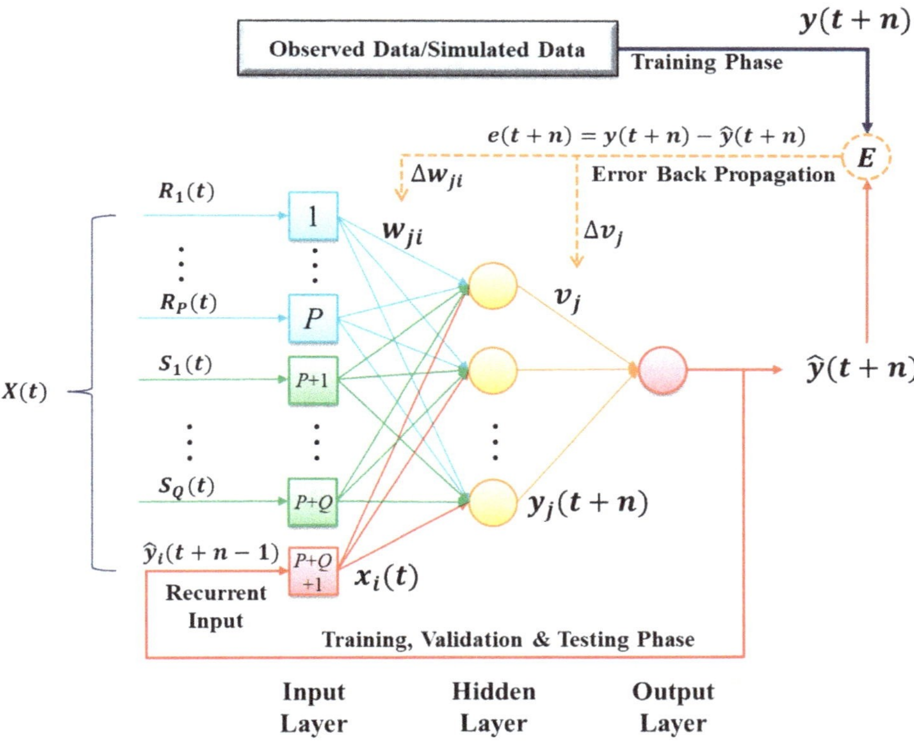

Figure 1. The architecture of the recurrent nonlinear autoregressive with exogenous inputs (RNARX) network.

The network contained P rainfall inputs (R), Q flood sensor inputs (S), and one recurrent input from the previous output. Let $X(t)$ denote the $(P + Q + 1) \times 1$ input vector, $\hat{y}(t + n)$ denote the n-step-ahead network output, and $y_j(t + n)$ denote the output vector of the jth layer. The network multiplied the input factors by weights and forwarded them to the next neuron. The neurons were summed up and outputted through the activation function $f(\cdot)$, which was continuously forwarded to the final output layer. The output values of each neuron were defined as Equations (1)–(4).

$$net_j(t + n) = \sum_i w_{ji} x_i(t) \tag{1}$$

$$y_j(t+n) = f\big(net_j(t+n)\big) \tag{2}$$

$$net(t+n) = \sum_j v_j y_j(t+n) \tag{3}$$

$$\hat{y}(t+n) = f(net(t+n)) \tag{4}$$

Let W denote the $N \times (P+Q+1)$ weight matrix of the hidden layer. Let V denote the $N \times 1$ weight matrix of the output layer. RNARX uses the conjugate gradient back-propagation learning algorithm to adjust the average regional inundation depth (ARID) in a specific time during the training phase. The target output value at time $t+n$ and its error vector were defined as $e(t+n)$, as shown in Equation (5), and the instantaneous error function $E(t+n)$ was as in Equation (6).

$$e(t+n) = y(t+n) - \hat{y}(t+n) \tag{5}$$

$$E(t+n) = \frac{1}{2}e^2(t+n) \tag{6}$$

The correction of v_j was defined as the partial differential of the error function $E(t+n)$ as Equation (7). After the chain-rule process, Equation (8) could be obtained, so v_j was updated in Equations (9) and (10).

$$\frac{\partial E(t+n)}{\partial v_j} = \frac{\partial\left(\frac{1}{2}e^2(t+n)\right)}{\partial v_j} \tag{7}$$

$$\frac{\partial E(t+n)}{\partial v_j} = (y(t+n) - \hat{y}(t+n))(-f'(net(t+n))) \times \left[y_j(t+n) + \sum_j v_j\left(f'\big(net_j(t+n)\big)w_{ji}\frac{\partial \hat{y}(t+n-1)}{\partial v_j} \right) \right] \tag{8}$$

$$\Delta v_j = -\eta_2 \frac{\partial E(t+n)}{\partial v_j} \tag{9}$$

$$v_j(p) = v_j(p-1) + \Delta v_j \quad p = epochs \tag{10}$$

where η_2 is the learning rate. Similarly, we could partially differentiate w_{ji} of the $E(t+n)$, and the results are shown in Equations (11)–(14).

$$\frac{\partial E(t+n)}{\partial w_{ji}} = \frac{\partial\left(\frac{1}{2}e^2(t+n)\right)}{\partial w_{ji}} \tag{11}$$

$$\frac{\partial E(t+n)}{\partial w_{ji}} = (y(t+n) - \hat{y}(t+n))(-f'(net(t+n))) \times \left[\sum_j v_j f'\big(net_j(t+1)\big)\left(w_{ji}\frac{\partial(\hat{y}(t+n-1))}{\partial w_{ji}} + \delta_{ji}x_i(t) \right) \right] \tag{12}$$

$$\Delta w_{ji} = -\eta_1 \frac{\partial E(t+n)}{\partial w_{ji}} \tag{13}$$

$$w_{ji}(p) = w_{ji}(p-1) + \Delta w_{ji} \quad p = epochs \tag{14}$$

where η_1 is the learning rate; δ_{ji} is Kronecker delta with value 1 if and only if $i = j$. Through the above weight adjusting process, the error, because of the complicated interaction between the inputs (rainfall, sensor data, feedback flood depth) and the outputs (flood depth at time T + 1–T + 3), would be computed, and the weights were systematically adjusted through continuous iterations so that the error would be gradually reduced until the satisfactory accuracy and/or the number of iterations was reached for the set requirements. Therefore, the RNARX model was used to predict the multi-step-ahead flood depth, and at the same time, the differences by using only rainfall and the extra sensors of inundation were compared.

3. Study Area and Datasets

This research area was the Erren River basin (Figure 2) located in southern Taiwan. The river length is 63.2 km with the upstream up to 460 m above sea level, and the catchment area is 350.4 km^2, which is further divided into 10,744 grids (40 m × 40 m). In this study, six actual storm events and twelve designed rainfall patterns with various return periods, as well as their correspondent 2D flood simulation results (simulated by the SOBEK software of Deltares), were collected from the Water Resources Agency. A total of 631 hourly datasets were used to configure the ML models. There were also 631 regional flood inundation maps generated by the SOBEK based on the rainfall patterns mentioned above, which were used as the virtual inundation maps for ML models' training and testing. These datasets were further divided into 391 datasets (10 events) for training, 144 datasets (4 events) for validation, and 96 datasets (4 events) for testing (Table 1).

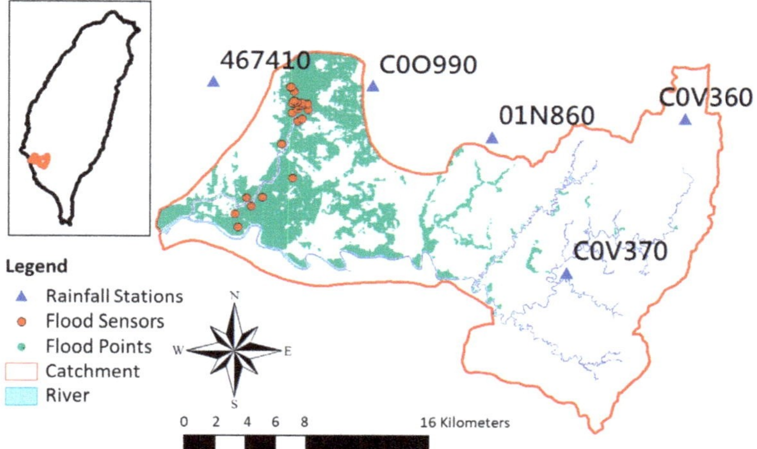

Figure 2. Study area of Erren river catchment.

Table 1. Simulation events used for model training, validation, and testing.

Event Number	Event Name	Average Rainfall (mm/h)	Total Rainfall (mm)	Average Inundation (m)	Max Inundation (m)
		Training phase (number of data: 391 h)			
1	24H800 mm[1]	33.3	800	0.46	0.89
2	24H450 mm	18.8	450.1	0.17	0.34
3	24H200 mm	8.3	200	0.05	0.10
4	24H10 y	17.4	417.6	0.14	0.32
5	24H500 y[2]	33.5	803.7	0.49	0.91
6	20190719[3]	3.2	99.7	0.01	0.02
7	20170601	1.2	86.7	0.01	0.02
8	20160925	7.6	550.1	0.16	0.34
9	20160912	2.7	128.7	0.03	0.10
10	20190813	4.7	226.2	0.07	0.20

Table 1. *Cont.*

Event Number	Event Name	Average Rainfall (mm/h)	Total Rainfall (mm)	Average Inundation (m)	Max Inundation (m)
		Training phase (number of data: 391 h)			
		Validation phase (number of data: 144 h)			
11	24H50 y	24.1	578	0.22	0.43
12	24H400 mm	16.7	400	0.16	0.29
13	24H5 y	14.4	344.8	0.11	0.27
14	20170613	1.4	100.2	0.01	0.03
		Testing phase (number of data: 96 h)			
15	24H100 y	26.9	646.1	0.26	0.52
16	24H2 y	9.8	234.7	0.06	0.11
17	24H500 mm	20.9	500.5	0.17	0.34
18	24H300 mm	12.5	300	0.11	0.25

[1,2] 24H800 mm and 24H500 y are the design storm events of 800 mm and 500-year for 24 h rainfall, respectively. [3] 20190719 (i.e., 2019, 07, 19) means the real event happened on the 19 July 2019.

Figure 3 shows a graph of rainfall and average regional flooding depth (ARID) history of 10 training events, which included three quantitative rainfalls (200, 450, and 800 mm), two different return period rainfalls (10 and 500 years), and five actual rainfall events. The rainfall pattern of quantitative rainfalls and two return period rainfalls adopted the centrally concentrated rain pattern, where the maximum rainfall is placed in the center and then sorted according to the amount of rainfall to the right and left, respectively. The average regional inundation depth (ARID) in a specific time was calculated by summarizing all the inundation depths in each grid and then dividing by the number of grids. The time series of ARID shown in Figure 3 presents a complete hydrograph for the correspondent rainfall event (pattern), with the ARID value along with the time from rising limb to the peak and then to the recession limb. We noticed that the actual rainfall events show more irregular rainfall patterns and produce a variant ARID hydrograph. According to the design of training cases, the model can learn the relationship between the different magnitudes of rainfall (large or small) and rainfall types (design rainfall or actual rainfall) with their correspondent average flooding depths during the training process.

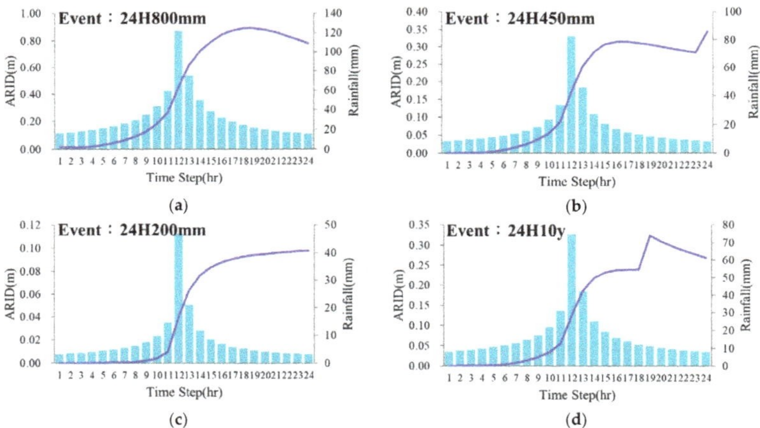

Figure 3. *Cont.*

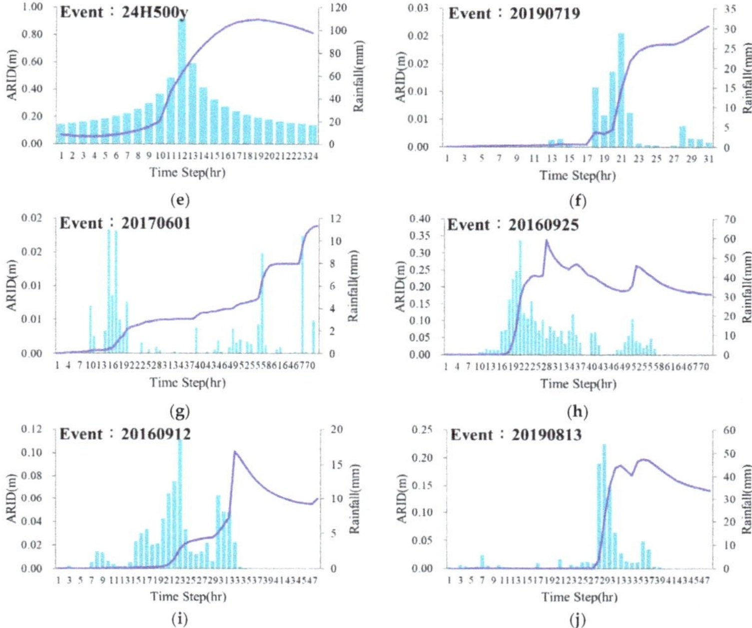

Figure 3. Rainfall histogram and average regional inundation depth hydrograph for training events.
(**a**) 24H800 mm; (**b**) 24H450 mm; (**c**) 24H200 mm; (**d**) 24H10 y; (**e**) 24H500 y; (**f**) 20190719; (**g**) 20170601;
(**h**) 20160925; (**i**) 20160912; (**j**) 20190813.

4. Model Construction

The research process was divided into three stages as shown in Figure 4. The first stage was to collect data, which includes two-dimensional flood simulation data, five stations of rainfall data, and 25 stations of flood sensor datasets. The total datasets were divided into three independent datasets for model training, validation, and testing. The second stage was to construct the forecast models. Three models were built to make multi-step-ahead forecasts of the average regional inundation depth (ARID). The models' parameters are shown in Table 2. The input factors of Model 1 were 5 rainfall stations' data ($R_1 \sim R_5$) and one model's self-feedback value $A\hat{R}ID(t + n - 1)$, a total of 6 input factors, and the number of weights was 41. Model 2 used 5 rainfall stations' data ($R_1 \sim R_5$), 25 flood sensors' data ($S_{01} \sim S_{25}$), and one model's self-feedback value $A\hat{R}ID(t + n - 1)$, a total of 31 input factors, the number of weights was 166. Model 3 used 5 rainfall stations' data ($R_1 \sim R_5$), 7 stations of flood representative sensor data (i.e., $S_{02}, S_{04}, S_{05}, S_{09}, S_{13}, S_{15}, S_{22}$) selected through correlation analysis, and one model's feedback value $A\hat{R}ID(t + n - 1)$, a total of 13 input factors, the number of weights was 76. The third stage was to assess the results of the three models using evaluation indicators.

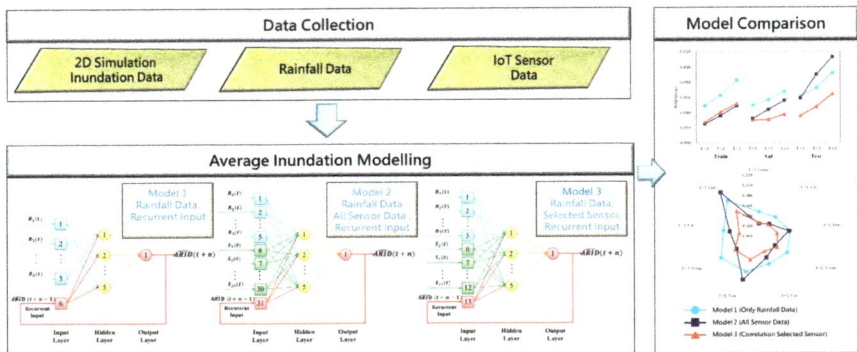

Figure 4. Research framework.

Table 2. Parameters of the proposed models.

Parameters	Model 1	Model 2	Model 3
Input	1. Rainfall Data 2. Recurrent Model Output	1. Rainfall Data 2. All Sensor Data 3. Recurrent Model Output	1. Rainfall Data 2. Correlation Selected Sensor 3. Recurrent Model Output
Input Dimension Counts	Rainfall = 5 Sensor = 0 Recurrent = 1	Rainfall = 5 Sensor = 25 Recurrent = 1	Rainfall = 5 Sensor = 7 Recurrent = 1
Hidden Neuron Counts	5	5	5
Weights Counts	41	166	76

Input variable selection is an essential step in the development of machine learning models. In recent years, various input selection methods have been satisfactorily used to improve prediction accuracy and produce parsimonious models in numerous applications [42–45]. For instance, in hydrological issues, Taormina and Chau [46] used binary-coded particle swarm optimization and extreme learning machines for rainfall–runoff modeling; Chang et al. used the Gamma test to identify the most suitable input variables for multi-step-ahead water level forecasting [26] and estimating stream total phosphate concentration [47]. While new methods continue to emerge, each has its own advantages and limitations, and there is no best method for all modeling purposes. In this study, we aimed learn whether the new implemented 25 sensors in the study area could be beneficial to model the regional flood inundation forecast. The input variable selection was mainly based on correlation analysis between the sensors. As shown above, the main difference between Model 2 and Model 3 was that Model 2 used all sensors (25) as the model input, while Model 3 used 7 selected sensors as input. A 25 × 25 correlation matrix was constructed. The selection process is shown in Figure 5 and explained as follows.

1. Calculate the correlation matrix between sensors.
2. Count the number of highly correlated ($R^2 > 0.9$) sensors and sorting their number from large to small.
3. Select a representative sensor from those sensors that have the highest number of correlated sensors; if the number is the same, compare their ARID and then select the sensor with the largest ARID for priority selection (as the representative sensor).

4. Remove those highly correlated sensors with the above representative sensor to avoid repeated selection. For example, in the first round, S_{22} was selected as the representative sensor (Figure 5). Before moving to the next round of selection, those sensors highly correlated with S_{22} sensor (i.e., $R^2 > 0.9$) would be removed from the correlation matrix. In this round, 10 sensors (including S_{22}) were removed.

5. If the selection has not been completed, return to step 1 and recalculate the correlation matrix of the remaining sensors until the selection is completed. That is, all the sensors will either be selected as the representative sensors or be removed during the selection process.

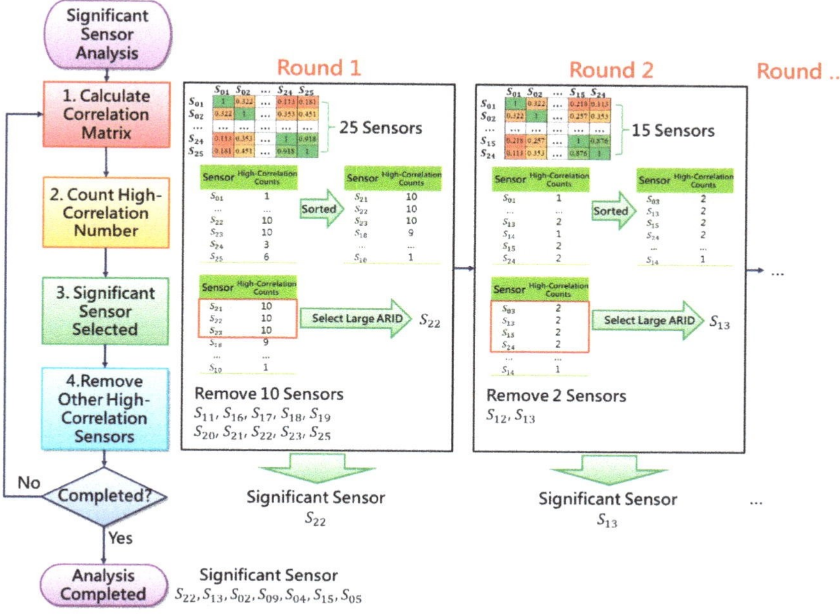

Figure 5. The procedure of selecting representative sensors.

In this way, one of the most representative sensors will be selected for each round until all sensors have been picked or removed. We noticed that this study conducted a total of 7 rounds and picked out $S_{22}, S_{13}, S_{02}, S_{09}, S_{04}, S_{15}, S_{05}$, a total of 7 representative sensors.

The hydrographs show 25 flood sensors and their correspondent ARID during the 500 year event used in Model 2 (Figure 6a) and 7 representative sensors used in Model 3 (Figure 6b), representatively. We noticed that many of these 25 sensors showed the same trend (such as $S_{11}, S_{12}, S_{13}, S_{14}, S_{15}, S_{16}, S_{17}, S_{19}, S_{20}, S_{21}, S_{23}, S_{24}, S_{25}$, etc.), which will result in additional parameters in the model and cause noise. In Model 3, which used only 7 representative sensors, the number of parameters and noise were both greatly reduced. Thus, the model could more accurately describe the relationship between rainfall, flooding sensor, and model output.

The evaluation indicators used in this study were RMSE (root-mean-square error), R^2, and Nash–Sutcliffe coefficient (NSE). The formulae are shown in Equations (15)–(17).

$$RMSE = \sqrt{\frac{\sum_{i=1}^{N}(d_i - y_i)^2}{N}} \tag{15}$$

$$R^2 = \left(\frac{\sum_{i=1}^{N}\left(d_i - \bar{d}\right) \times \left(y_i - \bar{y}\right)}{\sqrt{\sum_{i=1}^{N}\left(d_i - \bar{d}\right)^2 \times \sum_{i=1}^{N}\left(y_i - \bar{y}\right)^2}} \right)^2 \tag{16}$$

$$NSE = 1 - \frac{\sum_{i=1}^{N}\left(d_i - y_i\right)^2}{\sum_{i=1}^{N}\left(d_i - \bar{d}\right)^2} \tag{17}$$

where d_i and y_i are the observed value and the forecasted value of the *i*th data, respectively; $\bar{d}$ and $\bar{y}$ are the means of the observed and forecasted values.

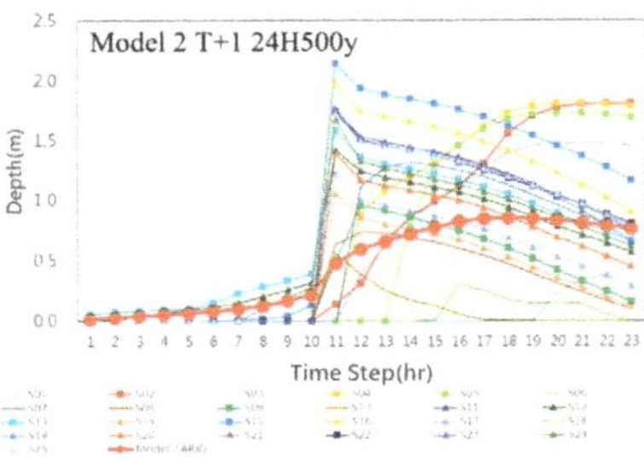

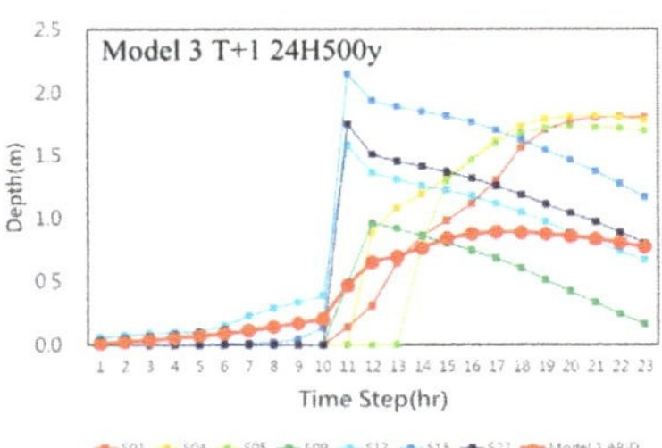

(**a**) Forecasted average regional inundation depth by Model 2 vs. 25 sensors' data

(**b**) Forecasted average regional inundation depth by Model 3 vs. 7 sensors' data

Figure 6. The sensor data vs. T + 1 forecasted average regional inundation depths by Models. (**a**) 25 sensors' data vs. T+1 forecasted ARIDs of Model 2; (**b**) 7 sensors' data vs. T+1 forecasted ARIDs of Model 3.

5. Results

As known, there is a temporal relationship between observations of the rainfall and IoT sensors' inundation level of the study area and hence the value of observation at a particular time depends to some extent on past values. This study sought to model the temporal relationship and forecast the average regional inundation depth by using the machine learning model. Our reason for employing IoT sensor data was that the sensors could provide monitored (real) inundation values for model's training as well as testing and thus improve the multi-step forecast accuracy. Moreover, the real-time monitoring datasets could be used to on-line adjust the model's parameters and visually assess the model's performance, which could largely promote the model's applicability and reliability. The results of the three models are shown in Table 3. There were two very large flood events (water depth > 0.5 m) in the training case. Model 1 results indicated those high flood events were underestimated, but its performances, in general, was good, where the RMSE could reach 0.049, 0.050, and 0.061 m in training, validation and testing cases, respectively. The small values of RMSE and high R^2 and NSE values indicate that Model 1, which was solely based on rainfall information as input, could provide suitable 1–3 h ahead forecasts, while it would largely underestimate the peak flood inundation depth.

Model 2 used the data of 5 rainfall stations and 25 flood sensors. In the training phase, its errors in T + 1–T + 3 were significantly reduced as compared with those of Model 1. For instance, the RMSE values of Model 1 and Model 2 at T + 3 were 0.083 and 0.049, respectively. We notice that Model 2 had much better performances in multi-step-ahead forecasts (T + 1–T + 3) than those of Model 1 in the training phase, while that was not the case in both the validation and testing phases. In order to investigate the inconsistency problem, we explored the relationship between sensors' data with the models' forecast values in the event (24H100y) of the testing case. Figure 7a shows that the time series of 25 sensors were inconsistent, especially during the 12th–18th hours, where some of them were ascending, while the rest of them were descending, and the time series of the 3 h ahead forecasted ARID made by Model 2 was in sharp descent. We noticed that the inconsistency between sensors' hydrographs might have been due to the propagation time between the upstream and downstream locations of the sensors resulting in the different lag time between the increasing and decreasing limb of their associated hydrograph. The inconsistency between the sensors' datasets as well as the forecasted ARID values, however, could result in a large error (i.e., RMSE = 0.113 shown in Table 3). The results indicated that adding the IoT sensors information could, in general, reduce the forecast error of the ARID in the training phase due to using more information (25 sensors' data) as input. Nevertheless, using a large number of inconsistent sensors values along the time series as input might introduce too many parameters in the model and result in an overfitting (overtraining) problem.

Table 3. Performance of one- to three-hour-ahead forecasts of the Model 1, Model 2, and Model 3.

Model	Time Step	RMSE (m)			R^2			NSE		
		Train	Val	Test	Train	Val	Test	Train	Val	Test
Model 1	T + 1	0.049	0.050	0.061	0.940	0.920	0.860	0.943	0.867	0.858
	T + 2	0.063	0.057	0.073	0.910	0.930	0.900	0.917	0.805	0.792
	T + 3	0.083	0.068	0.092	0.860	0.870	0.830	0.859	0.723	0.710
Model 2	T + 1	0.025	0.032	0.059	0.991	0.953	0.873	0.986	0.939	0.864
	T + 2	0.036	0.044	0.090	0.981	0.899	0.696	0.971	0.887	0.685
	T + 3	0.049	0.056	0.113	0.965	0.828	0.539	0.947	0.822	0.510
Model 3	T + 1	0.027	0.030	0.036	0.986	0.968	0.967	0.984	0.947	0.951
	T + 2	0.041	0.031	0.048	0.964	0.954	0.931	0.963	0.946	0.913
	T + 3	0.052	0.038	0.065	0.941	0.922	0.900	0.939	0.916	0.839

RMSE: root-mean-square error, NSE: Nash–Sutcliffe coefficient.

To reduce the input variables (sensors), correlation analysis was used to select a limited number of sensors as the representative sensors. Model 3 used the data of 5 rainfall stations and 7 representative sensors as inputs. Figure 7b shows the time series of 3 h forecasted ARID with the inundation depths of 7 representative sensors. As shown, the time series of 7 selected sensors as well as the forecasted ARID were relatively consistent, as compared with those of Figure 7a. The results show that Model 3, in general, was superior to the Model 1 and Model 2 in all the validation and testing phases, and its R^2 values in T + 1–T + 3 were consistent and higher than 0.9 in all the phases (Table 3). Thus, the Model 3 forecasts maintained fairly high accuracy (very small RMSE and high R^2 and NSE).

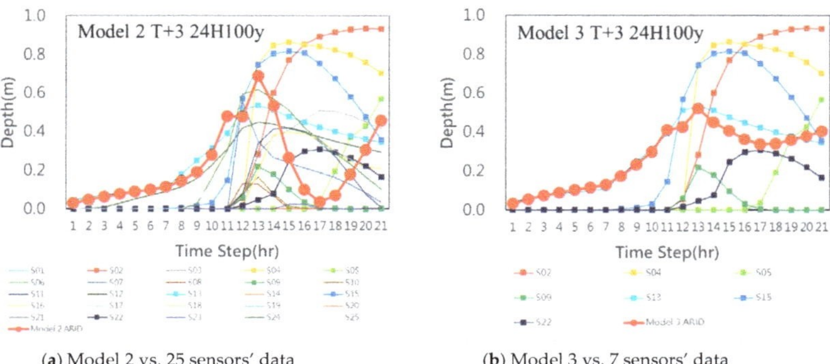

(**a**) Model 2 vs. 25 sensors' data   (**b**) Model 3 vs. 7 sensors' data

Figure 7. The sensor data vs. T + 3 forecasted average regional inundation depths by Models. (**a**) 25 sensors' data vs. T+3 forecasted ARIDs of Model 2; (**b**) 7 sensors' data vs. T+3 forecasted ARIDs of Model 3.

Figure 8 presents the forecast results of Model 1 and Model 3 at T + 1 in three phases. It shows that Model 1 underestimated the peak flow in the first and fifth events of the training session (i.e., 24H800mm and 24H500y), while they overestimated the peak flow in the validation and testing phases. In contrast, the results of Model 3 indicated that the problems of overestimated or underestimated peak values of ARIDs were significantly mitigated, its estimating errors (RMSE), in general, were also much smaller than those of Model 1 in all three phases.

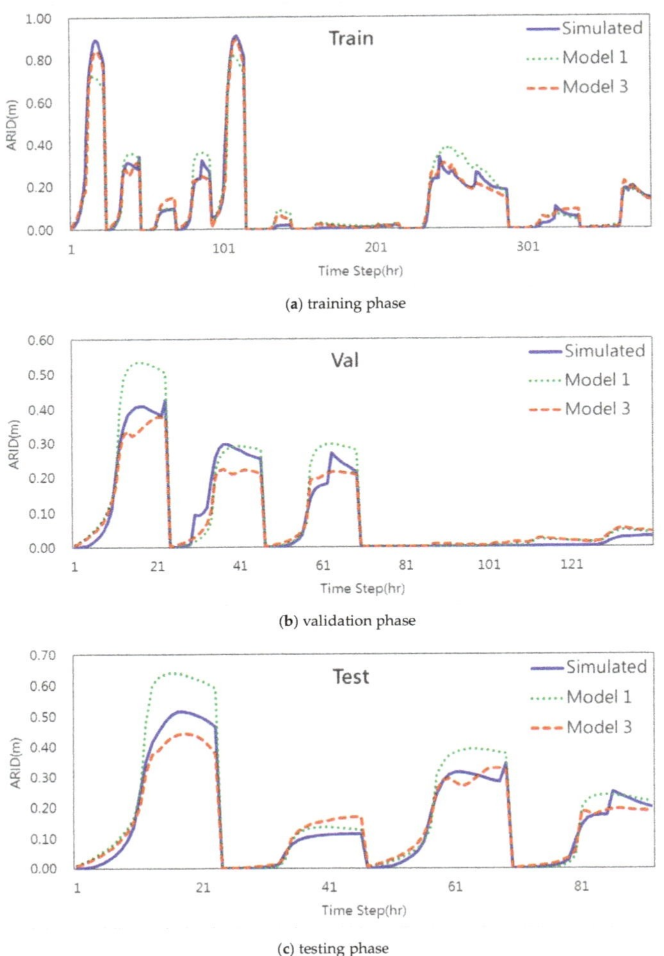

(a) training phase

(b) validation phase

(c) testing phase

Figure 8. Comparison of simulation, Model 1, and Model 3 in three phases. (a) training phase; (b)validation phase; (c)testing phase.

Figure 9 shows the RMSE charts in the training, validation, and testing phases of the three models. It is clearly shown that Model 3 had the most reliable and accurate performance compared to Models 1 and 2. Model 2 could be very well trained and was superior to Model 1 in all cases except in the cases of T + 2 and T + 3 in the testing phase, which was mainly caused by an inconsistency issue in the event of 24H100y. Model 3 had better performance than Model 2 in the validation and testing phases, where the average improvement rates were 22.65% and 42.71%, respectively. Moreover, the model's weights were reduced from 166 (Model 2) to 76 (Model 3), when the number of parameters was significantly reduced by 54.21%. The analyzed results provide an extra evidence and demonstrate that using the selected sensors as model input not only produces a parsimonious model but also improves prediction accuracy in multi-step-ahead flood inundation forecasts. These results represent the great value and benefit of IoT sensors, which fuse as inputs into the machine learning models.

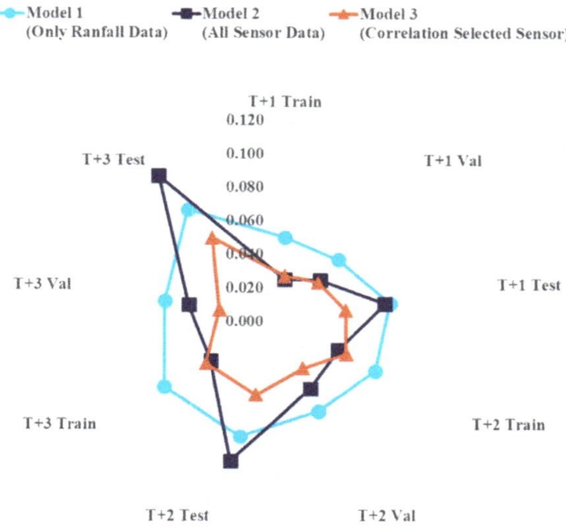

Figure 9. The radar chart of models' performance (RMSE) in three phases.

6. Conclusions

This study sought to model the temporal relationship and forecast the average regional inundation depth by using the IoT-based machine learning model. The datasets obtained from rainfall stations and IoT flood sensor data were used to model the average regional flood forecasts and explore the effectiveness and usefulness of the IoT sensors data in the model's reliability and accuracy. The results show that adding IoT sensor data as a model input can reduce the model error, especially for those of long horizontal (T + 2 and T + 3) forecasts in high-flood-depth conditions, where their underestimations are significantly mitigated. For instance, by adding 25 IoT sensors of inundation depth as extra inputs of Model 2, an average error improvement rate up to 18.49% could be reached as compared with Model 1 (only used rainfall datasets). While we also noticed that the inconsistent relationship between the 25 sensors datasets could over train the model and result in a chaotic issue in late application. For instance, Model 2 does provide the worst performances in the testing phase of T + 3 case. On the other hand, Model 3 used 7 IoT representative sensors, selected by the principle of correlation, as the extra inputs to model the multi-step flood forecasts. The number of parameters of Model 3 were greatly reduced (over 50%) as compared with Model 2. Furthermore, Model 3 was superior to Model 2, with an average improvement rate up to 18%, and it also provided much better forecast performances (small RMSE and high R^2 and NSE values) than Model 1 in all the T + 1–T + 3 cases. Therefore, these results give very promising evidence and demonstrate that using the IoT representative sensors as the model input to configure the machine learning models not only can produce a parsimonious model but also significantly improve the models' reliability and accuracy in multi-step-ahead regional flood inundation forecasts. The most fascinating achievement of this study is that the constructed model now can be on-line adjusted and its real-time forecast can be visually assessed and numerically evaluated based on the current implemented IoT inundation levels in the studied areas. Thus, the IoT-based machine learning model performance could be continuously assessed and adjusted, and its reliability and accuracy could be consistent.

Author Contributions: Conceptualization, L.-C.C.; Methodology, L.-C.C. and S.-N.Y.; Software, Validation, Data Curation and Visualization, S.-N.Y.; Formal Analysis and Investigation, L.-C.C. and S.-N.Y.; Resources, Writing-Original Draft Preparation, Writing-Review and Editing, Supervision, Project Administration and Funding Acquisition, L.-C.C. All authors have read and agreed to the published version of the manuscript.

Funding: This research was funded by the Water Resource Agency, Ministry of Economic Affairs, Taiwan, R.O.C. (grant number: MOEAWRA1060468).

Acknowledgments: The authors gratefully acknowledge the Water Resources Agency (WRA), Taiwan for the financial support on this research and for providing the investigative data. The authors would like to thank the Editors and anonymous Reviewers for their valuable and constructive comments related to this manuscript.

Conflicts of Interest: The authors declare no conflict of interest.

References

1. Kalyanapu, A.J.; Shankar, S.; Pardyjak, E.R.; Judi, D.R.; Burian, S.J. Assessment of GPU computational enhancement to a 2D flood model. *Environ. Model. Softw.* **2011**, *26*, 1009–1016. [CrossRef]
2. Nayak, P.C.; Sudheer, K.P.; Rangan, D.M.; Ramasastri, K.S. Short-term flood forecasting with a neurofuzzy model. *Water Resour. Res.* **2005**, *41*. [CrossRef]
3. Mai, D.T.; De Smedt, F. A combined hydrological and hydraulic model for flood prediction in Vietnam applied to the Huong river basin as a test case study. *Water* **2017**, *9*, 879. [CrossRef]
4. Adamowski, J.; Chan, H.F.; Prasher, S.O.; Ozga-Zielinski, B.; Sliusarieva, A. Comparison of multiple linear and nonlinear regression, autoregressive integrated moving average, artificial neural network, and wavelet artificial neural network methods for urban water demand forecasting in Montreal, Canada. *Water Resour. Res.* **2012**, *48*. [CrossRef]
5. Rezaeianzadeh, M.; Tabari, H.; Yazdi, A.A.; Isik, S.; Kalin, L. Flood flow forecasting using ANN, ANFIS and regression models. *Neural Comput. Appl.* **2014**, *25*, 25–37. [CrossRef]
6. Zare, M.; Koch, M. An analysis of MLR and NLP for use in river flood routing and comparison with the Muskingum method. In Proceedings of the ICHE 2014—11th International Conference on Hydroscience & Engineering, Hamburg, Germany, 28 September–2 October 2014; pp. 505–514. Available online: https://henry.baw.de/bitstream/handle/20.500.11970/99469/06_16.pdf?sequence=1&isAllowed=y (accessed on 31 May 2020).
7. Valipour, M.; Banihabib, M.E.; Behbahani, S.M.R. Comparison of the ARMA, ARIMA, and the autoregressive artificial neural network models in forecasting the monthly inflow of Dez dam reservoir. *J. Hydrol.* **2013**, *476*, 433–441. [CrossRef]
8. Huang, Y.F.; Mirzaei, M.; Yap, W.K. Flood analysis in Langat river basin using stochastic model. *Int. J. Geomate.* **2016**, *11*, 2796–2803.
9. Ab Razak, N.H.; Aris, A.Z.; Ramli, M.F.; Looi, L.J.; Juahir, H. Temporal flood incidence forecasting for Segamat River (Malaysia) using autoregressive integrated moving average modelling. *J. Flood Risk Manag.* **2018**, *11*, S794–S804. [CrossRef]
10. Machekposhti, K.H.; Sedghi, H.; Telvari, A.; Babazadeh, H. Flood Predicting in Karkheh River Basin Using Stochastic ARIMA Model. *Int. J. Agric. Biosyst. Eng.* **2018**, *12*, 89–96.
11. Kan, G.; He, X.; Ding, L.; Li, J.; Liang, K.; Hong, Y. Study on applicability of conceptual hydrological models for flood forecasting in humid, semi-humid semi-arid and arid basins in China. *Water* **2017**, *9*, 719. [CrossRef]
12. Chiang, Y.M.; Chang, L.C.; Tsai, M.J.; Wang, Y.F.; Chang, F.J. Dynamic neural networks for real-time water level predictions of Sewerage systems-covering gauged and unguaged sites. *Hydrol. Earth Syst. Sci.* **2010**, *14*, 1309–1319. [CrossRef]
13. Chang, F.J.; Chen, P.A.; Liu, C.W.; Liao, V.H.C.; Liao, C.M. Regional estimation of groundwater arsenic concentrations through systematical dynamic-neural modeling. *J. Hydrol.* **2013**, *499*, 265–274. [CrossRef]
14. Ruslan, F.A.; Samad, A.M.; Zain, Z.M.; Adnan, R. Flood water level modeling and prediction using NARX neural network: Case study at Kelang river. In Proceedings of the 2014 IEEE 10th International Colloquium

on Signal Processing and its Applications IEEE, Kuala Lumpur, Malaysia, 7–9 March 2014; pp. 204–207. [CrossRef]

15. Mosavi, A.; Ozturk, P.; Chau, K.W. Flood prediction using machine learning models: Literature review. *Water* **2018**, *10*, 1536. [CrossRef]

16. Noymanee, J.; Nikitin, N.O.; Kalyuzhnaya, A.V. Urban pluvial flood forecasting using open data with machine learning techniques in pattani basin. *Procedia Comput. Sci.* **2017**, *119*, 288–297. [CrossRef]

17. Puttinaovarat, S.; Horkaew, P. Flood Forecasting System Based on Integrated Big and Crowdsource Data by Using Machine Learning Techniques. *IEEE Access* **2020**, *8*, 5885–5905. [CrossRef]

18. PB, L.P.; Bhakthavathsalam, R.; Vishruth, K. Urban Flood Forecast using Machine Learning on Real Time Sensor Data. *Trans. Mach. Learn. Artif. Intell.* **2017**, *5*, 69. Available online: https://journals.scholarpublishing. org/index.php/TMLAI/article/view/3552/2104 (accessed on 31 May 2020). [CrossRef]

19. Tayfur, G.; Singh, V.P.; Moramarco, T.; Barbetta, S. Flood hydrograph prediction using machine learning methods. *Water* **2018**, *10*, 968. [CrossRef]

20. Leontaritis, I.J.; Billings, S.A. Input-output parametric models for non-linear systems Part I: Deterministic non-linear systems, Part II: Stochastic non-linear systems. *Int. J. Control* **1985**, *41*, 303–344. [CrossRef]

21. Tian, C.; Horne, R.N. Recurrent neural networks for permanent downhole gauge data analysis. In *SPE Annual Technical Conference and Exhibition*; Society of Petroleum Engineers: Calgary, AB, Canada, 2017. [CrossRef]

22. Koschwitz, D.; Frisch, J.; Van Treeck, C. Data-driven heating and cooling load predictions for non-residential buildings based on support vector machine regression and NARX Recurrent Neural Network: A comparative study on district scale. *Energy* **2018**, *165*, 134–142. [CrossRef]

23. Abou Rjeily, Y.; Abbas, O.; Sadek, M.; Shahrour, I.; Hage Chehade, F. Flood forecasting within urban drainage systems using NARX neural network. *Water Sci. Technol.* **2017**, *76*, 2401–2412. [CrossRef]

24. Khalid, E.G.; Jamal, E.K.; Isam, S.; Aziz, S. Comparison of M5 Model Tree and Nonlinear Autoregressive with eXogenous inputs (NARX) Neural Network for urban stormwater discharge modelling. In *MATEC Web of Conferences*; EDP Sciences: Les Ulis, France, 2019; Volume 295, p. 02002. [CrossRef]

25. Zainorzuli, S.M.; Abdullah, S.A.C.; Adnan, R.; Ruslan, F.A. Comparative Study of Elman Neural Network (ENN) and Neural Network Autoregressive with Exogenous Input (NARX) For Flood Forecasting. In Proceedings of the 2019 IEEE 9th Symposium on Computer Applications & Industrial Electronics (ISCAIE), Kota Kinabalu, Sabah, Malaysia, 27–28 April 2019; IEEE: Piscataway, NJ, USA, 2019; pp. 11–15. [CrossRef]

26. Chang, F.J.; Chen, P.A.; Lu, Y.R.; Huang, E.; Chang, K.Y. Real-time multi-step-ahead water level forecasting by recurrent neural networks for urban flood control. *J. Hydrol.* **2014**, *517*, 836–846. [CrossRef]

27. Shen, H.Y.; Chang, L.C. Online multistep-ahead inundation depth forecasts by recurrent NARX networks. *Hydrol. Earth Syst. Sci.* **2013**, *17*, 935–945. [CrossRef]

28. Chang, L.C.; Shen, H.Y.; Chang, F.J. Regional flood inundation nowcast using hybrid SOM and dynamic neural networks. *J. Hydrol.* **2014**, *519*, 476–489. [CrossRef]

29. Chang, L.C.; Amin, M.Z.M.; Yang, S.N.; Chang, F.J. Building ANN-based regional multi-step-ahead flood inundation forecast models. *Water* **2018**, *10*, 1283. [CrossRef]

30. Bande, S.; Shete, V.V. Smart flood disaster prediction system using IoT & neural networks. In Proceedings of the 2017 International Conference On Smart Technologies For Smart Nation (SmartTechCon), Bengaluru, India, 17–19 August 2017; IEEE: Piscataway, NJ, USA; pp. 189–194. [CrossRef]

31. Babu, V.; Rajan, V. Flood and Earthquake Detection and Rescue Using IoT Technology. In Proceedings of the 2019 International Conference on Communication and Electronics Systems (ICCES), Coimbatore, India, 17–19 July 2019; IEEE: Piscataway, NJ, USA; pp. 1256–1260. [CrossRef]

32. Wang, Y.; Chen, X.; Wang, L.; Min, G. Effective IoT-Facilitated Storm Surge Flood Modeling Based on Deep Reinforcement Learning. In *IEEE Internet of Things Journal*; IEEE: Piscataway, NJ, USA, 2020. [CrossRef]

33. Han, K.; Zhang, D.; Bo, J.; Zhang, Z. Hydrological monitoring system design and implementation based on IOT. *Phys. Procedia* **2012**, *33*, 449–454. [CrossRef]

34. Wang, J.; Yang, X. An Automatic Online Disaster Monitoring Network: Network Architecture and a Case Study Monitoring Slope Stability. *Int. J. Online Biomed. Eng.* **2018**, *14*, 4–19. [CrossRef]

35. Puttinaovarat, S.; Horkaew, P. Application Programming Interface for Flood Forecasting from Geospatial Big Data and Crowdsourcing Data. *Int. J. Interact. Mob. Technol.* **2019**, *13*, 137–156. [CrossRef]

36. Basha, E.A.; Ravela, S.; Rus, D. Model-based monitoring for early warning flood detection. In *SenSys '08: Proceedings of the 6th ACM Conference on Embedded Network Sensor Systems*; ACM: New York, NY, USA, 2008; pp. 295–308. [CrossRef]

37. Mitra, P.; Ray, R.; Chatterjee, R.; Basu, R.; Saha, P.; Raha, S.; Barman, R.; Patra, S.; Biswas, S.S.; Saha, S. Flood forecasting using Internet of things and artificial neural networks. In Proceedings of the 2016 IEEE 7th Annual Information Technology, Electronics and Mobile Communication Conference (IEMCON), Vancouver, BC, Canada, 13–15 October 2016; IEEE: Piscataway, NJ, USA; pp. 1–5. [CrossRef]

38. Pasi, A.A.; Bhave, U. Flood detection system using wireless sensor network. *Int. J. Adv. Res. Comput. Sci. Softw. Eng.* **2015**, *5*. [CrossRef]

39. Chang, L.C.; Chang, F.J.; Yang, S.N.; Kao, I.F.; Ku, Y.Y.; Kuo, C.L.; bin Mat, M.Z. Building an intelligent hydroinformatics integration platform for regional flood inundation warning systems. *Water* **2019**, *11*, 9. [CrossRef]

40. Sood, S.K.; Sandhu, R.; Singla, K.; Chang, V. IoT, big data and HPC based smart flood management framework. *Sustain. Comput. Inform. Syst.* **2018**, *20*, 102–117. [CrossRef]

41. Mishra, B.K.; Thakker, D.; Mazumdar, S.; Neagu, D.; Simpson, S. Using Deep Learning for IoT-enabled Smart Camera: A Use Case of Flood Monitoring. In Proceedings of the 2019 10th International Conference on Dependable Systems, Services and Technologies (DESSERT), Leeds, UK, 5–7 June 2019. [CrossRef]

42. Bowden, G.J.; Dandy, G.C.; Maier, H.R. Input determination for neural network models in water resources applications. Part 1—background and methodology. *J. Hydrol.* **2005**, *301*, 75–92. [CrossRef]

43. Galelli, S.; Humphrey, G.B.; Maier, H.R.; Castelletti, A.; Dandy, G.C.; Gibbs, M.S. An evaluation framework for input variable selection algorithms for environmental data-driven models. *Environ. Model. Softw.* **2014**, *62*, 33–51. [CrossRef]

44. Hu, J.H.; Tsai, W.P.; Cheng, S.T.; Chang, F.J. Explore the relationship between fish community and environmental factors by machine learning techniques. *Environ. Res.* **2020**, *184*, 109262. [CrossRef] [PubMed]

45. Muttil, N.; Chau, K.W. Machine-learning paradigms for selecting ecologically significant input variables. *Eng. Appl. Artif. Intell.* **2007**, *20*, 735–744. [CrossRef]

46. Taormina, R.; Chau, K.W. Data-driven input variable selection for rainfall–runoff modeling using binary-coded particle swarm optimization and extreme learning machines. *J. Hydrol.* **2015**, *529*, 1617–1632. [CrossRef]

47. Chang, F.J.; Chen, P.A.; Chang, L.C.; Tsai, Y.H. Estimating spatio-temporal dynamics of stream total phosphate concentration by soft computing techniques. *Sci. Total Environ.* **2016**, *562*, 228–236. [CrossRef]

Article

Improving the Reliability of Probabilistic Multi-Step-Ahead Flood Forecasting by Fusing Unscented Kalman Filter with Recurrent Neural Network

Yanlai Zhou [1,2], Shenglian Guo [1,*], Chong-Yu Xu [2,*], Fi-John Chang [3,*] and Jiabo Yin [1]

1 State Key Laboratory of Water Resources and Hydropower Engineering Science, Wuhan University, Wuhan 430072, China; yanlai.zhou@whu.edu.cn (Y.Z.); jboyn@whu.edu.cn (J.Y.)
2 Department of Geosciences, University of Oslo, P.O. Box 1047 Blindern, N-0316 Oslo, Norway
3 Department of Bioenvironmental Systems Engineering, National Taiwan University, Taipei 10617, Taiwan
* Correspondence: slguo@whu.edu.cn (S.G.); c.y.xu@geo.uio.no (C.-Y.X.); changfj@ntu.edu.tw (F.-J.C.)

Received: 7 January 2020; Accepted: 17 February 2020; Published: 20 February 2020

Abstract: It is fundamentally challenging to quantify the uncertainty of data-driven flood forecasting. This study introduces a general framework for probabilistic flood forecasting conditional on point forecasts. We adopt an unscented Kalman filter (UKF) post-processing technique to model the point forecasts made by a recurrent neural network and their corresponding observations. The methodology is tested by using a long-term 6-h timescale inflow series of the Three Gorges Reservoir in China. The main merits of the proposed approach lie in: first, overcoming the under-prediction phenomena in data-driven flood forecasting; second, alleviating the uncertainty encountered in data-driven flood forecasting. Two commonly used artificial neural networks, a recurrent and a static neural network, were used to make the point forecasts. Then the UKF approach driven by the point forecasts demonstrated its competency in increasing the reliability of probabilistic flood forecasts significantly, where predictive distributions encountered in multi-step-ahead flood forecasts were effectively reduced to small ranges. The results demonstrated that the UKF plus recurrent neural network approach could suitably extract the complex non-linear dependence structure between the model's outputs and observed inflows and overcome the systematic error so that model reliability as well as forecast accuracy for future horizons could be significantly improved.

Keywords: probabilistic forecast; Unscented Kalman Filter; artificial neural networks; Three Gorges Reservoir

---

1. Introduction

Reliable and accurate flood forecasting is one of the most important tasks of operational hydrology, while it is also very challenge due to the inordinately non-linear hydro-geological features and dynamic nature of climate conditions. High uncertainty encountered in the occurrence and magnitudes of future flood event stimulates the demands for probabilistic flood forecasting. The goal of probabilistic forecasting is to provide information about the uncertainty of the forecast [1]. Most hydrological forecast models produce deterministic forecasts, which provide the best point-value estimates rather than quantify the predictive uncertainty [2]. Nevertheless, when a deterministic forecast turns out to be far from what has taken place, the consequences will probably be worse than a situation where no forecast is available [3]. Probabilistic hydro-meteorological forecasts have been used frequently to communicate forecast uncertainty over the last few decades [4–6]. The transformation from a deterministic approach to a probabilistic approach is a development trend of flood forecasting around the world [6,7]. The optimal mega-reservoir operation for live-saving and resources utilization creates outreach demands for probabilistic flood forecasting; consequently, scientific research should focus on quantifying and mitigating the uncertainty of probabilistic flood forecasts [8,9]. The reliability of hydrologic forecasts can be affected by input uncertainty, meteorological uncertainty, and hydrologic uncertainty of model structure and parameters. One of the primary techniques to reflect different uncertainties in hydrologic forecasts is to create a probabilistic forecast [10,11]. Probabilistic forecasts can be made using three approaches: a probabilistic pre-processing approach plus a deterministic forecast model; a probabilistic forecast model; and a deterministic forecast model plus a probabilistic post-processing approach [12–14]. The first two approaches quantify uncertainties in inputs and model structure while the third quantifies the overall uncertainty in model structure and parameters. Our study would concentrate on improving hydrologic forecasts using deterministic models plus probabilistic post-processing technique.

Probabilistic post-processing techniques are commonly introduced to complement point-value estimations offered by the deterministic forecast model [15,16]. The Kalman filter (KF) proposed by Kalman [5] provides a theoretical post-processing framework based on model point estimation for reducing forecast uncertainty through recursively calculating a statistically optimal estimate of the prediction. The KF post-processing is a component of the probabilistic post-processing techniques, and is a recursive state estimator for a process that is assumed to be affected by stochastic interference and by stochastic noise [5]. The KF family consists of the linear KF (LKF) and non-linear KF (NKF) [17]. The LKF approach can only identify the linear error estimation whereas the NKF approach can quantify the non-linear error estimation. As is known, the NKF is widely used for extracting non-linear dependence of forecast errors and conquering the white noise with systematic over/under-predicting characteristics [17]. Furthermore, the extended KF (EKF) and the unscented KF (UKF) [17] are two common usages of NKF. Most importantly, the UKF approach has not yet been employed to lessen the uncertainty of multi-step-ahead flood forecasting driven by a recurrent neural network (RNN) according to a review of literature [18–21]. Despite there are several researches associated with the combination of UKF/EKF and hydrological models [22–24] on hydrological domain, all of them concentrate on quantifying the uncertainty of hydrological forecast driven by static (i.e., non-recurrent) artificial neural networks (ANNs), e.g., feed-forward neural network and local linear models as well as the hydraulic model. Bearing this in mind as motivation, for the first time, the UKF is introduced to quantify the uncertainty of multi-step-ahead flood forecasts driven by the RNN (i.e., RNN is more complicated than the static ANNs). Therefore, it is interesting to explore UKF for modeling and lowering the uncertainty appeared in RNN-driven flood forecasts.

Machine-learning techniques have developed fast during the last few decades, and they have been adopted as data-driven methods to model hydrological systems [11,25,26]. For instance, the back-propagation neural network (BPNN), the radial basis function (RBF), the support vector machine, the quantile regression neural network (QRNN), the recurrent neural network (RNN), the long-short term memory (LSTM) and the non-linear auto-regressive with exogenous inputs neural network (NARX) have been widely applied to modeling hydrologic and meteorological time series [27–38]. A number of recent studies indicate that ensemble artificial neural network can improve the probabilistic forecast skill for hydrological events [39–41]. The main advantage of ANN is owing to its ability to discern linear or non-linear relationships even with very limited data inputs and being able to recognize even complex patterns in a data set without a priori understating of the underlying mechanism. The major drawbacks, on the other hand, are that they are prone to under-predict flood series for extreme flood events. Therefore, it is essential to conduct in-depth research on machine-learning models for enhancing model accuracy and reliability through converting deterministic flood forecasts into probabilistic ones using a stochastic post-processing technique.

This study proposes a probabilistic forecasting approach to reduce the prediction intervals of multi-step-ahead flood forecasts, which consists of two parts: the deterministic forecast model and the probabilistic post-processing technique. First, the recurrent neural network (NARX) is introduced to make multi-step-ahead point forecasts. Then, the UKF technique driven by point forecasts is employed to create the prediction intervals of flood forecasts. We concentrate on hydrological uncertainty only (i.e., the uncertainty resulting from imperfect rainfall-runoff modeling), considering "perfect" rainfall as inputs. A static BPNN and a recurrent NARX are used to construct flood forecast models, and the model that produces more accurate point estimations will be employed to carry out probabilistic forecasting. The UKF approach is implemented separately to transform point flood forecasts into probabilistic flood forecasts. The rainfall and inflow datasets of the Three Gorges Reservoir (TGR) in China are used to demonstrate the reliability and applicability of the approach.

2. Methods

Figure 1 illustrates the probabilistic forecast architecture that separately integrates the NARX (Figure 1a) with the UKF approach (Figure 1b). The related methods are briefly described below.

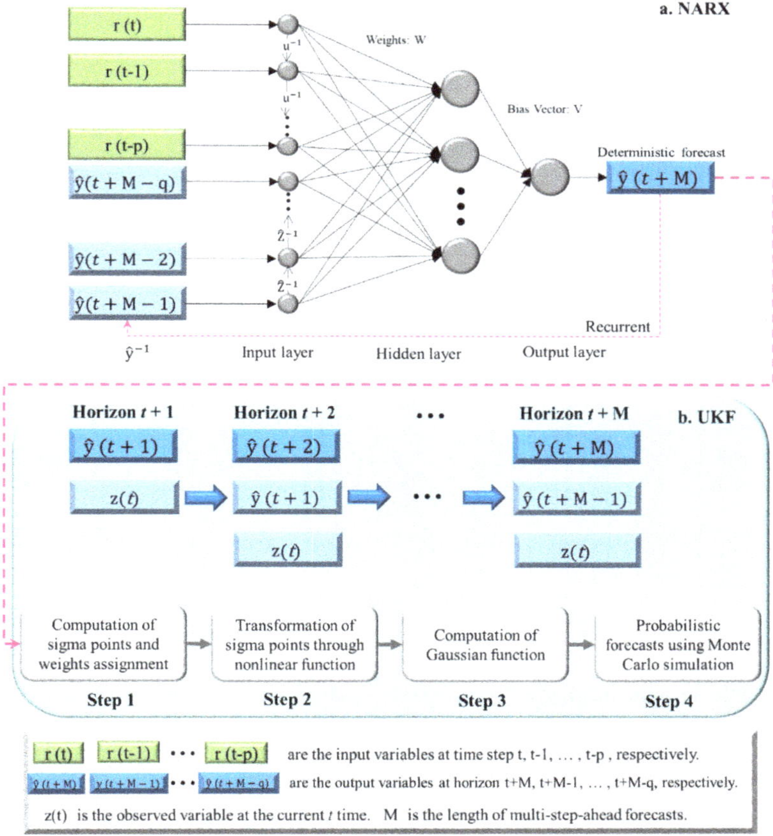

Figure 1. Probabilistic forecast architecture. (**a**) The non-linear auto-regressive with exogenous inputs (NARX) neural network model. (**b**) The unscented Kalman filter (UKF) post-processing approach.

2.1. Deterministic Flood Forecast Models Based on Artificial Neural Network (ANN)

The NARX is a recurrent neural network suitable for time-series prediction [42,43]. This study uses it to carry out deterministic flood forecasting because its recurrent mechanism can effectively integrate rainfall and discharge data with the latest outputs of the model to alleviate the time shift phenomenon and improve model reliability. The NARX network produces recurrent connections from the outputs, which could delay several unit times to produce new inputs (Figure 1a). This nonlinear system for M-step-ahead forecasting (M $\geq$ 1) can be described mathematically below:

$$\hat{y}(t+M) = f(\hat{y}(t+M-1), \ldots, \hat{y}(t+M-q), y(t), r(t), \ldots, r(t-p)) \tag{1}$$

where $R(t) = [y(t), r(t), \ldots, r(t-p)]$ and $\hat{y}(t+M)$ denote the input vector and the output value at the time step t and t + M, respectively. $f(\cdot)$ is the nonlinear function. p and q are the input-memory and output memory orders. Two inputs: $\hat{y}(t+M-i)$ $(i = 1, 2, \ldots, q)$ serves as an autoregressive variable (e.g., forecasted runoff), $R(t)$, serves as an implicit exogenous variable (e.g., observed rainfall and runoff) in a time series.

The BPNN, a static neural network, is implemented for comparison purposes. Both BPNN and NARX have the same model architecture of one input layer, one hidden layer and one output layer. The NARX model is a dynamic ANN model with the recurrent mechanism whereas the BPNN is a non-recurrent ANN model. The mathematical equation of the BPNN model is described as follows:

$$\hat{y}(t+M) = f(y(t), r(t), \ldots, r(t-p)) \tag{2}$$

In this study, the Levenberg–Marquardt back-propagation algorithm is used to train both ANN models [44]. The transfer functions of hidden and output layers are of a sigmoid type and a linear type, respectively, owing to their practicability and good performance in flood forecasting [45].

The parameters of both models consist of: the maximal epoch = 1000, the initial learning rate (mu) = 0.001, the increasing factor of mu = 10, the decreasing factor of mu = 0.1 and the maximal value of mu = 1000 respectively. Besides, the two ANN models are configured to have one hidden layer with 8–10 nodes in the perspective of different forecast horizons. The value of output memory q is set as one whereas the value of input memory p should be determined by using correlation analysis methods (presented in the section of study area and materials).

2.2. Probabilistic Forecasting Based on the Unscented Kalman Filter (UKF)

The UKF is an optimal recursive data processing algorithm and can be introduced to find the optimal estimation error for each state in multi-step-ahead flood forecasts [17]. The unscented Kalman filter is applied to model a nonlinear flood forecasting system and described as below:

$$x(t+1) = g(x(t), u(t), v(t), t) \tag{3}$$

$$z(t) = h(x(t), u(t), t) + w(t) \tag{4}$$

where $x(t)$ is the n-dimensional state of the flood forecasting system at time-step t, $u(t)$ is the input vector at time-step t, $v(t)$ is the k-dimensional state noise process vector owing to disturbances and model errors, $z(t)$ and $w(t)$ are the observation vector and noise, $g(\cdot)$ and $h(\cdot)$ are the distributions of forecasted and observed datasets, both are assumed following Gaussian distribution.

The implementation procedures for the UKF (Figure 1b) contain the following four steps [17].

Step 1: Computation of set of sigma points and assignment of weights to all sigma points. The n-dimensional random variable x with mean $\bar{x}$ and covariance P_{xx} is transformed to $2n + 1$ weighted points described below:

$$\chi_0 = \bar{x}, \ W_0 = \lambda/(n+\lambda) \tag{5a}$$

$$\chi_i = \bar{x} + \sqrt{(n+1)P_{xx}}, \ W_i = 1/2(n+\lambda) \tag{5b}$$

$$\chi_{i+n} = \bar{x} - \sqrt{(n+1)P_{xx}}, \ W_{i+n} = 1/2(n+\lambda) \tag{5c}$$

where χ_0, χ_i and χ_{i+n} are the $2n + 1$ sigma points, W_0, W_i and W_{i+n} are the $2n + 1$ weight coefficients, λ is the spreading parameter.

Step 2: Transformation of the points through non-linear function. The transformed set is produced by using a nonlinear function and the predicted mean and covariance are described below:

$$\chi_i(t+1|t) = g\big(\chi_i^{n+k}(t|t), u(t), t\big) \tag{6}$$

$$\hat{x}(t+1|t) = \sum_{i=0}^{2(n+k)} W_i \cdot \chi_i^{n+k}(t+1|t) \tag{7a}$$

$$\mathbf{P}(t+1|t) = \sum_{i=0}^{2(n+k)} W_i \cdot [\chi_i(t+1|t) - \hat{x}(t+1|t)] \cdot [\chi_i(t+1|t) - \hat{x}(t+1|t)]^{\mathrm{T}} \tag{7b}$$

where $\chi_i(t+1|t)$ is the transformed set. $\hat{x}(t+1|t)$ and $\mathbf{P}(t+1|t)$ are the predicted mean and covariance of the transformed set.

Step 3: Computation of Gaussian function from weighted and transformed points.

$$Z_i(t+1|t) = h(\chi_i(t+1|t), u(t), t) \tag{8}$$

where $Z_i(t+1|t)$ is the observed dataset computed by using the transformed point.

Step 4: Probabilistic forecasts. A Monte Carlo simulation is conducted to create probabilistic forecasts. A realization of observation h_m at the horizon m can be simulated according to the Gaussian function (Equation (7)) and the Monte Carlo simulation would be repeated for K times. K is the number of Monte Carlo samples and set as 1000 in this study; 90% confidence intervals are employed to reveal the uncertainty of probabilistic flood forecasts. Then, both observed and forecasted datasets are transformed into the real space for evaluating the performance of UKF probabilistic forecasts.

2.3. Evaluation Indicators

To evaluate the deterministic forecast accuracy and predictability of flood peak and flood volume, the mean absolute error (MAE), the peak percent threshold statistics (PPTS) [46] and the Nash–Sutcliffe efficiency (NSE) coefficient [47] were introduced accordingly. Their mathematical expressions are described below:

$$\mathrm{MAE} = \frac{1}{N} \sum_{t=1}^{N} |\hat{Z}(t) - Z(t)|, \ \mathrm{MAE} \geq 0 \tag{9}$$

$$\mathrm{PPTS}(l, u) = \frac{1}{(k_l - k_u + 1)} \sum_{t=1}^{N} \left(\left| \frac{\hat{Z}(t) - Z(t)}{Z(t)} \right| \right), \ \mathrm{PPTS} \geq 0 \tag{10}$$

$$\mathrm{NSE} = 1 - \frac{\sum_{t=1}^{N} \big(\hat{Z}(t) - Z(t)\big)^2}{\sum_{t=1}^{N} \big(Z(t) - \overline{Z}(t)\big)^2}, \ \mathrm{NSE} \leq 1 \tag{11}$$

where $k_l = \frac{l \times N}{100}$ and $k_u = \frac{u \times N}{100}$. N is the number of observed data while l and u are the lower and higher limits in percentage, respectively. For instance, PPTS(l, 10%) denotes the flood percentage threshold statistic of larger than 10% data whereas PPTS(90%, u) denotes the flood percentage threshold statistic of smaller than 90% data. $\hat{Z}(t)$, $Z(t)$ and $\overline{Z}(t)$ are the forecasted data (i.e., model output), observed data and the average of observed data at the t time, respectively.

The containing ratio (CR), the average relative bandwidth (RB) and the continuous ranked probability score (CRPS) were used for assessing the performance of probabilistic forecasts [48,49]. Their mathematical formulas are described below.

$$N(t) = \begin{cases} 1, & \text{if}\big(q_l(t) \leq \hat{Z}(t) \leq q_u(t)\big) \\ 0, & \text{else} \end{cases} \tag{12a}$$

$$CR = \frac{\sum_{t=1}^{N} N(t)}{N} \times 100\% \tag{12b}$$

$$RB = \frac{1}{N} \sum_{t=1}^{N} \left(\frac{q_u(t) - q_l(t)}{Z(t)} \right) \tag{13}$$

$$CRPS = \int_{-\infty}^{+\infty} \left[F^f(x) - F^o(x) \right]^2 dx \tag{14}$$

where $q_l(t)$ and $q_u(t)$ are the lower and upper limitation of the forecasted value at the t time, $F^f(x)$ and $F^o(x)$ are the cumulative distribution functions (CDF) of the forecast and observation distributions, respectively, x is the variable of the CDF.

From the standpoint of model performance, the indicators of MAE, PPTS and NSE are employed to evaluate the accuracy of deterministic flood forecasts while the indicators of CR, RB and CRPS are employed to evaluate the correctness and sharpness of probabilistic flood forecasts. The general implementation programming of the NARX and UKF can be downloaded from the Statistics and Machine Learning Toolbox of the Matlab software (website: https://ww2.mathworks.cn/products/statistics.html#machine-learning).

3. Study Area and Materials

Figure 2a illustrates the study area. The Yangtze River is famous for the longest river of China, and its length and drainage area are 6300 km and 1.80 million km^2 accordingly. The Three Gorges Reservoir (TGR) is a pivotal hydraulic facility that serves many purposes consisting of flood control, hydropower production and navigation etc. We notice that the upper Yangtze River' tributaries have notoriously complex hydro-geological characteristics. Plenty of studies were devoted to developing reliable and accurate short-term (less than 24-h ahead) flood forecast models for the Yangtze River due to the extremely non-linear relationship between rainfall and runoff over this basin during storm events [50,51]. Besides, a small improvement in the reliability and accuracy of short-term flood forecasts could be critical and beneficial to flood prevention as well as the dynamic management of the TGR. Heavy rainfalls usually cause floods and further induce downstream flooding within one day. In consequence, rational reservoir operation as well as river basin management require reliable and accurate flood forecasting so that they can adequately handle the high uncertainty of river discharge ranging from 8000 up to 70,000 m^3/s, according to the observed reservoir inflows (Figure 2b).

The inflow of the TGR contains three parts: the upstream inflow from the discharge of the XJB reservoir, the inflows from 8 tributaries (flow stations F1–F8), and rainfall (aggregated into two variables: Rainfall-I and Rainfall-II) monitored by 67 rain gauge stations spreading over the TGR intervening basin. Inflow and rainfall data for use in this study were gathered from 2003 to 2018 at a temporal scale of 6 h. When model training completes (2003–2010, 8 years), two ANN models (BPNN and NARX) are constructed. Then the trained ANN that creates the best forecast accuracy in the validation period (2011–2014, 4 years) is identified as the final model to be evaluated upon model reliability using test datasets (2015–2018, 4 years). The Kendall tau coefficient analysis [38,52] is employed to extract the highest correlation of lag-times between model input and output values. According to the highest correlation coefficients, lag-times between the inflow of the TGR and flow/rainfall at various gauge stations are set as 6 h (TGR), 48 h (XJB reservoir), 48 h (F_1), 48 h (F_2), 42 h (F_3), 42 h (F_4), 24 h (F_5), 18 h (F_6), 18 h (F_7), 12 h (F_8), 42 h (Rainfall-I) and12 h (Rainfal-II) [36]. To reduce the adverse effect of the distinct scales of input data on model performance, all 12 input variables (one autoregressive variable plus 11 exogenous variables) were transformed into the same scale during data preprocessing.

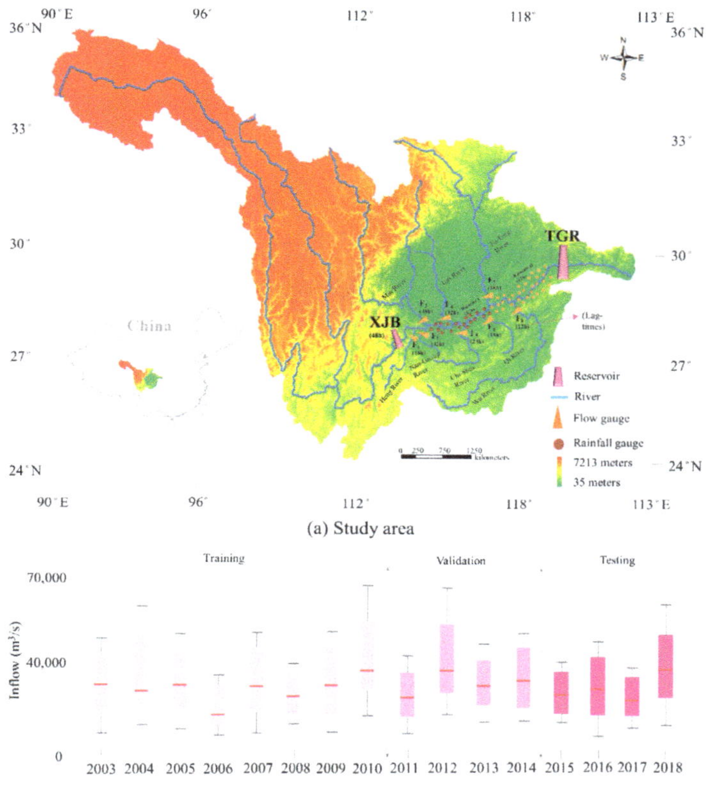

(a) Study area

(b) Boxplot of TGR inflows in flood seasons

Figure 2. Study area and the statistical characteristics of reservoir inflows. (**a**) Locations of the Xiang-Jia-Ba (XJB) Reservoir and the Three Gorges Reservoir (TGR), river flow as well as rain gauging stations; (**b**) The boxplot of TGR inflows collected in flood seasons (from 1 June to 30 September) during 2003 and 2018 at a temporal scale of 6 h.

4. Results and Discussion

This study intends to promote the predictability of the deterministic flood forecast model that is integrated with a UKF for probabilistic forecasting at different time horizons.

4.1. Performance of Deterministic Flood Forecasts

The short-term (one-day-ahead) forecast not only provides a crucial guideline in reservoir operation but also offers a warning to residents at inundation prone areas. Consequently, a 24 h lead time at a temporal scale of 6 h is suggested to evaluate the performance of two deterministic flood forecast models (NARX and BPNN). Four horizons (t + 1 to t + 4) are specified that 6 h (t + 1) is the first prediction, 12 h (t + 2) is the second prediction, 18 h (t + 3) is the third prediction and 24 h (t + 4) is the fourth prediction.

Table 1 summarizes the results of the NARX and BPNN models. It indicates that the BPNN model creates much lower NSE values but much higher PPTS and MAE values than the NARX model at all three stages. The PPTS indicator is able to show model forecast accuracy in different flood magnitudes. For instance, PPTS (l, 10%) denotes the flood percentage threshold statistic of larger than 10% data. The results in Table 1 demonstrate that the PPTS indicator raises gradually if the lead time increases from t +

1 to t + 4. The NARX model makes the smallest increments in PPTS and MAE values along the lead time while the BPNN model makes the largest ones. In addition, the NARX model displays its superiority compared with the BPNN model in predicting high-magnitude floods regarding frequencies of 2% and 5%. Given a lead time of one day (24 h), the NARX model can increase the NSE by 14.5% while decreasing the PPTS (l, 2%) by 35.8% and the MAE by 20.1% at the testing stage, in comparison to the BPNN model. The results demonstrate that the NARX model performs much better than the BPNN model as the forecast horizon increases. Therefore, it is obvious that the NARX model incorporated with a recurrent mechanism can improve forecast accuracy at longer horizons by feeding itself with the forecasted inflows attained from previous horizons.

Table 1. Performances (peak percent threshold statistics (PPTS), mean absolute error (MAE) and Nash–Sutcliffe efficiency (NSE)) of NARX and back-propagation neural network (BPNN) models in three stages.

Stage	Model	Indicator	Horizon			
			t + 1	t + 2	t + 3	t + 4
Training	NARX	PPTS (l, 2%)	0.0293	0.0310	0.0332	0.0349
		PPTS (l, 5%)	0.0314	0.0330	0.0355	0.0383
		PPTS (l, 10%)	0.0334	0.0356	0.0393	0.0450
		PPTS (l, 20%)	0.0366	0.0388	0.0429	0.0488
		MAE (m^3/s)	689	750	849	1037
		NSE	0.980	0.969	0.963	0.942
	BPNN	PPTS (l, 2%)	0.0293	0.0346	0.0364	0.0398
		PPTS (l, 5%)	0.0314	0.0447	0.0483	0.0512
		PPTS (l, 10%)	0.0334	0.0546	0.0606	0.0616
		PPTS (l, 20%)	0.0366	0.0601	0.0661	0.0672
		MAE (m^3/s)	689	951	1024	1295
		NSE	0.980	0.935	0.907	0.858
Validation	NARX	PPTS (l, 2%)	0.0295	0.0314	0.0340	0.0361
		PPTS (l, 5%)	0.0315	0.0334	0.0364	0.0397
		PPTS (l, 10%)	0.0335	0.0354	0.0395	0.0438
		PPTS (l, 20%)	0.0372	0.0399	0.0445	0.0491
		MAE (m^3/s)	716	782	919	1079
		NSE	0.978	0.965	0.957	0.936
	BPNN	PPTS (l, 2%)	0.0295	0.0432	0.0472	0.0493
		PPTS (l, 5%)	0.0315	0.0472	0.0515	0.0547
		PPTS (l, 10%)	0.0335	0.0551	0.0606	0.0643
		PPTS (l, 20%)	0.0372	0.0621	0.0690	0.0732
		MAE (m^3/s)	716	964	1088	1317
		NSE	0.980	0.931	0.904	0.853
Testing	NARX	PPTS (l, 2%)	0.0305	0.0312	0.0332	0.0349
		PPTS (l, 5%)	0.0320	0.0339	0.0365	0.0418
		PPTS (l, 10%)	0.0343	0.0368	0.0411	0.0456
		PPTS (l, 20%)	0.0374	0.0401	0.0448	0.0497
		MAE (m^3/s)	793	931	893	1006
		NSE	0.978	0.967	0.961	0.940
	BPNN	PPTS (l, 2%)	0.0305	0.0488	0.0536	0.0544
		PPTS (l, 5%)	0.0320	0.0511	0.0562	0.0571
		PPTS (l, 10%)	0.0343	0.0582	0.0647	0.0686
		PPTS (l, 20%)	0.0374	0.0635	0.0705	0.0749
		MAE (m^3/s)	793	992	1079	1259
		NSE	0.978	0.929	0.872	0.821

To distinguish between the BPNN and NARX models on deterministic forecasting capabilities in the testing stage, three flood events with maximal peak-flows reaching 35,000 m^3/s (small), 50,000 m^3/s (medium) and 60,000 m^3/s (large) are specified to test both models by evaluating the goodness-of-fit between observed and forecasted values at time-step t + 3 and t + 4 (Figure 3). The results appear to show that the NARX model is able to significantly mitigate the time gap between the observed and forecasted flood peaks. Furthermore, the NARX model can produce good forecast results at time-step t + 3 and t + 4, whereas the BPNN model creates a noticeable time-lag problem and fairly big gaps between observed and forecasted values at time-step t + 3 and t + 4. This demonstrates that the

NARX model is able to effectively reduce time-lag phenomena and provide accurate deterministic flood forecasting results.

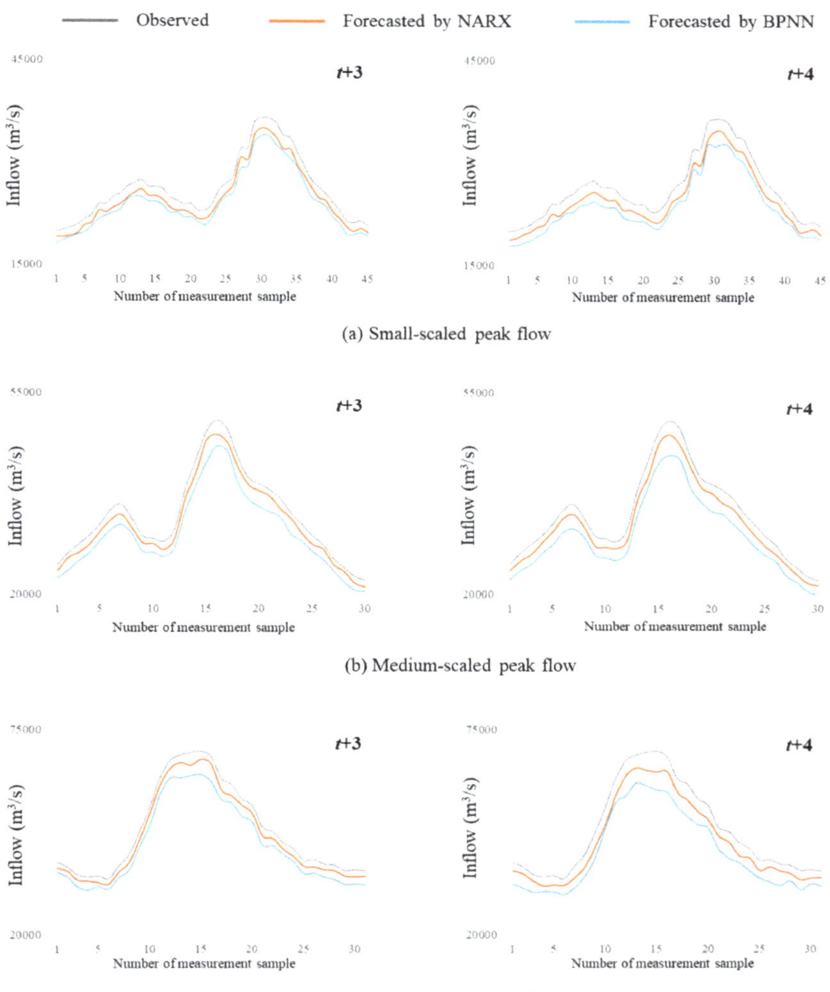

Figure 3. TGR inflow forecasts using the NARX and the BPNN models. Three flood events with maximal peak-flow exceeding (**a**) 35,000 m^3/s (small-scaled), (**b**) 50,000 m^3/s (medium-scaled) and (**c**) 60,000 m^3/s (large-scaled).

Although the NARX model provides substantial evidence of a good performance in flood forecasting, it is easy to produce systematic under-prediction results for extreme flood events (Figure 3). In addition, hydrologic uncertainties raised in model inputs (e.g., precipitation), as well as the structure and parameters, can be the drivers and causes of time-lag problems occurring in flood forecasting. Therefore, we next adopt a processing approach (i.e., UKF) to decrease the hydrological uncertainty based on the presumption that no uncertainty encounters in model input data.

4.2. Probabilistic Flood Forecasting Performance

Table 2 illustrates the results of the CR, RB and CRPS corresponding to the UKF plus NARX and the UKF plus BPNN approaches in all three stages at forecast horizons (t + 1 to t + 4). It supports the superiority of the UKF plus NARX approach in all three stages, whereas the UKF plus BPNN approach performs almost as well as the UKF plus NARX one only in the training stages at forecast horizons up to t + 2 (RB is lower than 0.15, CRPS is lower than 1200 m^3/s and CR is higher than 90%). The results demonstrate that the UKF plus NARX approach achieves higher reliability in probabilistic flood forecasting than the UKF plus BPNN one. For horizon t + 4 (one day ahead) in the testing stage, the UKF plus NARX approach is able to increase the CR value by 8.48% while reducing the RB and the CRPS values by 22.73% and 20.31%, respectively, in comparison to the UKF plus BPNN one. In other words, the UKF plus NARX approach is able to significantly improve probabilistic forecast accuracy by producing the narrower prediction bound.

Table 2. Probabilistic forecasting performance (containing ratio (CR), average relative bandwidth (RB) and continuous ranked probability score (CRPS)) regarding TGR reservoir inflow.

Stage	Model	Indicator	Horizon			
			t + 1	t + 2	t + 3	t + 4
Training	UKF plus NARX	CR (%)	98.23	96.37	95.06	94.53
		RB	0.08	0.10	0.13	0.17
		CRPS (m^3/s)	754	911	1092	1253
	UKF plus BPNN	CR (%)	98.23	94.11	92.25	89.03
		RB	0.08	0.12	0.18	0.21
		CRPS (m^3/s)	754	1092	1361	1517
Validation	UKF plus NARX	CR (%)	98.21	96.32	95.00	94.37
		RB	0.08	0.11	0.14	0.18
		CRPS (m^3/s)	789	934	1112	1274
	UKF plus BPNN	CR (%)	98.21	93.07	91.21	88.06
		RB	0.08	0.13	0.19	0.23
		CRPS (m^3/s)	789	1118	1395	1576
Testing	UKF plus NARX	CR (%)	98.18	96.29	95.02	94.41
		RB	0.09	0.10	0.13	0.17
		CRPS (m^3/s)	776	922	1104	1267
	UKF plus BPNN	CR (%)	98.18	92.04	89.25	87.03
		RB	0.09	0.12	0.18	0.22
		CRPS (m^3/s)	776	1127	1412	1590

For obviously differentiating the capabilities of the UKF plus NARX and the UKF plus BPNN approaches in the testing stage, the three aforementioned flood events are still specified to test both approaches by evaluating whether the observations fall in the 90% confidence interval at lead time t + 4 (Figure 4). It appears that most of the observations lie within the 90% confidence intervals created by both probabilistic forecasting approaches while the UKF plus NARX approach offers a narrower prediction range. Gneiting et al. [1] advocated that the maximization of the sharpness of the predictive distribution is the goal of probabilistic flood forecasts. Therefore, the UKF plus NARX approach is considered superior to the UKF plus BPNN one.

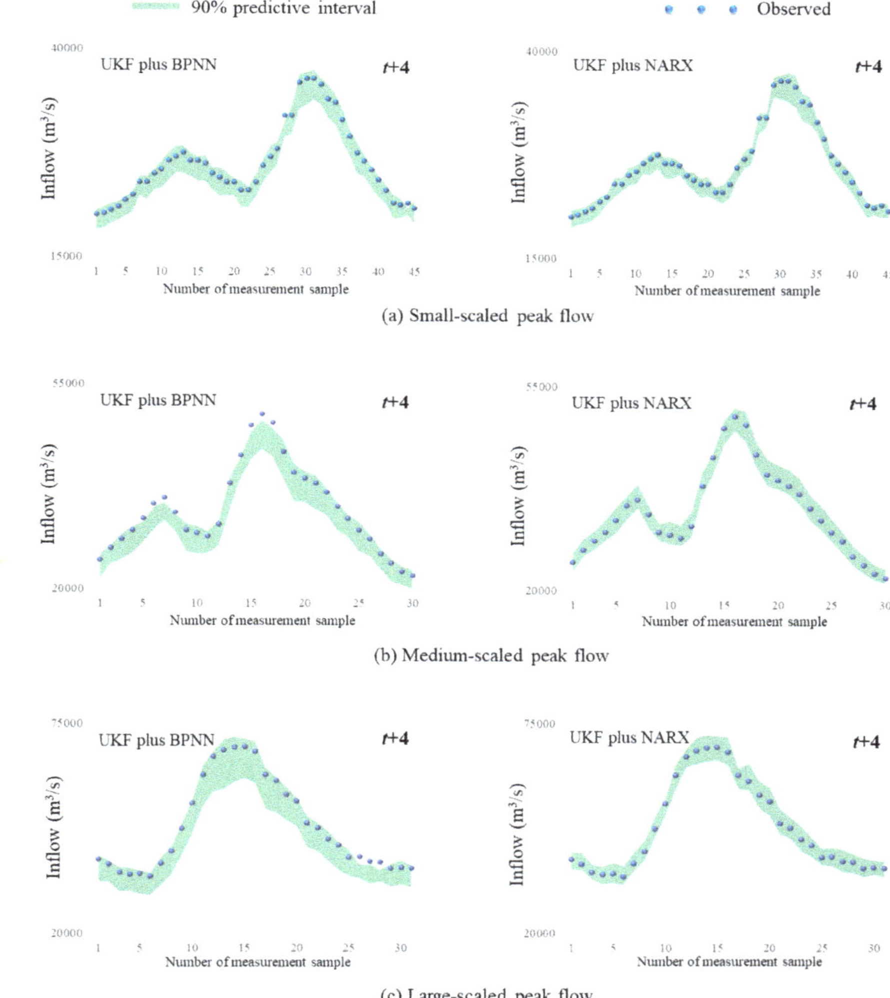

Figure 4. Probabilistic flood forecasts for TGR of the testing stage at lead time t + 4. Three flood events with maximal peak-flow exceeding (**a**) 35,000 m³/s (small-scaled); (**b**) 50,000 m³/s (medium-scaled); and (**c**) 60,000 m³/s (large-scaled). The 90% confidence bound is corresponding to the lower (5%) and upper (95%) limitation of the forecasted value at time t.

For median forecasts at horizons from t + 1 up to t + 4 in three stages, the indicators of MAE and CRPS closely associated with the deterministic forecast model (NARX) and the UKF plus NARX approach are listed in Table 3. It is revealed that the median forecasts of the UKF plus NARX approach would output lower values of MAE and CRPS indicators than those of the deterministic forecast model NARX after the horizon t + 2. That is to say, from the perspective of the median forecast, the probabilistic forecast approach also can mitigate the drawback of under-prediction.

Table 3. Results of median forecasts regarding inflows of Three Gorges Reservoir.

Stage	Model	Indicator	Horizon			
			$t+1$	$t+2$	$t+3$	$t+4$
Training	NARX	MAE (m^3/s)	453	750	849	1037
		CRPS (m^3/s)	781	1025	1189	1306
	UKF plus NARX	MAE (m^3/s)	412	702	811	954
		CRPS (m^3/s)	724	894	1064	1202
Validation	NARX	MAE (m^3/s)	474	782	919	1079
		CRPS (m^3/s)	813	1106	1216	1390
	UKF plus NARX	MAE (m^3/s)	439	736	851	975
		CRPS (m^3/s)	767	910	1097	1243
Testing	NARX	MAE (m^3/s)	465	931	893	1006
		CRPS (m^3/s)	798	1035	1194	1354
	UKF plus NARX	MAE (m^3/s)	428	719	838	960
		CRPS (m^3/s)	749	901	1082	1224

In summary, the UKF plus NARX approach not only can create more accurate and robust probabilistic forecasting results but can also mitigate the problems of systematic under-prediction of extreme flood events. Despite UKF indeed correcting for bias, it also produces a probabilistic forecast which we believe is more important. Furthermore, the reasons that we do not conduct a comparative analysis between LKF (including the autoregressive model, AR(1)) and NKF consist of: first, the AR(1) is a simple formation of LKF. The LKF approach can only identify the linear error estimation whereas the NKF approach can quantify the non-linear error estimation. Second, the rainfall-runoff process is intrinsically complex and non-linear and demands non-linear techniques for quantifying predictive uncertainty.

We want to mention that the probabilistic forecasting approach only spent around 60-s calculation temporal cost to provide the deterministic (within 50 s) and the probabilistic (within 10 s) flood forecasting with respect to the inflows of the TGR. A DELL computer conducted the computation (Intel® CoreTM i5, 7th Generation CPU @ 2.50 GHz, RAM 8 GB and 1 TB Hard Disk).

5. Conclusions

We adopt a probabilistic forecasting methodology that hybridises UKF and ANN to generate posterior distributions from observed and forecasted inflows for effectively reducing the predictive distributions occurring in data-driven flood forecasting to small ranges. The contribution of the UKF approach depends on modeling the non-linear correlation among hydrologic variables and on reducing the uncertainty arosing in flood forecasting. The results demonstrated that the recurrent neural network (NARX model) produced more accurate and stable deterministic flood forecasting on the inflows as longer lead time and effectively mitigated time-lag effects as compared with the static neural network (BPNN model). The reason could be due to a key strategy: the recurrent mechanism drives the integration of the antecedent observations and forecasts of input variables into the next forecasting step for alleviating the accumulation and propagation of multi-step-ahead forecast errors.

Nevertheless, the NARX model also suffered the technical barrier of systematic under-prediction of extreme flood events. Therefore, the UKF was applied to the post-processing of deterministic forecasts obtained from the NARX model. The results demonstrated that the UKF plus NARX approach not only can provide a narrower predictive distribution on the inflow series at longer forecast horizons but also can significantly alleviate under-prediction phenomena. The reason can be due to the key strategy: the effective quantification of the non-linear correlation among variables for lessening hydrologic uncertainty. Finally, it is worth noting that the computational time of the UKF plus NARX approach is extremely short (less than 60 s); therefore, it can be applied with success to real-time flood forecasting.

Author Contributions: Y.Z. carried out the analysis and wrote the article; S.G., C.-Y.X. and F.-J.C. developed the methodology; C.-Y.X., F.-J.C. and J.Y. provided technical assistance and contributed in writing the article. All authors have read and agreed to the published version of the manuscript.

Funding: This work was supported by the National Natural Science Foundation of China (No. 51539009 and No. U1865201), the National Key Research and Development Program of China (2018YFC0407904) and the Research Council of Norway (FRINATEK Project 274310).

Acknowledgments: We thank the Changjiang Water Resources Commission of China for providing the monitoring data, and the data can be downloaded from this website (http://www.cjh.com.cn, Chinese). The authors would like to thank the editors and anonymous reviewers for their constructive comments that greatly contributed to improving the manuscript.

Conflicts of Interest: The authors declare no conflict of interest.

References

1. Gneiting, T.; Balabdaoui, F.; Raftery, A.E. Probabilistic forecasts, calibration and sharpness. *J. R. Stat. Soc. Ser. B (Stat. Methodol.)* **2007**, *69*, 243–268. [CrossRef]
2. Han, S.; Coulibaly, P. Bayesian flood forecasting methods: A review. *J. Hydrol.* **2017**, *551*, 340–351. [CrossRef]
3. Wetterhall, F.; Pappenberger, F.; Alfieri, L.; Cloke, H.L.; Thielen-del Pozo, J.; Balabanova, S.; Danhelka, J.; Vogelbacher, A.; Salamon, P.; Carrasco, I.; et al. HESS Opinions: Forecaster priorities for improving probabilistic flood forecasts. *Hydrol. Earth Syst. Sci.* **2013**, *17*, 4389–4399. [CrossRef]
4. Arnal, L.; Ramos, M.H.; Perez, E.C.D.; Cloke, H.L.; Stephens, E.; Wetterhall, F.; van Andel, S.J.; Pappenberger, F. Willingness-to-pay for a probabilistic flood forecast: A risk-based decision-making game. *Hydrol. Earth Syst. Sci.* **2016**, *20*, 3109–3128. [CrossRef]
5. Kalman, R.E. A new approach to linear filtering and prediction problems. *J. Basic Eng.* **1960**, *82*, 35–45. [CrossRef]
6. Laio, F.; Tamea, S. Verification tools for probabilistic forecasts of continuous hydrological variables. *Hydrol. Earth Syst. Sci.* **2007**, *11*, 1267–1277. [CrossRef]
7. Herr, H.D.; Krzysztofowicz, R. Ensemble Bayesian forecasting system Part I: Theory and algorithms. *J. Hydrol.* **2015**, *524*, 789–802. [CrossRef]
8. Zhou, Y.; Guo, S. Risk analysis for flood control operation of seasonal flood-limited water level incorporating inflow forecasting error. *Hydrol. Sci. J.* **2014**, *59*, 1006–1019. [CrossRef]
9. Zhou, Y.; Guo, S.; Chang, F.J.; Liu, P.; Chen, A.B. Methodology that improves water utilization and hydropower generation without increasing flood risk in mega cascade reservoirs. *Energy* **2018**, *143*, 785–796. [CrossRef]
10. Madadgar, S.; Moradkhani, H. Improved Bayesian multimodeling: Integration of copulas and Bayesian model averaging. *Water Resour. Res.* **2014**, *50*, 9586–9603. [CrossRef]
11. Papacharalampous, G.A.; Tyralis, H.; Koutsoyiannis, D. Comparison of stochastic and machine learning methods for multi-step ahead forecasting of hydrological processes. *Stoch. Environ. Res. Risk Assess.* **2019**, *33*, 481–514. [CrossRef]
12. Engeland, K.; Steinsland, I. Probabilistic postprocessing models for flow forecasts for a system of catchments and several lead times. *Water Resour. Res.* **2014**, *50*, 182–197. [CrossRef]
13. Krapu, C.; Borsuk, M. Probabilistic programming: A review for environmental modellers. *Environ. Model. Softw.* **2019**, *114*, 40–48. [CrossRef]
14. Todini, E. From HUP to MCP: Analogies and extended performances. *J. Hydrol.* **2013**, *477*, 33–42. [CrossRef]
15. Siccardi, F.; Boni, G.; Ferraris, L.; Rudari, R.O.B.E.R.T.O. A hydrometeorological approach for probabilistic flood forecast. *J. Geophys. Res.: Atmos.* **2005**, *110*. [CrossRef]
16. Liu, Z.; Guo, S.; Xiong, L.; Xu, C.Y. Hydrological uncertainty processor based on a copula function. *Hydrol. Sci. J.* **2018**, *63*, 74–86. [CrossRef]
17. Julier, S.J.; UHlmann, J.K. A new extension of the Kalman filter to nonlinear systems. *Aerosense* **1997**, *97*, 182–193.
18. Bosov, A.V.; Miller, G.B. Conditionally minimax nonlinear filter and Unscented Kalman filter: Empirical analysis and comparison. *Autom. Remote. Control.* **2019**, *80*, 1230–1251. [CrossRef]
19. Jiang, P.; Sun, Y.; Bao, W. State estimation of conceptual hydrological models using unscented Kalman filter. *Hydrol. Res.* **2019**, *50*, 479–497. [CrossRef]

20. Kanakaraj, S.; Nair, M.S.; Kalady, S. Adaptive Importance Sampling Unscented Kalman Filter based SAR image super resolution. *Comput. Geosci.* **2019**, *133*, 104310. [CrossRef]

21. Fu, X.; Yu, Z.; Ding, Y.; Qin, Y.; Luo, L.; Zhao, C.; Yang, C. Unscented weighted ensemble Kalman filter for soil moisture assimilation. *J. Hydrol.* **2019**, *580*, 124352. [CrossRef]

22. Wu, X.; Wang, Y. Extended and Unscented Kalman filtering based feedforward neural networks for time series prediction. *Appl. Math. Model.* **2012**, *36*, 1123–1131. [CrossRef]

23. Wang, X.; Babovic, V. Application of hybrid Kalman filter for improving water level forecast. *J. Hydroinf.* **2016**, *18*, 773–790. [CrossRef]

24. Wu, X.-L.; Xiang, X.-H.; Wang, C.-H.; Chen, X.; Xu, C.-Y.; Yu, Z. Coupled hydraulic and Kalman Filter model for real-time correction of flood forecast in the Three Gorges interzone of Yangtze River, China. *J. Hydro. Eng.* **2013**, *18*, 1416–1425. [CrossRef]

25. Zhang, X.; Liang, F.; Yu, B.; Zong, Z. Explicitly integrating parameter, input, and structure uncertainties into Bayesian Neural Networks for probabilistic hydrologic forecasting. *J. Hydrol.* **2011**, *409*, 696–709. [CrossRef]

26. Zhu, F.; Zhong, P.A.; Sun, Y.; Yeh, W.W.G. Real-time optimal flood control decision making and risk propagation under multiple uncertainties. *Water Resour. Res.* **2017**, *53*, 10635–10654. [CrossRef]

27. Abrahart, R.J.; Anctil, F.; Coulibaly, P.; Dawson, C.W.; Mount, N.J.; See, L.M.; Wilby, R.L. Two decades of anarchy? Emerging themes and outstanding challenges for neural network river forecasting. *Prog. Phys. Geogr.* **2012**, *36*, 480–513. [CrossRef]

28. Shen, C.; Laloy, E.; Elshorbagy, A.; Albert, A.; Bales, J.; Chang, F.J.; Ganguly, S.; Hsu, K.L.; Kifer, D.; Fang, Z.; et al. HESS Opinions: Incubating deep-learning-powered hydrologic science advances as a community. *Hydrol. Earth Syst. Sci.* **2018**, *22*, 5639–5656. [CrossRef]

29. Asanjan, A.A.; Yang, T.; Hsu, K.; Sorooshian, S.; Lin, J.; Peng, Q. Short-term precipitation forecast based on the PERSIANN system and the long short-term memory (LSTM) deep learning algorithm. *J. Geophys. Res.: Atmos.* **2018**, *123*, 12543–12563.

30. Bui, D.T.; Panahi, M.; Shahabi, H.; Singh, V.P.; Shirzadi, A.; Chapi, K.; Ahmad, B.B. Novel hybrid evolutionary algorithms for spatial prediction of floods. *Sci. Rep.* **2018**, *8*, 15364. [CrossRef]

31. Cannon, A.J. Quantile regression neural networks: Implementation in R and application to precipitation downscaling. *Comput. Geosci.* **2011**, *37*, 1277–1284. [CrossRef]

32. Chang, F.J.; Chen, P.A.; Lu, Y.R.; Huang, E.; Chang, K.Y. Real-time multi-step-ahead water level forecasting by recurrent neural networks for urban flood control. *J. Hydrol.* **2014**, *517*, 836–846. [CrossRef]

33. Chang, F.J.; Tsai, M.J. A nonlinear spatio-temporal lumping of radar rainfall for modeling multi-step-ahead inflow forecasts by data-driven techniques. *J. Hydrol.* **2016**, *535*, 256–269. [CrossRef]

34. Chen, P.A.; Chang, L.C.; Chang, F.J. Reinforced recurrent neural networks for multi-step-ahead flood forecasts. *J. Hydrol.* **2013**, *497*, 71–79. [CrossRef]

35. Chen, L.; Sun, N.; Zhou, C.; Zhou, J.; Zhou, Y.; Zhang, J.; Zhou, Q. Flood forecasting based on an improved extreme learning machine model combined with the backtracking search optimization algorithm. *Water* **2018**, *10*, 1362. [CrossRef]

36. Hu, C.; Wu, Q.; Li, H.; Jian, S.; Li, N.; Lou, Z. Deep learning with a long short-term memory networks approach for rainfall-runoff simulation. *Water* **2018**, *10*, 1543. [CrossRef]

37. Zhou, Y.; Chang, F.J.; Chang, L.C.; Kao, I.F.; Wang, Y.S. Explore a deep learning multi-output neural network for regional multi-step-ahead air quality forecasts. *J. Clean. Prod.* **2019**, *209*, 134–145. [CrossRef]

38. Zhou, Y.; Guo, S.; Chang, F.J. Explore an evolutionary recurrent ANFIS for modelling multi-step-ahead flood forecasts. *J. Hydrol.* **2019**, *570*, 343–355. [CrossRef]

39. Kasiviswanathan, K.S.; Sudheer, K.P.; He, J. Probabilistic and ensemble simulation approaches for input uncertainty quantification of artificial neural network hydrological models. *Hydrol. Sci. J.* **2018**, *63*, 101–113. [CrossRef]

40. Kumar, S.; Tiwari, M.K.; Chatterjee, C.; Mishra, A. Reservoir inflow forecasting using ensemble models based on neural networks, wavelet analysis and bootstrap method. *Water Resour. Manag.* **2015**, *29*, 4863–4883. [CrossRef]

41. Zhong, Y.; Guo, S.; Ba, H.; Xiong, F.; Chang, F.J.; Lin, K. Evaluation of the BMA probabilistic inflow forecasts using TIGGE numeric precipitation predictions based on artificial neural network. *Hydrol. Res.* **2018**, *49*, 1417–1433. [CrossRef]

42. Shen, H.Y.; Chang, L.C. Online multistep-ahead inundation depth forecasts by recurrent NARX networks. *Hydrol. Earth Syst. Sci.* **2013**, *17*, 935–945. [CrossRef]

43. Wunsch, A.; Liesch, T.; Broda, S. Forecasting groundwater levels using nonlinear autoregressive networks with exogenous input (NARX). *J. Hydrol.* **2018**, *567*, 743–758. [CrossRef]

44. Lourakis, M.I. A brief description of the Levenberg-Marquardt algorithm implemented by Levmar. *Found. Res. Technol.* **2005**, *4*, 1–6.

45. Nanda, T.; Sahoo, B.; Beria, H.; Chatterjee, C. A wavelet-based non-linear autoregressive with exogenous inputs (WNARX) dynamic neural network model for real-time flood forecasting using satellite-based rainfall products. *J. Hydrol.* **2016**, *539*, 57–73. [CrossRef]

46. Lohani, A.K.; Goel, N.K.; Bhatia, K.K.S. Improving real time flood forecasting using fuzzy inference system. *J. Hydrol.* **2014**, *509*, 25–41. [CrossRef]

47. Nash, J.E. River flow forecasting through conceptual models, I: A discussion of principles. *J. Hydrol.* **1970**, *10*, 398–409. [CrossRef]

48. Gneiting, T.; Raftery, A.E. Strictly Proper Scoring Rules, Prediction, and Estimation. *J. Am. Stat. Assoc.* **2007**, *102*, 359–378. [CrossRef]

49. Xiong, L.; O'Connor, K.M. An empirical method to improve the prediction limits of the GLUE methodology in rainfall-runoff modeling. *J. Hydrol.* **2008**, *349*, 115–124. [CrossRef]

50. Bai, Y.; Chen, Z.; Xie, J.; Li, C. Daily reservoir inflow forecasting using multiscale deep feature learning with hybrid models. *J. Hydrol.* **2016**, *532*, 193–206. [CrossRef]

51. Chen, L.; Zhang, Y.; Zhou, J.; Singh, V.P.; Guo, S.; Zhang, J. Real-time error correction method combined with combination flood forecasting technique for improving the accuracy of flood forecasting. *J. Hydrol.* **2015**, *521*, 157–169. [CrossRef]

52. Maidment, D.R. *Handbook of Hydrology*; McGraw-Hill: New York, NY, USA, 1993.

Article

Modelling the Vegetation Response to Climate Changes in the Yarlung Zangbo River Basin Using Random Forest

Kaige Chi [1,2], Bo Pang [1,2,*], Lizhuang Cui [1,2], Dingzhi Peng [1,2], Zhongfan Zhu [1,2], Gang Zhao [1,3] and Shulan Shi [1,2]

[1] College of Water Sciences, Beijing Normal University, Beijing 100875, China; 15568297808@163.com (K.C.); 201921470003@mail.bnu.edu.cn (L.C.); dzpeng@bnu.edu.cn (D.P.); zhuzhongfan1985@bnu.edu.cn (Z.Z.); gang.zhao@bristol.ac.uk (G.Z.); 201821470021@mail.bnu.edu.cn (S.S.)
[2] Beijing Key Laboratory of Urban Hydrological Cycle and Sponge City Technology, Beijing 100875, China
[3] School of Geographical Science, University of Bristol, Bristol BS8 1SS, UK
* Correspondence: pb@bnu.edu.cn; Tel.: +86-135-2168-6445

Received: 31 March 2020; Accepted: 13 May 2020; Published: 18 May 2020

Abstract: Vegetation coverage variation may influence watershed water balance and water resource availability. Yarlung Zangbo River, the longest river on the Tibetan Plateau, has high spatial heterogeneity in vegetation coverage and is the main freshwater resource of local residents and downstream countries. In this study, we proposed a model based on random forest (RF) to predict the Normalized Difference Vegetation Index (NDVI) of the Yarlung Zangbo River Basin and explore its relationship with climatic factors. High-resolution datasets of NDVI and monthly meteorological observation data from 2000 to 2015 were used to calibrate and validate the proposed model. The proposed model was then compared with artificial neural network and support vector machine models, and principal component analysis and partial correlation analysis were also used for predictor selection of artificial neural network and support vector machine models for comparative study. The results show that RF had the highest model efficiency among the compared models. The Nash–Sutcliffe coefficients of the proposed model in the calibration period and verification period were all higher than 0.8 for the five subzones; this indicated that the proposed model can successfully simulate the relationship between the NDVI and climatic factors. By using built-in variable importance evaluation, RF chose appropriate predictor combinations without principle component analysis or partial correlation analysis. Our research is valuable because it can be integrated into water resource management and elucidates ecological processes in Yarlung Zangbo River Basin.

Keywords: NDVI; Yarlung Zangbo River; machine learning model; random forest

1. Introduction

Vegetation is produced as a result of the interactions among factors such as soil, atmosphere, and moisture [1]. Vegetation is affected by climate because of biophysical responses such as plant respiration, photosynthesis, and evapotranspiration [2]. Recent research found that vegetation plays a key role in future terrestrial hydrologic response, and understanding water stress is of the utmost importance for properly predicting future dryness and water resources [3]. Changes in global climate and associated effects on vegetation condition have received an increasing amount of attention [4]. Among such research, the Normalized Difference Vegetation Index (NDVI) is frequently used to monitor changes in vegetation conditions, because of its close relationship with photosynthetically active radiation, which is absorbed by photosynthesizing tissues [5,6]. With the improvement of remote sensors, the NDVI has been widely applied in continental and regional research [7]. Continuous NDVI datasets make it possible to trace vegetation conditions changes and explore the underlying climate factor-associated mechanisms [8,9]. The NDVI has been widely exploited to monitor and quantify drought disturbance in semiarid and arid regions with low values corresponding to stressed vegetation [10,11]. As a known covariate with other environmental variables, the NDVI was also applied to soil-loss-prone area identification [12,13], wetland delineation [14], irrigation and soil salinity management [15]. Therefore, quantifying the relationship between NDVI and climate factors, and predicting the NDVI trends will help effectively guide regional water resource managements [16,17].

Yarlung Zangbo River, the longest river on the Tibetan Plateau, has high spatial heterogeneity in vegetation conditions and is the main freshwater resource of local residents and downstream countries. As one of the most important ecosystems in the Tibetan Plateau, the vegetation conditions of the Yarlung Zangbo River Basin (YZRB) have a significant impact on the water balance and biological population of the Tibetan Plateau and surrounding areas [17]. Because of the influence of the plateau's high altitude, YZRB vegetation is extremely fragile and sensitive to global climate change. In recent years, statistically significant warming and intensive drought were observed in the YZRB [18], where the cultivated land accounts for about 62.89% of the area of the Tibet Autonomous Region [19]. Soil erosion is another water resources problem of YLZB, where the vegetation conditions play an important role [20]. Moreover, the changes in vegetation cover also influence the water availability of the YLZB [21,22]. Therefore, investigating and modelling the vegetation responses to climate changes is of great significance to the water resource management of YLRB and the water governance of the transboundary rivers [23]. Han et al. explore the relationship between the NDVI and the meteorological variables of the YZRB [24]. Liu et al. analyzed the spatiotemporal patterns of vegetation during 1998–2014 using the NDVI [25]. Sun et al. investigate the spatial heterogeneity of changes in vegetation growth and their driving forces using the NDVI of the YLZB [26]. Based on these researches, an NDVI prediction model that incorporates a comprehensive understanding of the climate–vegetation–hydrology relationships could be important for integrated water resource management.

A large amount of studies have been devoted to exploring the response of the NDVI to precipitation and temperature on regional and global scales, which are the most common climate factors [27,28]. Most of the studies adopted linear methods, such as partial correlation coefficient [29], complex correlation coefficient [30] and linear regression [31]. Due to the complexity of ecosystem and the uncertainties of vegetation dynamics, nonlinear modes, especially machine learning models, attached the attention of researchers [32–35]. Moreover, because the climate and topography show high heterogeneity from upstream to downstream regions [36,37], it puts forward higher requirements on the universal abilities of prediction models in the YZRB. Furthermore, because of the diversity of ecosystems and climate characters, the correlation between NDVI and climate are diverse in different regions [38]. Therefore, predictor selection is also a challenge for NDVI prediction models. Recently, random forest (RF) has received substantial attention in water resource research [39,40]. RF is advantageous because it can handle large datasets and undergoes predictor selection using a built-in variable importance evaluation method [41,42]. Therefore, RF should be highly suitable for the NDVI prediction of the YZRB. This is the first time RF has been applied to explore the complex relationship between the NDVI and climatic factors to the best of our knowledge.

The objective of this study was to propose feasible NDVI prediction models for the YZRB on the subzone scale. RF was adopted to simulate the relationships between NDVI and climatic factors. A comparison was then conducted between the RF and Artificial Neural Network (ANN) and Support Vector Machines (SVM) models. For comparative study, principal component analysis (PCA) and partial correlation analysis (PAR) were used for predictor selection of the models. This research will improve our knowledge on the climate–vegetation–hydrology relationships of the YZRB, which is an important high-altitude continental plateau basin.

2. Materials and Methods

2.1. Study Area

The Yarlung Zangbo River is the largest river on the Tibetan Plateau and one of the most important international rivers [43]. It originates from the Jemayangzong Glacier in southern Tibet and has a total length of 2229 km and its drainage area is 2.42×105 km^2 [44]. This river is one of the highest rivers in the world, with an average elevation of above 4600 m, and tilts from the west to the east, with an average slope of 2.6° [21]. The Yarlung Zangbo River has six major tributaries (the Dogxung Zangbo, Nyangqu, Lhasa, Nyang, Yigong Zangbo, and Purlung Zangbo Rivers). The locations of the YZRB and its tributaries are shown in Figure 1.

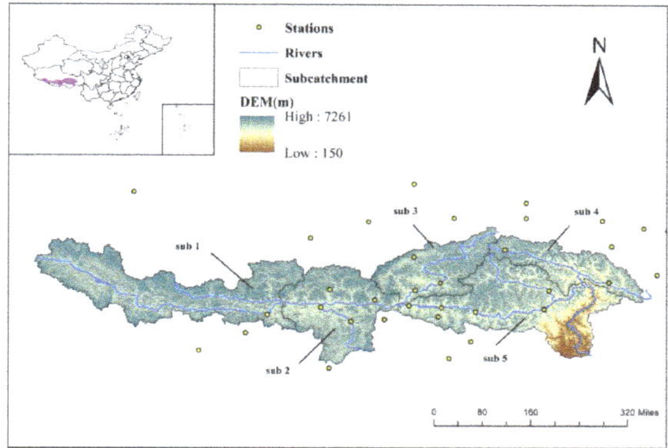

Figure 1. Location of the Yarlung Zangbo River Basin (YZRB) and its five subzones.

Because of the unique topographic characteristics and high altitude of the plateau, the vegetation and ecological environment of the YZRB are relatively fragile and complex [23] and show obvious changes from upstream to downstream [45]. According to the China Vegetation Atlas (Figure 2), the upstream region is located in an arid zone that is dominated by alpine grassland and meadows [46]. With decreasing elevation, the midstream transitions into a continental climate and is mainly covered by alpine grassland and meadows, and the cultivated vegetation slightly increases. The lower reaches of YZRB have a subtropical climate and are mainly covered by coniferous and broadleaf forests.

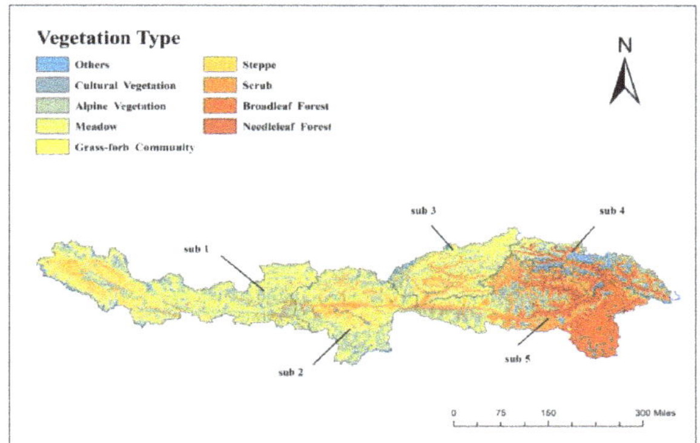

Figure 2. Vegetation conditions map of the YZRB.

Yarlung Zangbo river has more than 130 tributaries, which is larger than 100 km^2, and its major tributaries include Nianchu River, Lhasa River, Nyang River, and Parlung Tsangpo. By considering the hydrological and vegetation similarity, which is shown in Figures 1 and 2, the YLZB is divided into 5 subzones in the research. The area and vegetation conditions of the five subzones are shown in Tables 1 and 2.

Table 1. Name and area of the subzone.

Subzone	Watershed Name	Area (km^2)
1	Upper reaches of the Yarlung Zangbo River	70,048
2	Nianchu River	43,741
3	Lhasa River	31,571
4	Parlung Zangbo	26,574
5	Nyang River Lower reaches of the Yarlung Zangbo River	66,543

Table 2. Proportion of different vegetation types in each subzone.

Subzone	1	2	3	4	5
Cultural Vegetation	0.29	2.49	1.41	0.11	2.43
Alpine Vegetation	32.71	26.26	20.81	26.65	21.8
Broadleaf Forest	0	0	0	0.38	15.73
Needle leaf Forest	0	0	0	19.23	16.13
Meadow	49.23	48.82	53.86	11.48	8.85
Steppe	11.86	12.79	7.69	0	1.6
Scrub	3.44	8.34	13.97	23.69	30.47
Others	2.48	1.29	2.25	18.46	3
total	100	100	100	100	100

2.2. Data Description

A quality-controlled NDVI remote sensing product (MOD13A3) is selected in this study, obtained from the observation of MODIS (Moderate Resolution Image Spectroradiometer) data provided by NASA, spanning 16 years (February 2000 to December 2015). MOD13A3 is the third level product, based on the secondary product, corrected the edge distortion (Bowtie effect) produced by the sensor imaging process. The spatial resolution of the product is 1.1 km × 1.1 km, while the time resolution is monthly. The data were processed into the Geostationary Earth Orbit Tag Image File Format (GEO TIFF) by MODIS Reprojection Tool (MRT) software and processed by ArcGIS projection splicing. The monthly mean air temperature and precipitation data covering 2000–2015 from 30 meteorological stations located in the YZRB were collected from the China Meteorological Data Network. The locations of the meteorological stations are shown in Figure 1.

2.3. Methodology

2.3.1. Random Forest

RF was first proposed by Breiman (2001) [47] as an ensemble learning method that can be used in both classification and regression tasks. The model is considered capable of dealing with small sample sizes and high-dimensional correlation relationships [47]. RF is also advantageous because of its robustness; it does not easily lead to overfitting or provide biased estimates when predictors that do not add information are used [48].

RF includes a group of classification and regression decision trees (CARTs) that are unpruned and generated by bootstrap sampling and random variable selection. The RF algorithm can be divided into the following steps. First, the training dataset is randomly extracted from the original dataset by bootstrap resampling. Second, the CARTs are established for each training set. Compared with the traditional CART method, RF selects random feature combinations to split each node, and each CART grows to the maximum extent without any pruning. Finally, the RF output is obtained by voting in classification mode or averaging in the regression mode of all of the CART predictions.

RF provides a built-in cross-validation process that occurs in parallel with the training procedure for the out-of-bag (OOB) samples, which are not chosen by the bootstrap process. RF can evaluate variable importance by randomly permuting these variables and observing the difference in model performance using OOB samples. At the end of procedure, RF obtains variable importance by averaging these differences, which is then normalized by the standard deviation.

2.3.2. Model Implementation and Validation

The RF is utilized to simulate the nonlinear relationship between the NDVI and climate factors in the 5 subzones. The monthly area-averaged NDVI datasets for each subzone were obtained by calculating the average values of each pixel. The monthly area-averaged precipitation and temperature of 5 subzones were obtained from actual data. For model development, the monthly average NDVI datasets and monthly area precipitation and temperature datasets of each subzone are divided into two datasets. The datasets from 2000 to 2009 were used for model calibration, and those from 2010 to 2015 were used for model validation.

Two machine learning models, ANN and SVM, which were previously used for NDVI prediction [49,50], were selected to compare with RF performance. A three-layer back propagation (BP) ANN model was used in this research. The BP method has been the most widely used algorithm to design multiple layer neural networks, and has also been successfully used for NDVI prediction [16]. SVM was the first classification machine learning algorithm and was proposed by Vapnik, and then gradually derived to the regression algorithm [51]. SVM has been widely used for hydrological prediction, and most recently for NDVI prediction. In this research, SVM with linear kernel function was used, which has been widely used in former studies.

One of the most important steps in the development of machine learning prediction models is the choice of appropriate predictors. Due to the spatial heterogeneity of the vegetation in the YLZB, the relationships between the NDVI and climate factors are different in the 5 subzones. Therefore, it is important to choose appreciate climate factors as predictors in NDVI prediction. In previous studies, PCA and PAR were used for predictor selections of ANN and SVM models. For comparative study, here, PAR and PCA were both used for predictor selection of the ANN and SVM models [52]. For PCA, the climate factors are standardized by subtracting the mean from the original values and then dividing the results by the standard deviation of the original variables. The PCA method is then applied to the standardized climate factors to extract principal components (PCs) that are orthogonal. The obtained PCs preserve more than 90% of the variances that are selected as predictors. Then, the PCs are used in the ANN and SVM modeling, and these results are marked as, ANN-PCA and SVM-PCA. For PAR, climate factors with a partial correlation coefficient greater than 0.3 were selected as predictors, and the corresponding results are marked as ANN-PAR and SVM-PAR.

2.3.3. Model Evaluation Index

The mean absolute percentage error (MAE), Nash–Sutcliffe coefficient (NASH), root mean square errors (RMSE), and correlation coefficient (R) statistical indicators were used to assess the predictive performance of the ANN, SVM, and RF models. MAE, NASH, RMSE, and R were defined as:

$$NASH = 1 - \frac{\sum\limits_{i=1}^{n} \left(Y_{i,obs} - Y_{i,sim}\right)^2}{\sum\limits_{i=1}^{n} \left(Y_{i,obs} - \overline{Y_{obs}}\right)^2},$$

$$RMSE = \sqrt{\frac{\sum\limits_{i=1}^{n}\left(Y_{i,obs} - Y_{i,sim}\right)^2}{n}},$$

$$MAE = \frac{\sum\limits_{i=1}^{n}\left|Y_{i,obs} - Y_{i,sim}\right|}{n},$$

$$r = \frac{\sum\limits_{i=1}^{n}\left[\left(Y_{i,obs} - \overline{Y_{obs}}\right)\left(Y_{i,sim} - \overline{Y_{sim}}\right)\right]}{\sqrt{\left[\sum\limits_{i=1}^{n}\left(Y_{i,obs} - \overline{Y_{obs}}\right)^2\right]}\sqrt{\left[\sum\limits_{i=1}^{n}\left(Y_{i,sim} - \overline{Y_{sim}}\right)^2\right]}}$$

where $Y_{i,obs}$ is the measured NDVI value of the station, $\overline{Y_{obs}}$ is the mean of the observed NDVI value, $Y_{i,sim}$ is the vector of the simulated NDVI value, and $\overline{Y_{sim}}$ is the mean of the simulated NDVI value. In general, a higher NASH value indicates better model efficiency; in contrast, smaller RMSE, MAE, and R values indicate higher accuracy.

3. Results and Discussion

3.1. Spatial and Temporal Characteristics of the NDVI in the YZRB

The inter-annual variations of the NDVI, precipitation, and temperature on the subzone scale from 2000 to 2015 are shown in Table 3. The NDVI and temperature values showed a statistically insignificant increase, whereas the average precipitation of the Yarlung Zangbo River Basin significantly decreased from 528 mm in 2000 to 396 mm in 2015, with a total increase of 0.8 °C over the 16 years. This finding is consistent with the results of previous studies [26].

Table 3. Rainfall, temperature and the Normalized Difference Vegetation Index (NDVI) perennial change rate.

Subzone	Precipitation (mm)	Temperature (°C)	NDVI
Sub1	−3.9	0.02	0.1×10^{-3}
Sub2	−3.7	0.04	0.1×10^{-3}
Sub3	−9.86	0.07	0.4×10^{-3}
Sub4	−13.86	0.04	0.7×10^{-3}
Sub5	−12.8	0.01	0.2×10^{-3}
Total	−8.25	0.03	0.2×10^{-3}

In the five subzones, NDVI gradually increased from upstream to downstream. The average annual growth of NDVI in the five subzones was 0.1×10^{-3}, 0.1×10^{-3}, 0.4×10^{-3}, 0.7×10^{-3}, and 0.2×10^{-3}. The precipitation and temperature show similar trends. The average annual growth of precipitation was −3.9, −3.7, −9.86, −13.86, and −12.8; the average annual growth of temperature was 0.02, 0.04, 0.07, 0.04, and 0.01.

3.2. Predictors Selection

In order to determine the optimal predictors for NDVI prediction models, PCA and PAR were used to analyze the relationships between NDVI and precipitation/temperature at different lead times. The results are shown in Tables 4 and 5, where Pn represents the average precipitation with a lead time n month, and Tn represents the average precipitation with a lead time n month. With reference to similar studies and the meteorological cycles [32–35], the maximum lead times were set to 6 months.

As shown in Tables 4 and 5, the correlations between the NDVI and precipitation/temperature gradually decayed with the increase of lead time. The PCA results show that the precipitation and temperature whose lead time was shorter than 2 months had major impacts on the NDVI in these subzones. However, the PAR results varied in these subzones. In Sub1 and Sub5, the precipitation in the present month and temperature whose lead time was shorter than 2 months had major impacts on the NDVI. In Sub2, the precipitation whose lead time was shorter than 1 month and temperature whose lead time was shorter than 2 months had major impacts on the NDVI. In Sub3, the precipitation whose lead time shorter than 1 months and temperature whose lead time shorter than 3 months had major impacts on NDVI. In Sub4, the precipitation whose lead time was shorter than 2 months and temperature whose lead time was shorter than 3 months had major impacts on the NDVI. In general, the relationships between the NDVI and temperature were slightly closer than those between NDVI and precipitation in the five subzones.

RF evaluates the relative contribution of each predictor using a built-in variable importance evaluation process. The importance of the precipitation/temperature at different lead times in these subzones are calculated and indicated in Figure 3. As illustrated in Figure 3, although the importance of precipitation and temperature gradually decreased, the increase in lead time and the decreases were not as significant as in the PCA and PAR results. This finding may indicate that RF can use all predictors without overfitting. Thus, the precipitation and temperature whose lead time was shorter than 6 months were used for RF modeling of the five subzones.

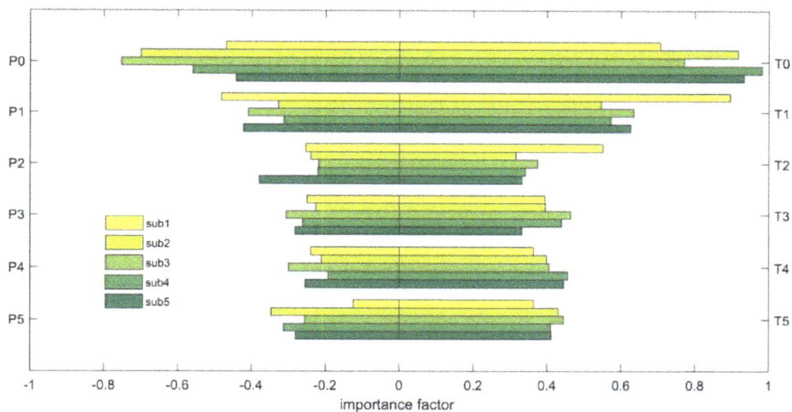

Figure 3. Important factors (rainfall and temperature) in different months.

Table 4. Contribution and cumulative contribution rates for selected principal components.

	Subzone PCA	T_0	P_0	T_1	P_1	T_2	P_2	T_3	P_3
sub1	Contribution rate	0.46	0.26	0.14	0.06	0.03	0.03	0.01	0.01
	Cumulative contribution rate	0.46	0.72	0.86	0.92	0.95	0.98	0.99	1.00
sub2	Contribution rate	0.56	0.27	0.05	0.05	0.03	0.02	0.01	0.01
	Cumulative contribution rate	0.56	0.83	0.88	0.93	0.96	0.98	0.99	1.00
sub3	Contribution rate	0.59	0.19	0.06	0.05	0.04	0.03	0.03	0.01
	Cumulative contribution rate	0.59	0.78	0.84	0.89	0.93	0.96	0.99	1.00
sub4	Contribution rate	0.58	0.24	0.05	0.05	0.03	0.03	0.01	0.01
	Cumulative contribution rate	0.58	0.82	0.87	0.92	0.95	0.98	0.99	1.00
sub5	Contribution rate	0.57	0.20	0.08	0.06	0.04	0.03	0.01	0.01
	Cumulative contribution rate	0.57	0.77	0.85	0.91	0.95	0.98	0.99	1.00

T_0: Temperature of the month, P_0: rainfall of the month, T_1: temperature of the previous month, P_1: rainfall of the previous month, T_2: temperature of the first 2 months, P_2: rainfall of the last 2 months, T_3: temperature of the first 3 months, P_3: rainfall in the first 3 months.

Table 5. Partial correlation calculation results.

PAR	Sub1	Sub2	Sub3	Sub4	Sub5
T_0	0.61	0.80	0.60	0.75	0.60
P_0	0.56	0.78	0.58	0.66	0.57
T_1	0.41	0.53	0.44	0.57	0.39
P_1	−0.06	0.50	0.42	0.50	−0.06
T_2	0.38	0.45	0.36	0.45	0.32
P_2	−0.16	0.11	0.28	0.35	−0.11
T_3	0.27	0.20	0.33	0.43	0.28
P_3	−0.25	−0.20	−0.15	0.22	−0.25

T_0: Temperature of the month, P_0: rainfall of the month, T_1: temperature of the previous month, P_1: rainfall of the previous month, T_2: temperature of the first 2 months, P_2: rainfall of the last 2 months, T_3: temperature of the first 3 months, P_3: rainfall in the first 3 months.

3.3. Comparative Study

The calibration and validation results of the RF and comparative models are summarized in Table 6.

Table 6. Machine learning calculation results.

Subzone	Model	Calibration				Validation			
		NASH	RMSE	MAEP	R	NASH	RMSE	MAEP	R
Sub1	ANN-PCA	0.68	0.03	0.02	0.84	0.67	0.03	0.03	0.86
	ANN-PAR	0.65	0.03	0.02	0.84	0.63	0.03	0.03	0.82
	SVM-PCA	0.90	0.02	0.01	0.95	0.87	0.02	0.01	0.95
	SVM-PAR	0.90	0.02	0.01	0.94	0.85	0.03	0.02	0.95
	RF	0.96	0.02	0.01	0.98	0.91	0.02	0.01	0.98
Sub2	ANN-PCA	0.74	0.03	0.03	0.91	0.73	0.04	0.03	0.92
	ANN-PAR	0.69	0.03	0.03	0.83	0.71	0.04	0.03	0.84
	SVM-PCA	0.90	0.02	0.02	0.95	0.91	0.02	0.01	0.95
	SVM-PAR	0.89	0.02	0.02	0.94	0.90	0.02	0.02	0.94
	RF	0.97	0.01	0.01	0.98	0.95	0.01	0.01	0.98
Sub3	ANN-PCA	0.78	0.05	0.05	0.94	0.77	0.05	0.04	0.94
	ANN-PAR	0.79	0.05	0.04	0.91	0.79	0.05	0.04	0.91
	SVM-PCA	0.91	0.04	0.03	0.95	0.89	0.04	0.03	0.95
	SVM-PAR	0.89	0.04	0.03	0.94	0.87	0.04	0.03	0.94
	RF	0.96	0.02	0.02	0.98	0.96	0.02	0.02	0.98
Sub4	ANN-PCA	0.75	0.05	0.04	0.90	0.75	0.05	0.04	0.89
	ANN-PAR	0.71	0.05	0.04	0.84	0.67	0.06	0.05	0.82
	SVM-PCA	0.85	0.04	0.03	0.92	0.82	0.04	0.03	0.92
	SVM-PAR	0.79	0.05	0.04	0.88	0.77	0.05	0.04	0.88
	RF	0.94	0.03	0.02	0.97	0.89	0.03	0.03	0.97
Sub5	ANN-PCA	0.78	0.04	0.03	0.89	0.72	0.05	0.04	0.87
	ANN-PAR	0.72	0.05	0.04	0.86	0.67	0.05	0.04	0.83
	SVM-PCA	0.84	0.04	0.03	0.92	0.73	0.05	0.03	0.92
	SVM-PAR	0.78	0.04	0.04	0.89	0.68	0.05	0.03	0.89
	RF	0.92	0.03	0.02	0.96	0.83	0.04	0.03	0.96

The results show that RF was superior to the comparative models in the calibration and validation periods. The NASH RF values for the five subzones were 0.96, 0.97, 0.96, 0.94, and 0.92 in the calibration period, and 0.91, 0.95, 0.96, 0.89, and 0.83 in the validation period. All of the measured criteria were superior to those of the compared models (ANN and SVM).

The results of the two-parameter selection were also compared between the ANN and SVM models. PCA was superior to PAR for both the ANN and SVM models. For the ANN models, the average RMSE and MAE were similar in both the calibration and validation periods. However, the average NASH and R of the results using PAR were superior to those of the PCA by 0.03 and 0.04 in the calibration period, and 0.03 and 0.05 in the validation period, respectively. For the SVM models, the average NASH and R increased by 0.03 and 0.02 in the calibration period, and 0.03 and 0.02 in the validation period, respectively. The average RMSE and MAE decreased by 0.002 and 0.004 in the calibration period, and 0.004 and 0.006 in validation period, respectively. Therefore, PCA was advantageous over PAR, with increases of NASH and R, and decreases of RMSE and MAE.

4. Conclusions

As a key component of ecohydrological processes, vegetation conditions influence the efficiency of plant water use and potentially affect water resources. Therefore, investing the changes of vegetation conditions and exploring the vegetation responses to climate changes will provide essential information for regional water resource management [53,54]. Combining with climate models, NDVI prediction models can assess the effects of future drought events [10]. As a covariate with other environmental variables, NDVI prediction models will also provide essential information for irrigation management [15] and soil-loss-prone area identification [12,13], etc. By exploring the vegetation condition changes of the YZRB and their relationship with climatic factors, we proposed an NDVI prediction model based on RF with area-averaged precipitation and temperature as predictors. The monthly rainfall and temperature observations from 30 meteorological stations in the YZRB and the MODIS NDVI datasets from 2000 to 2015 were selected to calibrate and validate the proposed model. The RF results were also compared with those of ANN and SVM models. The primary conclusions are as follows:

1. RF successfully simulated the relationship between NDVI and climatic factors. The NASH coefficients of the proposed model during the calibration period in the five subzones were all higher than 0.9, and those during the verification period were all higher than 0.8. Among the five tested models, RF showed the highest model efficiency in both the calibration and validation periods among all compared models.
2. RF showed advantages for predictor selection. The built-in variable importance evaluation allowed RF to select predictors without additional selection methods, such as PAR and PCA. Moreover, the numbers of predictors were greatest for RF among the compared models. RF showed robustness for modeling, because it could take full advantage of all predictor and avoid overfitting.
3. PCA and PAR were used to analyze the factors that affect the NDVI in YZRB subzones. The results show that the rainfall and temperature of the first 3 months had significant impacts on NDVI, and temperature had a greater influence than rainfall in most of the subzones.

Because of sparse meteorological networks, this research was conducted on a subzone scale. In the future, we will try to explore the relationships between NDVI and climatic factors at a higher resolution with gridded meteorological observations, which will be more applicable for integrated water resource management. The adoption of more vegetation indices, such as leaf area index (LAI), is another important direction.

Author Contributions: K.C. used RF, SVM, ANN model for simulation, calibration, and validation; B.P., L.C. and D.P. collected and processed data; K.C., B.P. and wrote the paper; Z.Z., G.Z. and S.S. supervised the research. All authors have read and agreed to the published version of the manuscript.

Funding: This research was funded by three research programs: (1) National Natural Science Foundation of China (91647202). (2) National Natural Science Foundation of China (51879008); (3) China Scholarship Council (No. 201906045024).

Conflicts of Interest: The authors declare no conflict of interest.

References

1. Che, M.; Chen, B.; Innes, J.L.; Wang, G.; Dou, X.; Zhou, T.; Zhang, H.; Yan, J.; Xu, G.; Zhao, H. Spatial and temporal variations in the end date of the vegetation growing season throughout the Qinghai–Tibetan Plateau from 1982 to 2011. *Agric. For. Meteorol.* **2014**, *189*, 81–90. [CrossRef]
2. Nouri, H.; Anderson, S.; Sutton, P.; Beecham, S.; Nagler, P.; Jarchow, C.J.; Roberts, D.A. NDVI, scale invariance and the modifiable areal unit problem: An assessment of vegetation in the Adelaide Parklands. *Sci. Total. Environ.* **2017**, *584*, 11–18. [CrossRef] [PubMed]
3. Lemordant, L.; Gentine, P.; Swann, A.L.S.; Cook, B.I.; Scheff, J. Critical impact of vegetation physiology on the continental hydrologic cycle in response to increasing CO2. *Proc. Natl. Acad. Sci. USA* **2018**, *115*, 4093–4098. [CrossRef] [PubMed]
4. Yang, Y.; Piao, S. Variations in grassland vegetation cover in relation to climatic factors on the Tibetan Plateau. *J. Plant Ecol.* **2006**, *30*, 1–8.
5. Zhong, L.; Ma, Y.; Salama, M.S.; Su, Z. Assessment of vegetation dynamics and their response to variations in precipitation and temperature in the Tibetan Plateau. *J. Clim. Chang.* **2010**, *103*, 519–535. [CrossRef]
6. Jiang, D.; Fu, X.; Wang, K. Vegetation dynamics and their response to freshwater inflow and climate variables in the Yellow River Delta, China. *J. Quatern Int.* **2013**, *304*, 75–84. [CrossRef]
7. Barbosa, H.A.; Kumar, T.L.; Silva, L.R.M. Recent trends in vegetation dynamics in the South America and their relationship to rainfall. *J. Nat. Hazards* **2015**, *77*, 883–899. [CrossRef]
8. Bao, G.; Qin, Z.; Bao, Y.; Zhou, Y.; Li, W.; Sanjjav, A. NDVI-Based Long-Term Vegetation Dynamics and Its Response to Climatic Change in the Mongolian Plateau. *Remote Sens.* **2014**, *6*, 8337–8358. [CrossRef]
9. Aguilar, C.; Zinnert, J.C.; Polo, M.J.; Young, D.R. NDVI as an indicator for changes in water availability to woody vegetation. *Ecol. Indic.* **2012**, *23*, 290–300. [CrossRef]
10. Dutta, D.; Kundu, A.; Patel, N. Predicting agricultural drought in eastern Rajasthan of India using NDVI and standardized precipitation index. *Geocarto Int.* **2013**, *28*, 192–209. [CrossRef]
11. Omute, P.; Corner, R.; Awange, J.; Corner, R. The use of NDVI and its Derivatives for Monitoring Lake Victoria's Water Level and Drought Conditions. *Water Resour. Manag.* **2012**, *26*, 1591–1613. [CrossRef]
12. Carvalho, D.F.D.; Durigon, V.L.; Antunes, M.A.H.; Almeida, W.S.D.; Oliveira, P.T.S.D. Predicting soil erosion using Rusle and NDVI time series from TM Landsat 5. *Pesq. Agropec. Bras.* **2014**, *49*, 215–224. [CrossRef]
13. Singh, D.; Herlin, I.; Berroir, J.; Silva, E.; Meirelles, M.S. An approach to correlate NDVI with soil colour for erosion process using NOAA/AVHRR data. *Adv. Space Res.* **2004**, *33*, 328–332. [CrossRef]
14. White, D.C.; Lewis, M.M.; Green, G.; Gotch, T.B. A generalizable NDVI-based wetland delineation indicator for remote monitoring of groundwater flows in the Australian Great Artesian Basin. *Ecol. Indic.* **2016**, *60*, 1309–1320. [CrossRef]
15. Fu, B.; Burgher, I. Riparian vegetation NDVI dynamics and its relationship with climate, surface water and groundwater. *J. Arid. Environ.* **2015**, *113*, 59–68. [CrossRef]
16. Huang, S.; Ming, B.; Leng, G.; Hou, B. A Case Study on a Combination NDVI Forecasting Model Based on the Entropy Weight Method. *Water Resour. Manag.* **2017**, *31*, 3667–3681. [CrossRef]
17. Piao, S.; Wang, T.; Ciais, P.; Zhu, B.; Liu, J. Changes in satellite?derived vegetation growth trend in temperate and boreal Eurasia from 1982 to 2006. *Glob. Chang. Boil.* **2011**, *17*, 3228–3239. [CrossRef]
18. Aldakheel, Y.Y. Assessing NDVI Spatial Pattern as Related to Irrigation and Soil Salinity Management in Al-Hassa Oasis, Saudi Arabia. *J. Indian Soc. Remote Sens.* **2011**, *39*, 171–180. [CrossRef]
19. Li, H.; Liu, L.; Shan, B.; Xu, Z.; Niu, Q.; Cheng, L.; Liu, X.; Xu, Z. Spatiotemporal Variation of Drought and Associated Multi-Scale Response to Climate Change over the Yarlung Zangbo River Basin of Qinghai–Tibet Plateau, China. *Remote Sens.* **2019**, *11*, 1596. [CrossRef]
20. Wang, X.; Zhong, X.; Fan, J. Assessment and spatial distribution of sensitivity of soil erosion in Tibet. *J. Geogr. Sci.* **2004**, *14*, 41–46. [CrossRef]
21. Li, F.; Zhang, Y.; Xu, Z.; Teng, J.; Liu, C.; Liu, W.; Mpelasoka, F. The impact of climate change on runoff in the southeastern Tibetan Plateau. *J. Hydrol.* **2013**, *505*, 188–201. [CrossRef]
22. Liu, Z.; Yao, Z.; Huang, H.; Wu, S.; Liu, G. Land use and climate changes and their impacts on runoff in The Yarlung Zangbo river basin. *J. Land Degrad. Dev.* **2014**, *25*, 203–215. [CrossRef]

23. Li, H.; Li, Y.; Shen, W.; Li, Y.; Lin, J.; Lu, X.; Xu, X.; DeAngelis, D. Elevation-Dependent Vegetation Greening of the Yarlung Zangbo River Basin in the Southern Tibetan Plateau, 1999–2013. *Remote Sens.* **2015**, *7*, 16672–16687. [CrossRef]

24. Han, X.; Zuo, D.; Xu, Z.; Cai, S.; Gao, X. Analysis of vegetation condition and its relationship with meteorological variables in the Yarlung Zangbo River Basin of China. *Proc. Int. Assoc. Hydrol. Sci.* **2018**, *379*, 105–112. [CrossRef]

25. Liu, X.; Xu, Z.; Peng, D. Spatio-Temporal Patterns of Vegetation in the Yarlung Zangbo River, China during 1998–2014. *J. China Rural Water Hydropower* **2019**, *11*, 1–11. (In Chinese) [CrossRef]

26. Sun, W.; Wang, Y.; Fu, Y.H.; Xue, B.; Wang, G.; Yu, J.; Zuo, D.; Xu, Z. Spatial heterogeneity of changes in vegetation growth and their driving forces based on satellite observations of the Yarlung Zangbo River Basin in the Tibetan Plateau. *J. Hydrol.* **2019**, *574*, 324–332. [CrossRef]

27. Di, L.; Rundquist, D.C.; Han, L. Modelling relationships between NDVI and precipitation during vegetative growth cycles. *Int. J. Remote Sens.* **1994**, *15*, 2121–2136. [CrossRef]

28. Braswell, B.; Schimel, D.S.; Linder, E.; Moore, B. The Response of Global Terrestrial Ecosystems to Interannual Temperature Variability. *Science* **1997**, *278*, 870–873. [CrossRef]

29. Tian, F.; Fensholt, R.; Verbesselt, J.; Grogan, K.; Horion, S.; Wang, Y. Evaluating temporal consistency of long-term global NDVI datasets for trend analysis. *Remote Sens. Environ.* **2015**, *163*, 326–340. [CrossRef]

30. Wang, J.; Rich, P.M.; Price, K.P. Temporal responses of NDVI to precipitation and temperature in the central Great Plains, USA. *Int. J. Remote Sens.* **2003**, *24*, 2345–2364. [CrossRef]

31. Iwasaki, H. NDVI prediction over mongolian grassland using Gsmap Precipitation Data and Jra-25/jcdas Temperature Data. *J. Arid Environ.* **2009**, *73*, 557–562. [CrossRef]

32. Meng, B.; Gao, J.; Liang, T.; Cui, X.; Ge, J.; Yin, J.; Feng, Q.; Xie, H. Modeling of Alpine Grassland Cover Based on Unmanned Aerial Vehicle Technology and Multi-Factor Methods: A Case Study in the East of Tibetan Plateau, China. *Remote Sens.* **2018**, *10*, 320–339. [CrossRef]

33. Kang, L.; Di, L.; Deng, M.; Yu, E.; Xu, Y. Forecasting vegetation index based on vegetation-meteorological factor interactions with artificial neural network. In Proceedings of the 2016 Fifth International Conference on Agro-Geoinformatics (Agro-Geoinformatics), Tianjin, China, 18–20 July 2016; pp. 1–6.

34. Nay, J.; Burchfield, E.; Gilligan, J.M. A machine-learning approach to forecasting remotely sensed vegetation health. *Int. J. Remote Sens.* **2017**, *39*, 1800–1816. [CrossRef]

35. Asoka, A.; Mishra, V.; Akarsh, A. Prediction of vegetation anomalies to improve food security and water management in India. *Geophys. Res. Lett.* **2015**, *42*, 5290–5298. [CrossRef]

36. Thukaram, D.; Khincha, H.; Vijaynarasimha, H. Artificial Neural Network and Support Vector Machine Approach for Locating Faults in Radial Distribution Systems. *IEEE Trans. Power Deliv.* **2005**, *20*, 710–721. [CrossRef]

37. Feng, X.W.; Hang, L.M.; Shen, B. Comparative study on multivariate linear regression and BP neural network model in the prediction of flood volume. *J. Water Resour. Water Eng.* **2017**, *28*, 123–126. (In Chinese)

38. Kaufmann, R.K.; Zhou, L.; Myneni, R.; Tucker, C.J.; Slayback, D.; Shabanov, N.V.; Pinzon, J. The effect of vegetation on surface temperature: A statistical analysis of NDVI and climate data. *Geophys. Res. Lett.* **2003**, *30*, 2147–2150. [CrossRef]

39. Ge, J.; Meng, B.; Liang, T.; Feng, Q.; Gao, J.; Yang, S.; Huang, X.; Xie, H. Modeling alpine grassland cover based on MODIS data and support vector machine regression in the headwater region of the Huanghe River, China. *Remote Sens. Environ.* **2018**, *218*, 162–173. [CrossRef]

40. Hsu, C.; Chang, C.; Lin, C. A practical guide to support vector classification. *J. Bju Int.* **2008**, *101*, 1396–1400.

41. Binren, X.; Yuanyuan, W. Spatial statistics of TRMM precipitation in the Tibetan Plateau using random forest algorithm. *J. Remote Sens. Land Resour.* **2018**, *30*, 181–188.

42. Tongtiegang, Z.; Dawen, Y.; Ximing, C.; Yong, C. Predict seasonal low flows in the upper Yangtze River using random forests model. *J. Hydroelectr. Eng.* **2015**, *31*, 19–27. (In Chinese)

43. Lehnert, L.; Meyer, H.; Wang, Y.; Miehe, G.; Thies, B.; Reudenbach, C.; Bendix, J. Retrieval of grassland plant coverage on the Tibetan Plateau based on a multi-scale, multi-sensor and multi-method approach. *Remote Sens. Environ.* **2015**, *164*, 197–207. [CrossRef]

44. Ma, Y.; Yang, Y.; Han, Z.; Tang, G.; Maguire, L.; Chu, Z.; Hong, Y. Comprehensive evaluation of Ensemble Multi-Satellite Precipitation Dataset using the Dynamic Bayesian Model Averaging scheme over the Tibetan plateau. *J. Hydrol.* **2018**, *556*, 634–644. [CrossRef]

45. Chen, B.; Li, H.; Cao, X.; Shen, W.; Jin, H. Vegetation Pattern and Spatial Distribution of NDVI in the Yarlung Zangbo River Basin of China. *J. Desert Res.* **2015**, *35*, 120–128. (In Chinese)

46. Shen, M.; Piao, S.; Cong, N.; Zhang, G.; Jassens, I.A. Precipitation impacts on vegetation spring phenology on the Tibetan Plateau. *Glob. Chang. Boil.* **2015**, *21*, 3647–3656. [CrossRef] [PubMed]

47. Breiman, L. Random forests. *Mach. Learn.* **2001**, *45*, 5–32. [CrossRef]

48. Biau, G.; Scornet, E. A random forest guided tour. *TEST* **2016**, *25*, 197–227. [CrossRef]

49. Baez-Villanueva, O.M.; Zambrano-Bigiarini, M.; Beck, H.E.; McNamara, I.; Ribbe, L.; Nauditt, A.; Birkel, C.; Verbist, K.; Giraldo-Osorio, J.D.; Thinh, N.X. RF-MEP: A novel Random Forest method for merging gridded precipitation products and ground-based measurements. *Remote Sens. Environ.* **2020**, *239*, 111606. [CrossRef]

50. Were, K.; Bui, D.T.; Øystein, B.D.; Singh, B.R. A comparative assessment of support vector regression, artificial neural networks, and random forests for predicting and mapping soil organic carbon stocks across an Afromontane landscape. *Ecol. Indic.* **2015**, *52*, 394–403. [CrossRef]

51. Pan, X.C. Application of SCA-SVM to annual runoff wet-dry identification. *J. China Three Gorges Univ.* **2016**, *38*, 6–11. (In Chinese)

52. Pang, B.; Yue, J.; Zhao, G.; Xu, Z. Statistical Downscaling of Temperature with the Random Forest Model. *Adv. Meteorol.* **2017**, *2017*, 1–11. [CrossRef]

53. Peng, D.; Du, Y. Comparative analysis of several Lhasa River basin flood forecast models in Yarlung Zangbo River. In Proceedings of the 2010 4th International Conference on Bioinformatics and Biomedical Engineering, Chengdu, China, 18–20 June 2010; pp. 1–4.

54. Zhang, J.; Ren, Z. Responses of vegetation changes in growing season to precipitationin Yarlung Zangbo River Basin. *J. Soil Water Conserv.* **2015**, *2*, 209–212.

 *water*

Article

Uncertainty Analysis of Spatiotemporal Models with Point Estimate Methods (PEMs)—The Case of the ANUGA Hydrodynamic Model

Maikel Issermann and Fi-John Chang *,†

Department of Bioenvironmental Systems Engineering, National Taiwan University, Taipei 10617, Taiwan; d04622003@ntu.edu.tw
* Correspondence: changfj@ntu.edu.tw
† Current address: National Taiwan University, Department of Bioenvironmental Systems Engineering, No. 1, Section 4, Roosevelt Rd, Da'an District, Taipei City 10617, Taiwan.

Received: 28 November 2019; Accepted: 9 January 2020; Published: 14 January 2020

Abstract: Practitioners often neglect the uncertainty inherent to models and their inputs. Point Estimate Methods (PEMs) offer an alternative to the common, but computationally demanding, method for assessing model uncertainty, Monte Carlo (MC) simulation. PEMs rerun the model with representative values of the probability distribution of the uncertain variable. The results can estimate the statistical moments of the output distribution. Hong's method is the specific PEM implemented here for a case study that simulates water runoff using the ANUGA model for an area in Glasgow, UK. Elevation is the source of uncertainty. Three realizations of the Sequential Gaussian Simulation, which produces the random error fields that can be used as inputs for any spatial model, are scaled according to representative values of the distribution and their weights. The output from a MC simulation is used for validation. A comparison of the first two statistical moments indicates that Hong's method tends to underestimate the first moment and overestimate the second moment. Model efficiency performance measures validate the usefulness of Hong's method for the approximation of the first two moments, despite the method suffering from outliers. Estimation was less accurate for higher moments but the moment estimates were sufficient to use the Grams-Charlier Expansion to fit a distribution to them. Regarding probabilistic flood-inundation maps, Hong's method shows very similar probabilities in the same areas as the MC simulation. However, the former requires just three 11-minute simulation runs, rather than the 500 required for the MC simulation. Hong's method therefore appears attractive for approximating the uncertainty of spatiotemporal models.

Keywords: flood-risk map; hydrodynamic modelling; Sequential Gaussian Simulation; urban stormwater

---

1. Introduction

Flood inundation, in its many forms, is one the most devastating types of natural disaster for civilization [1]. Floods are involved in the majority of fatalities associated with natural disasters [2] and cause severe economic damage by disrupting and destroying processes within the society and economy. The socio-economic impact of flood inundations has given rise to studies on the subject receiving widespread interest. Flood-risk assessments are in particular focus and are legally required in some parts of the world, for example in the European Directive on the Assessment and Management of Flood Risks [3].

Maps are one of the best ways to present the spatial distribution of uncertainty and risk. Flood-risk maps are event-based and show the probability of the occurrence and consequences of a flood [1]. The process of producing flood-risk maps involves floodplain mapping and uncertainty analysis.

Floodplain maps are generated by a deterministic approach using either empirical, hydrodynamic or simplified-conceptual models [4–9]. Historical flood data is employed to validate models and calibrate their parameters [10]. Despite a notable research effort in recent decades, deterministic models remain afflicted by uncertainty [9]. This is in part because the models are assumed to be a comprehensive representation of the relevant physical processes and their relationship to water flow, and in part because they are assumed to use an optimal set of parameters. The issue is further compounded because uncertainty is likely to be non-stationary. Practitioners such as environmental agencies, river basin authorities and engineering consultancies often fail to fully invest the effort required for the detailed setting up of model and lack the computational resources required to solve the physical models [1]. This means that, in practice, additional uncertainty can be introduced to flood inundation modelling by the application of simplified, conceptual models in a deterministic way.

Realistically evaluating the risk of flood events and planning countermeasures therefore requires probabilistic floodplain maps, which are produced from ensemble simulations, like sensitivity analysis, of deterministic models [1]. The probabilistic approach is also more robust concerning the non-stationary aspect of uncertainty because of the many simulations that use a diverse set of data and parameters. Despite awareness of the benefits of this approach, Di Baldassarre et al. [1] claims that practitioners rarely apply probabilistic methods because of a slow rate of knowledge transfer and their familiarity with deterministic results. The authors further conclude there is a lack of a coherent terminology, a systematic approach and mature guidance for the use of probabilistic methods in hydrological analyses.

Reliability engineering has made a major contribution to uncertainty analysis. For one-dimensional models, several approaches can be used to quantify uncertainty. These include the First-Order Second-Moment (FOSM) method [11], Monte Carlo (MC) simulations [12], the Moment method [13], the spectral method [14], the Mellin transformation [15,16] and Point Estimate Methods (PEMs) [17–22].

Two-dimensional models are complicated by the need to consider spatial variability, which is handled by random field theories. Random fields can be generated in many ways [23] and serve as input for MC simulations which are common in uncertainty analysis, though very demanding in terms of computational resources. The study by Aerts et al. [24] is a good example: the researchers attempt to optimally locate and design a ski run while considering the uncertainty arising from an uncertain Digital Elevation Map (DEM). The uncertainty is analyzed by a MC simulation, the distinct DEM realizations are derivatives of random error fields, and the error values are the result of measurements at ground control locations. The measured differences are then analyzed for their spatial structure in the form of a variogram. Kriging is used as part of a Sequential Gaussian Simulation (SGS) [25–27]. SGS is a method of generating random fields to interpolate the non-measured errors between the ground control locations.

The popularity of MC simulations is illustrated by their continuous development. This development has led to various stratified sampling methods [28,29] that reduce the computational burden of MC, for example, Latin Hypercube Sampling [30–32]. Another improvement was the combination of MC with the Markov Chain process [33–35]. A useful implementation of MC can be found in the work by Feinberg and Langtangen [36].

Suchomel et al. [37] present a hybrid method combining the FOSM with random fields. Although the method can only consider uncorrelated uncertain variables, spatial correlation is accounted for by spatial averaging, which thus implies a variance reduction. The method was implemented as a finite element model.

In hydrology, uncertainty is commonly quantified by sensitivity analysis [9]. This involves perturbing within feasible limits variables and parameters–either individually and gradually (local) or simultaneously and randomly–and then assessing the effects on the outputs and any interactions.

The Generalized Likelihood Uncertainty Estimation (GLUE) is based on a Bayesian approach [1]. Similar to MC, GLUE requires several model realizations using random samples of the parameter in focus. The realizations are subsequently distinguished into behavioral and non-behavioral groups.

The outputs (e.g., the extent of flood inundation) are merged using Bayesian equations to result in a distributed probability profile. GLUE is less rigorous and requires the user to make many choices, increasing the potential for methodological errors [9].

A probabilistic modelling approach usually involves repeated runs of deterministic models using samples from across the range of parameters and available data. The purpose is to approximate the probability distribution of the output. For users with scarce computational resources, simplified conceptual methods are often more convenient than physics-based models.

The purpose of this study is to expand the application of PEMs to uncertainty analysis of spatiotemporal models and to develop a convenient way to produce probabilistic inundation maps.

2. Materials and Methods

This section explains the modelling approach and data used in this study. First, the ANUGA inundation spatial model [38,39] is briefly explained. The MC approach is subsequently discussed and this is followed by an introduction to Hong's PEM [20]. Furthermore, the expansion of PEM to two-dimensional space is clarified. Finally, the section concludes by describing a case study and its data set.

2.1. ANUGA

ANUGA is developed by the Australian National University (ANU) and Geoscience Australia (GA). The fluid dynamics of ANUGA is implemented using a finite volume approach to solve the Shallow Water Wave Equations. The simulation domain is partitioned into a mesh of triangular cells. The governing Shallow Water Wave Equations are solved at the centroid of the cell and thus provide water depth and horizontal momentum over time. ANUGA is capable of simulating the wetting and drying process when water enters and leaves a cell. It can therefore approximate the flow onto a beach, dry land and around structures such as buildings. Discontinuities, such as hydraulic jumps, are accommodated by the finite-volume implementation. A scenario set-up requires the domain's geometry, initial water level and boundary conditions. The drivers of the system in the inundation model, like rainfall, abstraction and culverts, are designed as operators and are also required to set up the scenario. The mesh generator can be used interactively to help the user to tailor the domain's geometry. Mesh triangles are symbolically tagged to indicate boundary regions or regions with different parameters, such as the Manning friction coefficients. ANUGA is mainly written in the Python programming language, which allows for flexible usage [39]. Computationally intensive components are executed as C routines by the means of Python's numerical module Numpy. The approach has been validated by adequately modelling various inundation problems [40,41].

2.2. Monte Carlo Simulation

MC simulation involves running a simulation many times with different random samples from the probability distribution of an uncertain input variable to obtain a numerical probability distribution of the output variable [24]. MC is very commonly used in uncertainty quantification to capture a system's natural variability [12,42,43]. The MC simulation serves as benchmark in this study to test the accuracy of the spatially implemented PEM. The outcome of MC is currently the best approximation to the "ground truth" of a stochastic system.

Distinct spatial realizations of the input variable in question (ground elevation) are produced by a SGS. Random error fields induce noise into the DEM data to derive distinct realizations. The error values originate from the differences between the original data set and the ground control data set at 50 locations. Figure 1 shows the histogram of the elevation error, which displays a non-normal form. Error values were interpolated by Ordinary Kriging to obtain error values for the whole spatial domain. Because Kriging provides an estimate of a variable's mean and standard deviation, the variable can be represented as a random variable according to a Gaussian distribution. However, Figure 1 shows that the errors are not normally distributed, and thus need to be normalized by the means of normal score transformation [44].

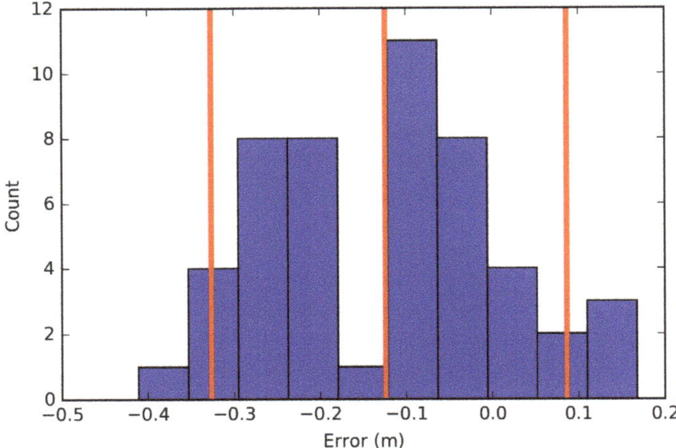

Figure 1. Histogram showing the elevation error and the location of the representative values (red): mean $= -0.12$, standard deviation $= 0.14$.

The MC approach requires many simulation runs. Accordingly, 500 were carried out for this study, requiring 500 distinct realizations of the random error field. The first step was to create the same number of random paths visiting every grid cell. Following these paths, the error value for each non-ground control cell was interpolated according to neighboring cells in relationship to the variogram. Because each path is different, the error value of each cell differed slightly from those around it. The mean and standard deviation of the Kriging enabled a random selection from a normal distribution. SGS was conducted by a Python library using the mean of the High Performance Geostatistics Library [45]. Subsequently, the error values were back-transformed into the original distribution [44]. The advantage of using Kriging was weighting the influence of the neighboring cells according to the semivariogram permitted the spatial structure to be respected.

2.3. Hong's Point Estimate Method

2.3.1. Theory

The core idea of the PEM is to calculate the statistical moments of the output variables. The moments are derived from several realizations of a deterministic model using representative values of the uncertain variable as an input. The purpose is to approximate the probability distribution of the model's output variable, which is related to the probability distribution of the uncertain input variables. The process only requires a few statistical moments of the input variables to be known. The derivation of PEM assumes that the rth moment of the output variable is the integral of the evaluation function with the random input variables over their domains with respect to their joint distribution functions. The equations of the representative values and weights are the products of expanding this function by Taylor series at various points of the variable distributions. The original PEM was proposed by Rosenblueth [21], but this has since proved to be too inefficient because it requires 2^m model evaluations, where m is the number of uncertain variables. Thus, the refinement suggested by Hong [20], which is a is a $2m + 1$ PEM scheme, was used.

Although this study considers only one variable (elevation), the set of random variables is defined as $X_i, i = 1, 2, \ldots, m$. Therefore, to compute the statistical moments of the output requires just three simulation runs, one for each representative input value, k. The first four statistical central moments (mean, variance, skewness and kurtosis) were used to approximate the distribution of the uncertain variable and thus compute the location of the representative values.

The locations of the representative values for each uncertain variable, i, in standardized space, $\tilde{\varsigma}_{i,k}$, were computed by Equation (1) based on the skewness, $\lambda_{i,3}$, and kurtosis, $\lambda_{i,4}$, of the uncertain variable.

$$\tilde{\varsigma}_{i,k} = \begin{cases} \frac{\lambda_{i,3}}{2} + (-1)^{3-k}\sqrt{\lambda_{i,4} - \frac{3}{4}\lambda_{i,3}^2} & ,k = 1,2 \\ 0 & ,k = 3 \end{cases} \tag{1}$$

Subsequently, the standardized location was transformed using the mean, μ_i, and standard deviation, σ_i, of the uncertain variable to its original space, $x_{i,k}$, by Equation (2). The representative value $k = 3$ was the mean value of the variable.

$$x_{i,k} = \mu_i + \sigma_i \tilde{\varsigma}_{i,k} \tag{2}$$

Based on the standardized values for location, skewness and kurtosis, the weights for the representative values, $\omega_{i,k}$, were then calculated by Equation (3).

$$\omega_{i,k} = \begin{cases} (-1)^{3-k}\frac{1}{\tilde{\varsigma}_{i,k}(\tilde{\varsigma}_{i,1} - \tilde{\varsigma}_{i,2})} & ,k = 1,2 \\ \frac{1}{m} - \frac{1}{\tilde{\varsigma}_{i,4} - \tilde{\varsigma}_{i,3}^2} & ,k = 3 \end{cases} \tag{3}$$

The weight determines the proportion of impact that each evaluation of the model $F(.)$ at the location of the representative value has on the statistical moment of the output variable. The rth statistical moment of a function of m random variables, Z, was then approximated by Equation (4).

$$E(Z^r) \approx \Sigma_{i=1}^m \omega_{i,3}(F(\mu_{x1}, \mu_{x2}, ..., \mu_{xi}, ..., \mu_{xm}))^r + \\ \Sigma_{i=1}^m \Sigma_{k=1}^2 \omega_{i,k}(F(\mu_{x1}, \mu_{x2}, ..., x_{i,k}, ..., \mu_{xm}))^r \tag{4}$$

2.3.2. Spatial Implementation of PEM

Similar to the MC approach, the spatial implementation also relies on random error fields. Instead of 500 error fields, this study only uses $2m + 1$ error fields, i.e., three, given that only one uncertain variable is considered. The three error fields were produced by the SGS process. Hong's method was then used to determine the representative values of the distribution as described by Equations (1) and (2). Figure 1 shows the histogram of the errors with the locations of the representative values.

The representative values and the corresponding weights were thereafter used to scale the three error fields, as can be seen in Equation (5).

$$sREF_{i,k} = x_{i,k} - (REF_i - \mu_{REF_i})\frac{\omega_{i,k}\sigma_i}{\sigma_{REF_i}} \tag{5}$$

where REF_i is the SGS-produced random error field of the uncertain variable, i. Figure 2 shows the realizations of the random error field, $sREF_{i,k}$, scaled by the representative values, $x_{i,k}$; their weights, $\omega_{i,k}$; and the standard deviation of the uncertain variable, σ_i. It is important that the range of values in three scaled random error fields is similar to that of the original random error field.

The simulation was run three times with the corresponding error fields. The statistical moments of the output variable were computed with Equation (4) as the weighted sum of the simulation results at the locations of the representative values to the power of the order of the statistical moment.

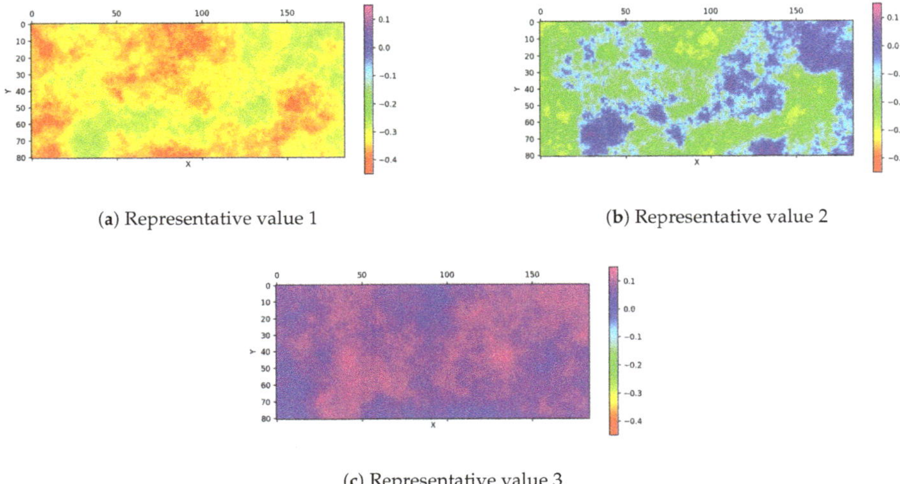

(**a**) Representative value 1

(**b**) Representative value 2

(**c**) Representative value 3

Figure 2. The random error field produced by SGS and scaled according to Equation (5) for three representative values.

2.4. Workflow

The Figure 3 illustrates the workflow of this study. The point of departure is the DEM and elevation of ground control locations. Both data sets combined result in error values. The error values need to be normalized because SGS requires that the input variable has a normal distribution. SGS generates 500 error fields for MC and three for PEM. The error values are back-transformed to the original distribution. Equation (5) scales the error fields for PEM. The error fields modify the DEM. The modified DEMs are supplied to ANUGA with the purpose to run an inundation scenario with a rainfall operator. The outputs are 500 raster time series data sets for MC and three for PEM. In the case of MC, the 500 raster time series data sets are reduced to four data sets, one for each statistical moment. In the case of PEM, Equation (4) estimates the raster time series data of the moments. The model efficiency performance measures finally compare the raster time series of each moment estimated by both methods. For each model efficiency performance measure, the results are one raster data set for each moment where each grid cell contains a model efficiency value. Each raster data set is subsequently summarized by mean, median, standard deviation, range and skew. The flowchart only displays the model efficiency performance assessment for the reason of readability.

2.5. Case Study

The location of the case study is Cockenzie Street and its surroundings in the city of Glasgow, UK. The data set was provided by the UK Environment Agency following their benchmark test No. 8 of two-dimensional hydraulic packages [41]. The data set contains the DEM, the geometry of buildings and roads, the Manning friction coefficients of paved and unpaved surfaces, the extent of the simulation domain, the location of the nine water level gauges and a rainfall curve. The study area is visualized in Figure 4. The DEM covers an area of approximately 0.4 km by 0.96 km. The original resolution of the DEM (0.5 m) was reduced to 5 m because of computational limitations. The range of the ground elevation is 21 m to 37 m. There are over 14.000 grid cells in the domain.

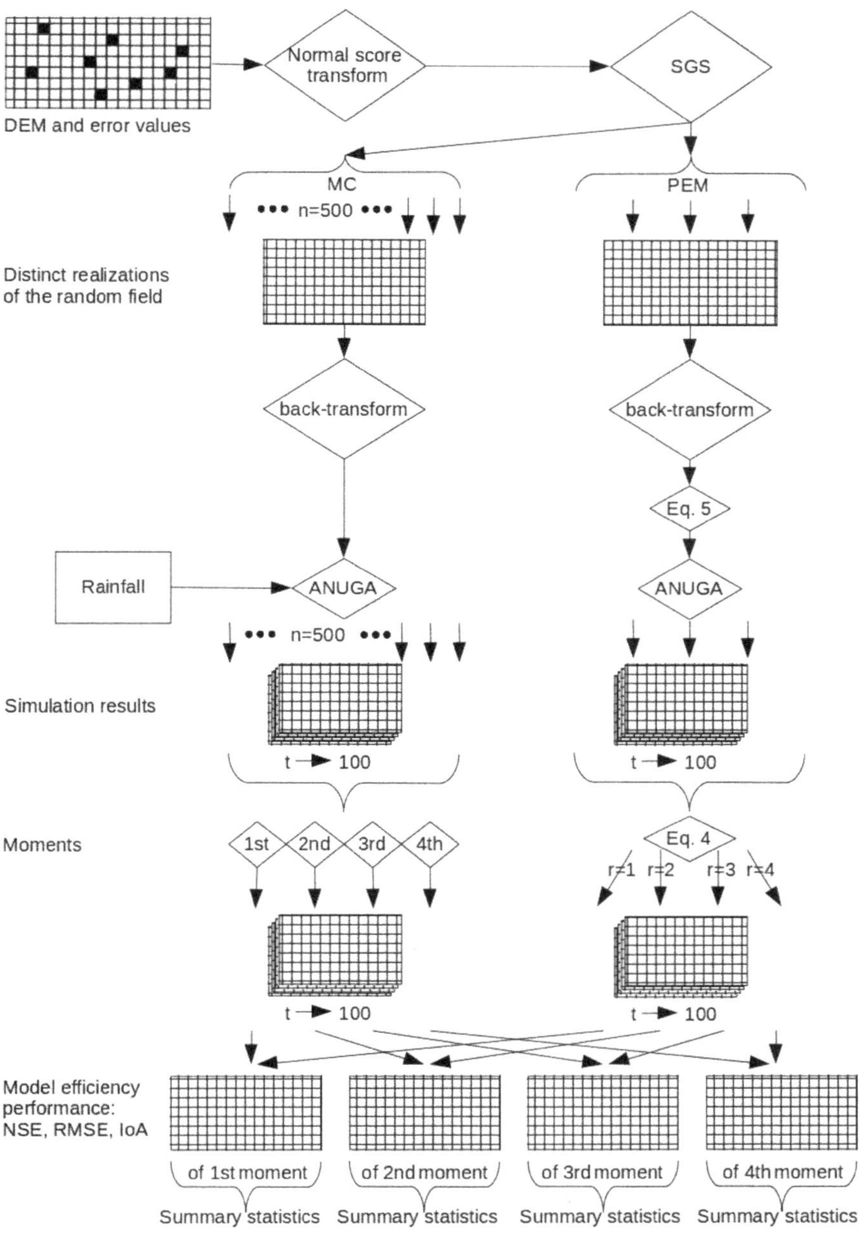

Figure 3. Workflow of the study.

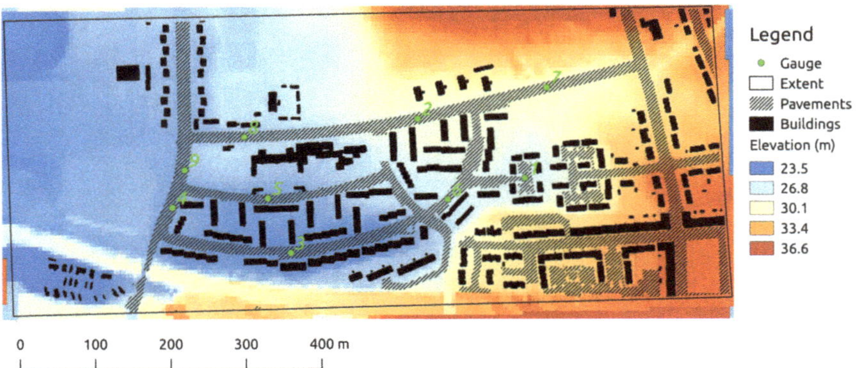

Figure 4. Digital Elevation Map with land cover of the area around Cockenzie Street in Glasgow, UK.

There are only two land cover categories with Manning friction coefficients: pavements (0.02) and unpaved surfaces (0.05). The simulation time was 1000 s with a 10-second time-step. The initial condition was a dry bed. The rainfall began after 1 min and lasted for 3 min with an intensity of 400 mm/h.

3. Results and Discussion

The model comparison below follows the recommendations in Bennett et al. [46] and Biondi et al. [47]. The times series at the nine gauge locations indicated in Figure 4 were visually compared. The similarities and differences of the spatial extents of the maximal inundations are then briefly assessed. The evaluation is completed by using the following model efficiency performance measures [47] on all time series over the whole spatial domain: Nash-Sutcliffe Model Efficiency (NSE), Root Mean Square Error (RMSE) and Index of Agreement (IoA). The statistical moments of MC should correspondingly be considered as "ground truth", i.e., akin to observed data. We focus on the first and second statistical moments because of their more important relationship with real-world applications, but also consider the third and fourth moments because they permit the fitting of a probability distribution to the output variable.

A comparison of the water depth values is shown using residual plots in Figures 5 and 6. The residual plots show the differences and similarities over time. Figure 6 groups the time series with noteworthy behavior. During rainfall (from time-step 61 to 240), PEM mostly underestimates the first and second moment, except for at gauges 4 and 6 where it overestimates both moments. Gauge 7 is particular as PEM underestimates the first moment here but then overestimates the second moment. PEM also appears to be less sensitive during the more dynamic period, i.e., when mass (rainfall) is added to the system. Therefore, the residuals are much higher during these periods. After the rainfall had stopped, the output from MC and PEM converged at most gauges, except at gauges 1 and 3. PEM continued to slightly underestimate the first moment at most gauge locations while also overestimating the second moment at many gauge locations. Concerning the first moment, there is a bias towards MC, except at gauge 4. The plots show a bias of the second moment towards PEM, which is especially visible for gauge 3. The estimations of the second moment by PEM are also marked by outliers, especially for gauges 2, 6, 7 and 9. However, it is very sensitive to the multitude of paths runoff can take when slightly changing elevation. This is especially the case for the gauges grouped in Figure 6.

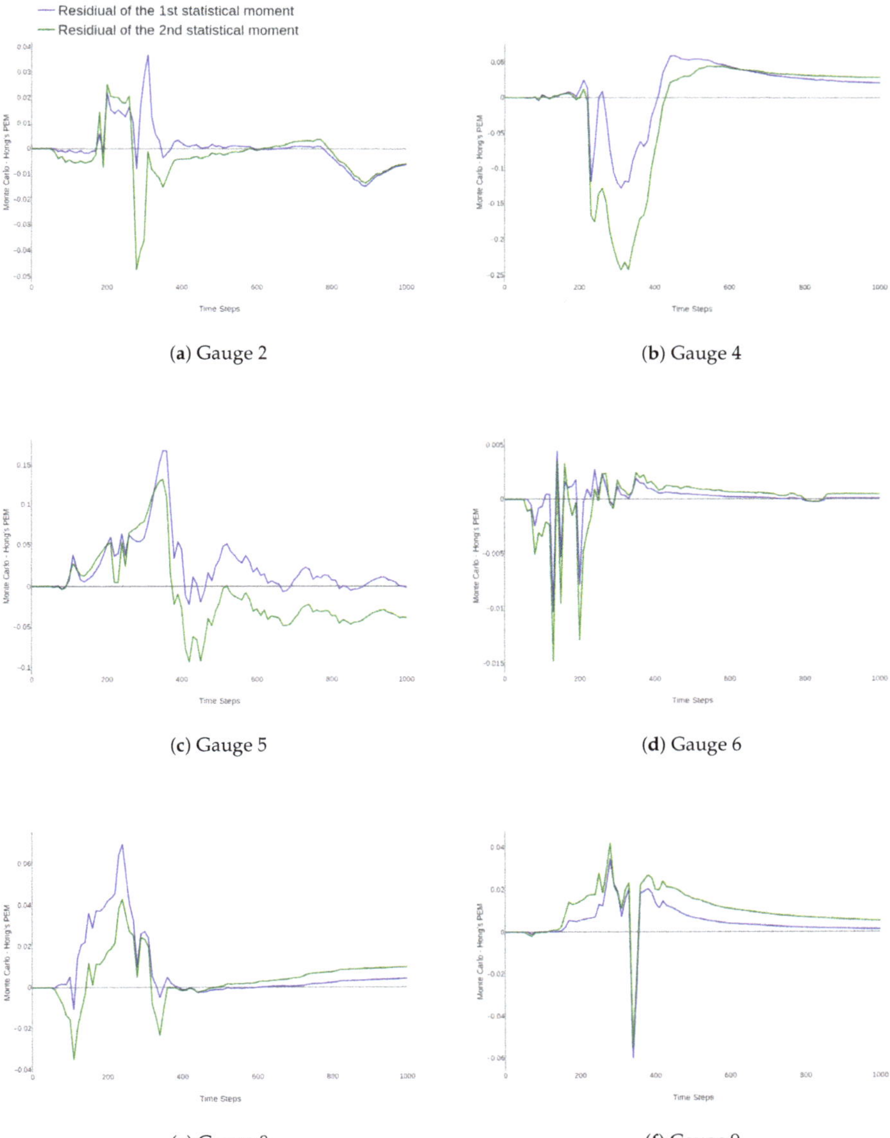

(**a**) Gauge 2

(**b**) Gauge 4

(**c**) Gauge 5

(**d**) Gauge 6

(**e**) Gauge 8

(**f**) Gauge 9

Figure 5. Residual plots showing similarities and differences between predictions made by MC simulation and PEM.

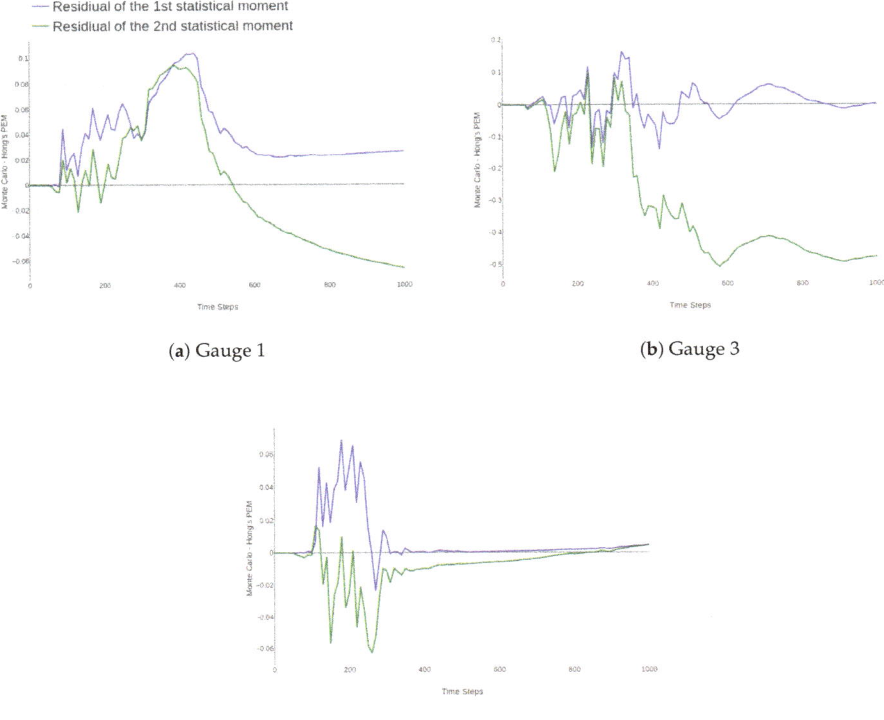

(**a**) Gauge 1

(**b**) Gauge 3

(**c**) Gauge 7

Figure 6. Residual plots showing similarities and differences between predictions made by MC simulation and PEM for gauges with noteworthy behaviour.

The spatial extent of the first two moments of the maximal flood inundation are compared in both Figures 7 and 8 because peak water depth is important for practical purposes. The mean estimations by MC and PEM in Figure 7 depict a very similar spatial structure and magnitude. Figure 8 confirms the previous statements about the second moment. Despite similar spatial structures, the second moment is systematically overestimated by PEM. This observation is useful as it seems that PEM sets the upper limit for the second moment.

Table 1 summarizes the model efficiency performance measures for all time series of the estimated statistical moments. The model efficiency performance measures consequently compare the similarity between the results of MC and PEM. The empirical distributions of the performance measures are described by mean, median, standard deviation, range and skew. The column NSE_{mean} characterizes the means of the Nash–Sutcliffe efficiency between MC and PEM of every estimated moment. The other columns follow the same logic. The NSE in Table 1 is the most commonly employed performance measure in hydrology. It ranges from 1 to negative infinity, where 1 indicates a perfect match. The RMSE in Table 1 ranges from 0 to positive infinity, where 0 indicates a perfect match. The NSE and RMSE suffer potential bias when comparing models with offsets [46], perhaps because of the usage of the square root. To complement the evaluation, the IoA in Table 1 was also calculated. The advantage of using the IoA is that it uses absolute differences, and thus might better avoid bias. It ranges from 0 to 1, where 1 indicates a perfect match. The mean values for the performance measures for the first and second moments show a good match. The situation is further improved when regarding the median owing to the impact on the mean values of far outliers, as indicated by the range and skew values. The outliers also prevent the performance measures being summarized in a box plot. Estimations of the second moment by PEM yielded more under-outliers, as is clearly observed when comparing the negative mean of the NSE with the median. The third and fourth moments perform less well. The reasons for this appear to be more systematic because range and skew indicate that outliers are less of a problem than for the first and second moments. Interestingly, the NSE and IoA suggest that the estimations of the fourth moment match the observations better than those of the third moment, but RMSE suggests the opposite. The decreasing performance of the PEM estimations with the order of moments might have been caused by the usage of the square function in the calculation of the moments in Equation (4).

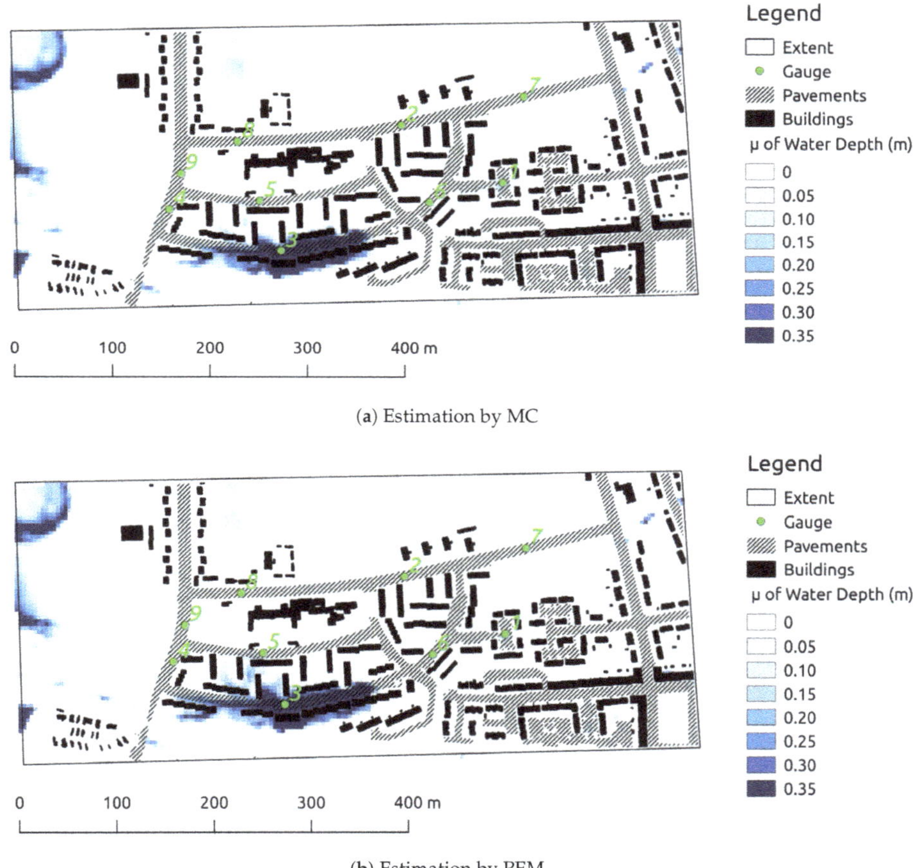

(a) Estimation by MC

(b) Estimation by PEM

Figure 7. Map of the 1st statistical moment of maximal water depth.

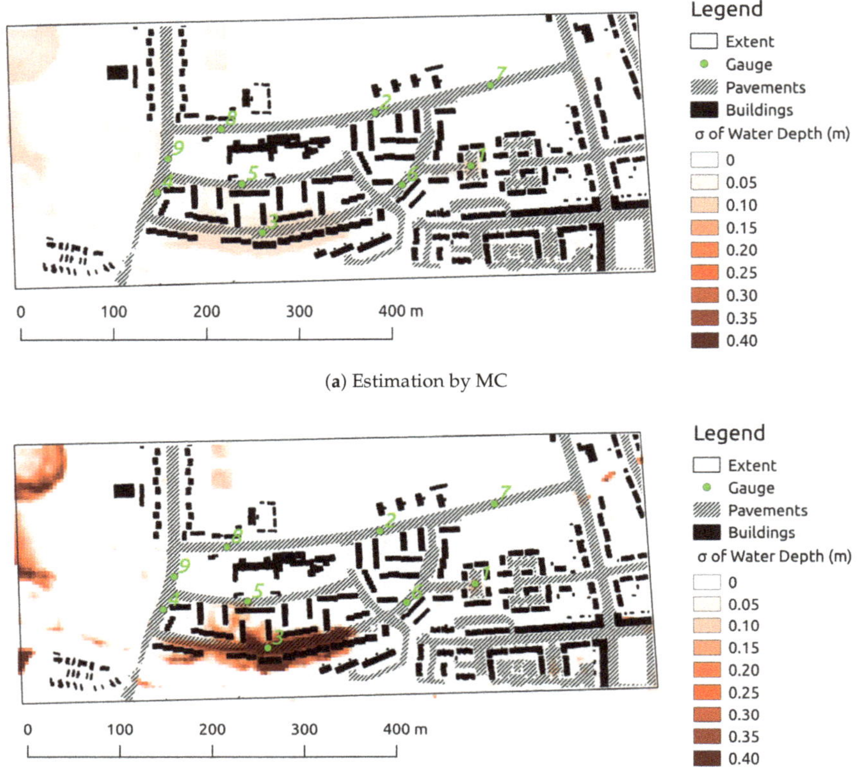

(**a**) Estimation by MC

(**b**) Estimation by PEM

Figure 8. Map of the 2nd statistical moment of maximal water depth.

Table 1. Summary statistics of model efficiency performance for all time series of the estimated statistical moments at each grid cell.

	NSE_{mean}	NSE_{median}	NSE_{SD}	NSE_{range}	NSE_{skew}
1st Moment	0.73	0.99	2.35	141.49	−34.87
2nd Moment	−1.95	0.96	38.10	2008.66	−35.86
3rd Moment	−0.78	−0.25	1.81	12.61	−1.72
4th Moment	0.07	−0.32	0.96	10.92	−1.74

	$RMSE_{mean}$	$RMSE_{median}$	$RMSE_{SD}$	$RMSE_{range}$	$RMSE_{skew}$
1st Moment	0.0074	0.0007	0.0164	0.1775	3.6896
2nd Moment	0.0179	0.0014	0.054	0.9263	6.1075
3rd Moment	5.092	3.6726	4.101	20.4579	1.0632
4th Moment	74.6667	33.7397	85.76	443.706	1.7032

	IoA_{mean}	IoA_{median}	IoA_{SD}	IoA_{range}	IoA_{skew}
1st Moment	0.96	0.99	0.09	0.96	−5.28
2nd Moment	0.88	0.99	0.18	0.99	−2.22
3rd Moment	0.55	0.64	0.35	0.99	−0.23
4th Moment	0.62	0.81	0.36	0.99	−0.51

The end-product of this study is the probabilistic inundation map. The Gram-Charlier expansion, which was implemented using the R package PDQutils [48], was used to fit the statistical moments to a probability distribution based on a normal distribution. Figure 9 displays the probability of a maximal inundation over 0.2 m. Despite the differences of the higher moments between MC and PEM, PEM predicts a similar risk of inundation of over 0.2 m in terms of probability values and spatial extent.

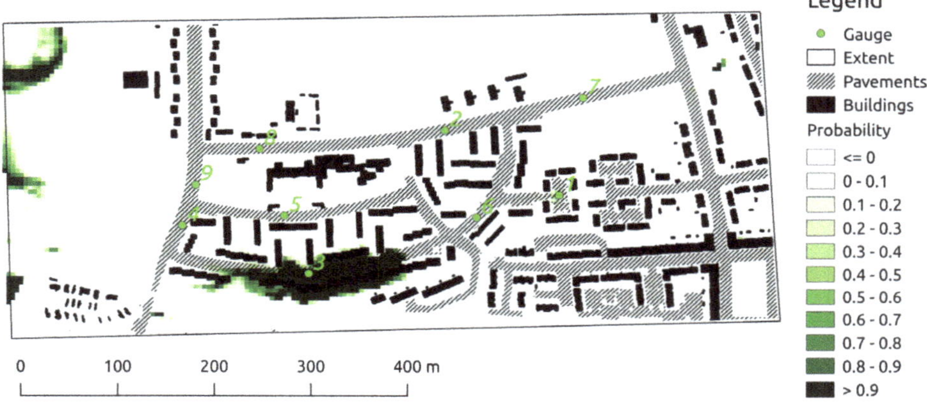

(**a**) Estimation by MC

(**b**) Estimation by PEM

Figure 9. Probabilistic inundation maps showing the probability of inundation over 0.2 m.

4. Conclusions

This study demonstrated that PEMs are capable of approximating the uncertainty of an output in spatiotemporal models. Using PEMs instead of the more commonly used MC simulations yields considerable savings of computational resources. Instead of 500 simulation runs, each lasting 11 min for the data used in this study, only 3 simulation runs were required for the PEM approach. This decrease in computational resource requirement is especially important for spatiotemporal models because of the amount of data and the complexity involved. Methodological advances like this are key to development because challenges arising from the use of Big Data are unlikely to be overcome by increases in computational power alone, despite its continually falling cost.

Uncertainty analysis is important to mitigate the impact of an overreliance by practitioners on deceptive deterministic solutions that arise because fundamental natural principles are less deterministic than we think and our continued lack of capacity to model phenomena in their totality. While MC simulations are computationally too demanding to be widely adopted outside of the research community, PEMs' ability to decrease computational costs could facilitate the uptake of uncertainty analysis in various fields. For example, hydrologists could combine water pollution models and PEMs to assess the risk of exceeding the legislated values (such as the U.S. EPA's Total Maximum Daily Loads). In a similar fashion, PEMs could help to determine the risk of exceeding health-related air pollution thresholds. Concerning flooding, flood-control authorities and insurance companies could better evaluate the potential risks for critical infrastructure and for insured objects, respectively.

Future studies should investigate different implementations of PEMs for spatiotemporal models and how to accommodate multivariate cases, where spatial cross-correlation is of tremendous importance. Current visualization techniques need to be improved to exploit the full potential of the information furnished by uncertainty analysis.

Author Contributions: Conceptualization, M.I.; methodology, M.I.; software, M.I.; validation, M.I.; formal analysis, M.I.; investigation, M.I.; resources, M.I.; data curation, M.I.; writing—original draft preparation, M.I.; writing—review and editing, M.I.; visualization, M.I.; supervision, F.-J.C.; project administration, F.-J.C.; funding acquisition, F.-J.C. All authors have read and agreed to the published version of the manuscript.

Funding: This research and the APC was funded by the Ministry of Science and Technology, Taiwan (Grant number: 107-2621-M-002-004-MY3).

Acknowledgments: We are grateful to the UK Environment Agency for providing data. We also wish to thank Sam Pickard, Rainer Wunderlich, Bryon Flowers, Hangyeh Lin and Pierre-Alexandre Château for their constructive comments on previous versions of this article. The authors would like to thank the Editors and anonymous Reviewers for their constructive comments that are greatly contributive to the revision of the manuscript.

References

1. Di Baldassarre, G.; Schumann, G.; Bates, P.D.; Freer, J.E.; Beven, K.J. Flood-plain mapping: A critical discussion of deterministic and probabilistic approaches. *Hydrol. Sci. J.* **2010**, *55*, 364–376. [CrossRef]
2. Daniell, J.; Wenzel, F.; McLennan, A.; Daniell, K.; Kunz-Plapp, T.; Khazai, B.; Schaefer, A.; Kunz, M.; Girard, T. The global role of natural disaster fatalities in decision-making: statistics, trends and analysis from 116 years of disaster data compared to fatality rates from other causes. In Proceedings of the EGU General Assembly 2016 Conference, Vienna, Austria, 17–22 April 2016; Volume 18, p. 2021.
3. Bubeck, P.; Kreibich, H.; Penning-Rowsell, E.C.; Botzen, W.; De Moel, H.; Klijn, F. Explaining differences in flood management approaches in Europe and in the USA–a comparative analysis. *J. Flood Risk Manag.* **2017**, *10*, 436–445. [CrossRef]
4. Almeida, G.A.; Bates, P.; Freer, J.E.; Souvignet, M. Improving the stability of a simple formulation of the shallow water equations for 2-D flood modeling. *Water Resour. Res.* **2012**, *48*. [CrossRef]
5. Bates, P.D. Integrating remote sensing data with flood inundation models: How far have we got? *Hydrol. Process.* **2012**, *26*, 2515–2521. [CrossRef]
6. Hapuarachchi, H.A.P.; Wang, Q.J.; Pagano, T.C. A review of advances in flash flood forecasting. *Hydrol. Process.* **2011**, *25*, 2771–2784. [CrossRef]
7. Merkuryeva, G.; Merkuryev, Y.; Sokolov, B.V.; Potryasaev, S.; Zelentsov, V.A.; Lektauers, A. Advanced river flood monitoring, modelling and forecasting. *J. Comput. Sci.* **2015**, *10*, 77–85. [CrossRef]
8. Neal, J.; Villanueva, I.; Wright, N.; Willis, T.; Fewtrell, T.; Bates, P. How much physical complexity is needed to model flood inundation? *Hydrol. Process.* **2012**, *26*, 2264–2282. [CrossRef]

9. Teng, J.; Jakeman, A.J.; Vaze, J.; Croke, B.F.; Dutta, D.; Kim, S. Flood inundation modelling: A review of methods, recent advances and uncertainty analysis. *Environ. Model. Softw.* **2017**, *90*, 201–216. [CrossRef]

10. Stephens, E.M.; Bates, P.; Freer, J.; Mason, D. The impact of uncertainty in satellite data on the assessment of flood inundation models. *J. Hydrol.* **2012**, *414*, 162–173. [CrossRef]

11. Wang, S.J.; Hsu, K.C. Dynamic interactions of groundwater flow and soil deformation in randomly heterogeneous porous media. *J. Hydrol.* **2013**, *499*, 50–60. [CrossRef]

12. Chang, C.H.; Tung, Y.K.; Yang, J.C. Monte Carlo simulation for correlated variables with marginal distributions. *J. Hydraul. Eng.* **1994**, *120*, 313–331. [CrossRef]

13. Graham, W.; McLaughlin, D. Stochastic analysis of nonstationary subsurface solute transport: 2. Conditional moments. *Water Resour. Res.* **1989**, *25*, 2331–2355. [CrossRef]

14. Li, S.G.; McLaughlin, D. A nonstationary spectral method for solving stochastic groundwater problems: Unconditional analysis. *Water Resour. Res.* **1991**, *27*, 1589–1605. [CrossRef]

15. Gires, A.; Onof, C.; Maksimovic, C.; Schertzer, D.; Tchiguirinskaia, I.; Simoes, N. Quantifying the impact of small scale unmeasured rainfall variability on urban runoff through multifractal downscaling: A case study. *J. Hydrol.* **2012**, *442*, 117–128. [CrossRef]

16. Tung, Y.K. Mellin transform applied to uncertainty analysis in hydrology/hydraulics. *J. Hydraul. Eng.* **1990**, *116*, 659–674. [CrossRef]

17. Che-Hao, C.; Yeou-Koung, T.; Jinn-Chuang, Y. Evaluation of probability point estimate methods. *Appl. Math. Model.* **1995**, *19*, 95–105. [CrossRef]

18. Franceschini, S.; Tsai, C.; Marani, M. Point estimate methods based on Taylor Series Expansion—The perturbance moments method—A more coherent derivation of the second order statistical moment. *Appl. Math. Model.* **2012**, *36*, 5445–5454. [CrossRef]

19. Harr, M.E. Probabilistic estimates for multivariate analyses. *Appl. Math. Model.* **1989**, *13*, 313–318. [CrossRef]

20. Hong, H.P. An efficient point estimate method for probabilistic analysis. *Reliab. Eng. Syst. Saf.* **1998**, *59*, 261–267. [CrossRef]

21. Rosenblueth, E. Point estimates for probability moments. *Proc. Natl. Acad. Sci. USA* **1975**, *72*, 3812–3814. [CrossRef]

22. Rosenblueth, E. Two-point estimates in probabilities. *Appl. Math. Model.* **1981**, *5*, 329–335. [CrossRef]

23. Christakos, G. *Random Field Models in Earth Sciences*; Courier Corporation: North Chelmsford, MA, USA, 2012.

24. Aerts, J.C.J.H.; Goodchild, M.F.; Heuvelink, G.B.M. Accounting for Spatial Uncertainty in Optimization with Spatial Decision Support Systems. *Trans. GIS* **2003**, *7*, 211–230. [CrossRef]

25. Ehlers, L.; Refsgaard, J.C.; Sonnenborg, T.O.; He, X.; Jensen, K.H. Using sequential Gaussian simulation to quantify uncertainties in interpolated gauge based precipitation. In Proceedings of the EGU General Assembly 2016 Conference, Vienna, Austria, 17–22 April 2016; Volume 18, p. 15751.

26. Gonçalvès, J.; Vallet-Coulomb, C.; Petersen, J.; Hamelin, B.; Deschamps, P. Declining water budget in a deep regional aquifer assessed by geostatistical simulations of stable isotopes: Case study of the Saharan "Continental Intercalaire". *J. Hydrol.* **2015**, *531*, 821–829. [CrossRef]

27. Varouchakis, E.A.; Hristopulos, D.T. Dynamic Modelling of Aquifer Level Using Space-Time Kriging and Sequential Gaussian Simulation. In Proceedings of the EGU General Assembly 2016 Conference, Vienna, Austria, 17–22 April 2016; Volume 18, p. 11694.

28. Shields, M.D.; Teferra, K.; Hapij, A.; Daddazio, R.P. Refined stratified sampling for efficient Monte Carlo based uncertainty quantification. *Reliab. Eng. Syst. Saf.* **2015**, *142*, 310–325. [CrossRef]

29. Wu, K.; Li, J. A surrogate accelerated multicanonical Monte Carlo method for uncertainty quantification. *J. Comput. Phys.* **2016**, *321*, 1098–1109. [CrossRef]

30. Gurdak, J.J.; Qi, S.L. *Vulnerability of Recently Recharged Ground Water in the High Plains Aquifer to Nitrate Contamination*; Technical Report; U. S. Geological Survey: Reston, VA, USA, 2006.

31. Niemunis, A.; Wichtmann, T.; Petryna, Y.; Triantafyllidis, T. Stochastic modelling of settlements due to cyclic loading for soil-structure interaction. In Proceedings of the International Conference on Structural Damage and Lifetime Assessment, Rome, Italy, 17–18 February 2005; Volume 18.

32. Tejchman, J.; Górski, J. Deterministic and statistical size effect during shearing of granular layer within a micro-polar hypoplasticity. *Int. J. Numer. Anal. Methods Geomech.* **2008**, *32*, 81–107. [CrossRef]

33. Dodwell, T.J.; Ketelsen, C.; Scheichl, R.; Teckentrup, A.L. A hierarchical multilevel Markov chain Monte Carlo algorithm with applications to uncertainty quantification in subsurface flow. *SIAM/ASA J. Uncertain. Quantif.* **2015**, *3*, 1075–1108. [CrossRef]

34. Moradkhani, H.; DeChant, C.M.; Sorooshian, S. Evolution of ensemble data assimilation for uncertainty quantification using the particle filter-Markov chain Monte Carlo method. *Water Resour. Res.* **2012**, *48*. [CrossRef]

35. Vrugt, J.A.; Ter Braak, C.J.; Clark, M.P.; Hyman, J.M.; Robinson, B.A. Treatment of input uncertainty in hydrologic modeling: Doing hydrology backward with Markov chain Monte Carlo simulation. *Water Resour. Res.* **2008**, *44*. [CrossRef]

36. Feinberg, J.; Langtangen, H.P. Chaospy: An open source tool for designing methods of uncertainty quantification. *J. Comput. Sci.* **2015**, *11*, 46–57. [CrossRef]

37. Suchomel, R.; Mašı, D. Comparison of different probabilistic methods for predicting stability of a slope in spatially variable c-phi soil. *Comput. Geotech.* **2010**, *37*, 132–140. [CrossRef]

38. Nielsen, O.; Roberts, S.; Gray, D.; McPherson, A.; Hitchman, A. Hydrodynamic modelling of coastal inundation. In Proceedings of the MODSIM 2005 International Congress on Modelling and Simulation, Melbourne, Austrilia, 12–15 December 2005; pp. 518–523.

39. Roberts, S.; Nielsen, O.; Gray, D.; Sexton, J. *ANUGA User Manual*; Geoscience Australia and Australian National University: Canberra, Austrilia, 2010.

40. Mungkasi, S.; Roberts, S. Validation of ANUGA hydraulic model using exact solutions to shallow water wave problems. *J. Phys. Conf. Ser.* **2013**, *423*, 012029. [CrossRef]

41. Néelz, S.; Pender, G. *Benchmarking the Latest Generation of 2D Hydraulic Modelling Packages*; Technical Report SC120002; UK Environment Agency: Rotherham, UK, 2013.

42. Fan, Y.; Huang, W.; Huang, G.; Huang, K.; Zhou, X. A PCM-based stochastic hydrological model for uncertainty quantification in watershed systems. *Stoch. Environ. Res. Risk Assess.* **2015**, *29*, 915–927. [CrossRef]

43. Miller, K.; Berg, S.; Davison, J.; Sudicky, E.; Forsyth, P. Efficient uncertainty quantification in fully-integrated surface and subsurface hydrologic simulations. *Adv. Water Resour.* **2018**, *111*, 381–394. [CrossRef]

44. Deutsch Clayton, V.; Journel André, G. *GSLIB Geostatistical Software Library and User's Guide*; Oxford University Press: Oxford, UK, 1998.

45. Savichev, V.; Bezrukov, A.; Muharlyamov, A.; Barskiy, K.; Rustam, S. High Performance Geostatistics Library, 2010. Available online: http://hpgl.github.io/hpgl/index.html (accessed on 13 October 2017).

46. Bennett, N.D.; Croke, B.F.; Guariso, G.; Guillaume, J.H.; Hamilton, S.H.; Jakeman, A.J.; Marsili-Libelli, S.; Newham, L.T.; Norton, J.P.; Perrin, C. Characterising performance of environmental models. *Environ. Model. Softw.* **2013**, *40*, 1–20. [CrossRef]

47. Biondi, D.; Freni, G.; Iacobellis, V.; Mascaro, G.; Montanari, A. Validation of hydrological models: Conceptual basis, methodological approaches and a proposal for a code of practice. *Phys. Chem. Earth Parts A/B/C* **2012**, *42*, 70–76. [CrossRef]

48. Pav, S. PDQ Functions via Gram Charlier, Edgeworth, and Cornish Fisher Approximations. 2017. Available online: https://cran.r-project.org/web/packages/PDQutils/index.html (accessed on 13 October 2017).

Article

Uncertainty Assessment of Urban Hydrological Modelling from a Multiple Objective Perspective

Bo Pang [1,2], **Shulan Shi** [1,2], **Gang Zhao** [1,2,3,*], **Rong Shi** [4], **Dingzhi Peng** [1,2] **and Zhongfan Zhu** [1,2]

[1]   College of Water Sciences, Beijing Normal University, Beijing 100875, China; pb@bnu.edu.cn (B.P.);
      201821470021@mail.bnu.edu.cn (S.S.); dzpeng@bnu.edu.cn (D.P.); zhuzhongfan1985@bnu.edu.cn (Z.Z.)
[2]   Beijing Key Laboratory of Urban Hydrological Cycle and Sponge City Technology, Beijing 100875, China
[3]   School of Geographical Science, University of Bristol, Bristol BS8 1SS, UK
[4]   Water in the Pearl River Planning Survey Design Co., Ltd. Beijing Branch, Beijing 100875, China;
      shirong610@126.com
*   Correspondence: gang.zhao@bristol.ac.uk; Tel.: +86-186-1289-3886

Received: 31 March 2020; Accepted: 12 May 2020; Published: 14 May 2020

Abstract: The uncertainty assessment of urban hydrological models is important for understanding the reliability of the simulated results. To satisfy the demand for urban flood management, we assessed the uncertainty of urban hydrological models from a multiple-objective perspective. A multiple-criteria decision analysis method, namely, the Generalized Likelihood Uncertainty Estimation-Technique for Order Preference by Similarity to Ideal Solution (GLUE-TOPSIS) was proposed, wherein TOPSIS was adopted to measure the likelihood within the GLUE framework. Four criteria describing different urban stormwater characteristics were combined to test the acceptability of the parameter sets. The TOPSIS was used to calculate the aggregate employed in the calculation of the aggregate likelihood value. The proposed method was implemented in the Storm Water Management Model (SWMM), which was applied to the Dahongmen catchment in Beijing, China. The SWMM model was calibrated and validated based on the three and two flood events respectively downstream of the Dahongmen catchment. The results showed that the GLUE-TOPSIS provided a more precise uncertainty boundary compared with the single-objective GLUE method. The band widths were reduced by 7.30 m^3/s in the calibration period, and by 7.56 m^3/s in the validation period. The coverages increased by 20.3% in the calibration period, and by 3.2% in the validation period. The median estimates improved, with an increase of the Nash–Sutcliffe efficiency coefficients by 1.6% in the calibration period, and by 10.0% in the validation period. We conclude that the proposed GLUE-TOPSIS is a valid approach to assess the uncertainty of urban hydrological model from a multiple objective perspective, thereby improving the reliability of model results in urban catchment.

Keywords: urban hydrological model; Generalized Likelihood Uncertainty Estimation (GLUE); Technique for Order Preference by Similarity to Ideal Solution (TOPSIS); uncertainty analysis

---

1. Introduction

Land use modifications accompanied by urbanization, including the decrease of vegetation cover, increase of impervious surfaces, and drainage channel modifications, result in changes in the characteristics of the surface runoff hydrograph [1,2]. Urban flooding and waterlogging are severe global environmental issues. Urban hydrological models play an important role in the planning and construction of urban drainage and flood control systems [3,4]. The Storm Water Management Model (SWMM), which was developed by the United States Environmental Protection Agency (EPA), has become one of the most popular urban hydrological models [5,6] in urban stormwater simulation.

Urban hydrological models typically require a large number of parameters, which are difficult or impossible to measure with sufficient accuracy, and must generally be estimated or evaluated from secondary information sources [7,8]. Hence, the obtained simulation results are typically laden with notable degrees of uncertainty [9]. Thus, various methods have been developed to assess the uncertainty in urban hydrological models, such as the Monte Carlo Markov chain [10,11], grey box model [12], Bayesian approach [13], and Generalized Likelihood Uncertainty Estimation (GLUE) method [14,15]. The selection of the objective function is critical for many of these uncertainty analysis methods. According to GLUE method, the objective function and acceptability threshold exert substantial influence on the final assessment results. The Nash–Sutcliffe Efficiency index (NSE) is widely used as the objective function in the GLUE method for urban hydrological models [16,17].

In 1998, Gupta et al. [18] proposed that hydrological model calibration inherently comprises multiple objectives, and that converting it into a single-objective model must involve some degree of subjectivity. The complexity of urban stormwater management comes from flood characteristics such as the flood volume, peak flood value, lag time, and overflow characteristics such as the overflow volume and duration. These characteristics concern different management system components [15,19] and are difficult to be captured by a single-objective function during model calibration and validation. Moreover, the adoption of the single-objective function introduces additional uncertainties in applications because it may consider unacceptable parameter sets [20,21]. The controversy can also be found in the use of informal likelihood measures based on the NSE, which involves subjective decisions [22].

This paper proposes a new framework for analyzing the uncertainty of an urban hydrological model. TOPSIS (evaluation technique for order performance by similarity to ideal solution), which is a multiple-criteria decision analysis (MCDA) method developed by Hwang and Yoon [23,24], was adopted as the likelihood criterion of the GLUE method using four important objective criteria in urban flood modelling. The proposed method was used to quantify the parameter uncertainties in the SWMM model and was applied to the Dahongmen (DHM) catchment located in Beijing, China. We attempted to improve the uncertainty assessment of an urban hydrological model by comprehensively investigating the performance of the parameter sets during simulation.

2. Methods

The proposed uncertainty analysis method is based on the GLUE framework, which was developed by Beven and Binley [25]. In the GLUE method, a likelihood measure is selected and calculated to reflect the goodness of fit between the model simulation and the observations. The model simulations are considered as non-behavioral when the values of the likelihood measures are lower than the cut-off threshold. The selection of the likelihood measure and cut-off threshold has been discussed in various papers, owing to the subjective decisions involved [22,26–28]. Because the single objective always involves some degree of subjectivity [18], the proposed method uses MCDA to carry out the likelihood measure; the critical points are summarized as follows:

- The performance of the parameter sets was comprehensively investigated by considering the threshold of each objective function. Only the parameter sets for which all objective function values exceeded their threshold were considered as behavioral.
- An integrated likelihood measure was carried out using TOPSIS, which was used as a weighting factor to derive the posterior probability density functions for both the parameters and the predictions. The prominent advantage of TOPSIS is that different types of objective functions can be easily integrated into a unified evaluation index by setting the benefit criteria and loss criteria.

The uncertainty assessment process can be described as follows:

1. Based on Monte Carlo sampling from feasible parameter spaces with a defined prior distribution, wherein uniform probability distributions are often adopted without prior knowledge [29,30].
2. Instead of a unique likelihood measure, multiple thresholds from multiple criteria are used to determine the behavioral parameter sets. In urban stormwater management, not only the precision of the flood process but also the precision of the flood volume and flood peak is considered by the

modeler [31]. A behavior parameter set should be able to achieve these objectives. The criteria typically used in flood prediction are as follows:

$$NSE = 1 - \frac{\sum_{i=1}^{n}\left(y_{obs,i} - y_{sim,i}\right)^2}{\sum_{i=1}^{n}\left(y_{obs,i} - \overline{y_{obs,i}}\right)^2} \tag{1}$$

$$VB = \frac{\left|\sum_{i=1}^{n} y_{obs,i} - \sum_{i=1}^{n} y_{sim,i}\right|}{\sum_{i=1}^{n} y_{obs,i}} \tag{2}$$

$$PB = \frac{\left|\max_{1\leq i\leq n}\{y_{obs,i}\} - \max_{1\leq i\leq n}\{y_{sim,i}\}\right|}{\max_{1\leq i\leq n}\{y_{obs,i}\}} \tag{3}$$

and

$$R = \frac{\sum\left(y_{obs,i} - \overline{y_{obs}}\right)\left(y_{sim,i} - \overline{y_{sim}}\right)}{\sqrt{\sum\left(y_{obs,i} - \overline{y_{obs}}\right)^2\left(y_{sim,i} - \overline{y_{sim}}\right)^2}} \tag{4}$$

where y_{obs} is the observed flow, y_{sim} is the simulated flow, $\overline{y_{obs}}$ is the average measured flow, $\overline{y_{sim}}$ is the average simulated flow, and n is the number of the observed flow points. In the above formula, NSE is the widely used Nash–Sutcliffe efficiency index [32]. The flood volume bias (VB) and flood peak bias (PB) are the modified expressions for the flood volume and flood peak deviation [18,33], respectively. The parameter R represents the consistency between the observed flow and the simulated flow [34]. The threshold of these criteria can be defined with reference to practical demand [35]. The reasonable range of the four criteria is between 0 to 1. The optimum value of NSE and R is 1, and optimum value of VB and PB is 0, which means the simulation results of the model completely fit the measured results. In this study, the behavioral parameter sets whose likelihood values of the four criteria were greater than the corresponding thresholds were chosen for further analysis.

3. TOPSIS, which is a well-known MCDA method and can provide the ranking order of all alternatives [36,37], was employed in the calculation of the aggregate likelihood value $L(\theta_i)$ of the behavioral parameter set θ_i. In the TOPSIS method, the four criteria of all parameter sets should be normalized by the classification of the benefit and cost criterion, where the benefit criterion means that a larger value is more valuable, and vice-versa for the cost criterion [37]. In this study, NSE and R are benefit criteria, while VB and PB are cost criteria; x_{ij} is the ith criterion of the jth parameter set. For the benefit criteria, the normalized value (r_{ij}) is calculated as follows:

$$r_{ij} = \frac{x_{ij}}{\sqrt{\sum_{i=1}^{n} x_{ij}^2}}. \tag{5}$$

For the cost criteria, the normalized value (r_{ij}) is calculated as follows:

$$r_{ij} = \frac{\frac{1}{x_{ij}}}{\sqrt{\sum_{i=1}^{n} \frac{1}{x_{ij}}^2}}. \tag{6}$$

In many situations, the criterion should be weighted according its importance [36]. Because it is difficult to identify which criterion is more important, we assume that all criteria are equally important. The ideal solution R_j^+ and negative-ideal solution R_j^- can be calculated as follows:

$$R_j^+ = \max\{x_{1j}, x_{2j}, \dots, x_{nj}\} \tag{7}$$

$$R_j^+ = \min\{x_{1j}, x_{2j}, \dots, x_{nj}\}. \tag{8}$$

Then, the separation of each alternative from the ideal and negative-ideal solutions are expressed, respectively, as follows:

$$D_i^+ = \sqrt{\sum_{j=1}^{n}\left(r_{ij} - R_j^+\right)^2} \tag{9}$$

$$D_i^- = \sqrt{\sum_{j=1}^{n}(r_{ij} - R_j^-)^2} \tag{10}$$

The aggregate likelihood value $L(\theta_i)$ is evaluated by comparing the distance from the ideal solution and the distance from the negative-ideal solution:

$$L(\theta_i) = \frac{D_i^+}{D_i^+ + D_i^-} \tag{11}$$

4. Finally, the predictions from the behavioral parameter sets are ranked in the order of the likelihood weights $W(i)$, which is defined as follows:

$$W(i) = \frac{L(\theta_i)}{\sum_{i=1}^{N} L(\theta_i)} \tag{12}$$

where N is the number of behavioral parameter sets. Additionally, the cumulative probability distribution for the ranked discharge predictions can be obtained as follows:

$$P(Q < Q_i) = \frac{\sum_{j=1}^{i} W(j)}{\sum_{j=1}^{n} W(j)} \tag{13}$$

where Q denotes the discharge, and Qi is the discharge prediction ranked at the ith position; n has the same meaning as in Equation (2). According to the cumulative probability distribution, the uncertainty bound can be obtained for a given certainty level.

3. Study Area and SWMM Model

3.1. Study Area

The Dahongmen catchment is a typical urbanized area in Beijing, PRC, with a high population density and heavily built-up underlying surface. The catchment covers an area of approximately 131.49 km^2 and is located upstream of the Liangshui River basin in Beijing, between 39°48′–39°55′ N and 116°90′–116°24′ E. In the catchment, the terrain exhibits a downward trend from the western mountains to the eastern plains. The annual average precipitation is 522.4 mm, and 80% of the precipitation occurs during the period from June to September. The river systems and hydro-meteorological stations of the Dahongmen catchment are shown in Figure 1.

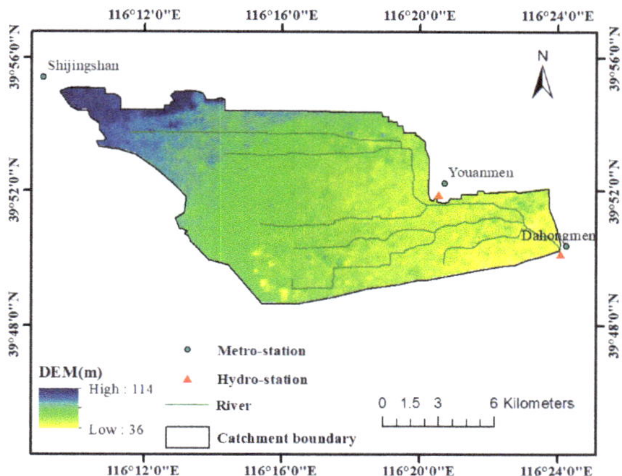

Figure 1. Station distribution at Dahongmen catchment.

3.2. SWMM Model

The SWMM is a dynamic hydrological simulation model used in single-event or long-term (continuous) simulations of runoff quantity and quality from primarily urban areas. The runoff component of SWMM operates on a collection of subcatchment areas that receive precipitation and generate runoff and pollutant loads. The routing portion of the SWMM transports this runoff through a system of pipes, channels, storage/treatment devices, pumps, and regulators [38,39]. Additionally, the SWMM tracks the quantity and quality of the runoff generated within each subcatchment, and the flow rate, flow depth, and quality of the water in each pipe and channel during a simulation period with multiple time steps. In this study, SWMM version 5.1 was used; its technical details have been reported by Rossman et al. [40]. The best performance of SWMM before and after urbanization in the Dahonmen catchment was addressed by Xu and Zhao (2017) [41].

In this study, five heavy storms (accumulated precipitation > 50 mm) occurred between 2011 and 2012 in the Dahongmen catchment were used for model calibration and verification. Precipitation on 23 June 2011, 26 July 2011, and 14 August 2011 were used to calibrate the model, and the other two on 24 June 2012 and 21 July 2012 were used for validation. The accumulated precipitation of the five events was 110.6, 66.8, 66.9, 63.9, and 197.4 mm, respectively. The rainfall duration of the five storms was14, 5, 3, 16, and 18 h, respectively. As shown in Figure 1, the hourly series from three precipitation stations and a DHM gauge station were obtained from the Hydrographic Station of Beijing, CHINA. The hourly inflow data of the Youanmen station were obtained from the Liangshui River Basin Authority. The floodwater from this gate only account for very small amount of streamflow thus will not have great impact on the results of uncertainly analysis. The Digital Elevation Model (DEM) and sewer system map were provided by the National Aeronautics and Space Administration (NASA, ASTER GDEM) and Beijing Municipal Institute of City Planning and Design. These datasets include the locations, section shapes, and conveyance capacities of the river and sewer system in the study area. The catchment was divided into 13 subcatchments, jointly controlled by the river and pipe network, wherein there existed a total of 351 drainage channels (282 watercourses and 69 road drainage channels), as shown in Figure 2.

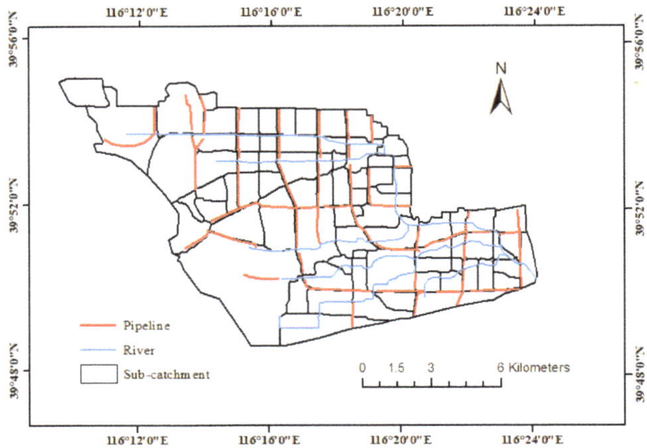

Figure 2. Structure of Storm Water Management Model (SWMM) model in Dahongmen catchment.

Table 1 lists the parameters used in the uncertainty analysis of this study, along with their units and distribution. The minimum and maximum values were obtained from the SWMM user's manual [40] and relevant literature [42,43]. The values for the selected model parameters were randomly selected from uniform probability distributions.

Table 1. Distribution of SWMM parameters.

Category	Parameter	Description	Units	Distribution
Basic characteristic	%Imperv	Percent of impervious area	%	60–80
Manning roughness	N-Imperv	Manning coefficient for impervious area		0.025–0.045
	N-perv	Manning coefficient for pervious area		0.1–0.5
	N-river	Manning coefficient for riverway		0.03–0.06
	N-conduit	Manning coefficient for conduit		0.01–0.03
Reservoir in depressions	D-imperv	Depth of depression storage on impervious area	mm	15–40
	D-perv	Depth of depression storage on pervious area	mm	20–50
Infiltration parameters	MaxRate	Maximum rate on Horton infiltration curve	mm/h	100–150
	MinRate	Minimum rate on Horton infiltration curve	mm/h	10–90
	Decay	Decay constant for the Horton infiltration curve	h^{-1}	0–50

3.3. Interval Evaluation Index

To analyze the effectiveness of the range of uncertainty, we selected three evaluation indices, namely, the average band width (B), coverage (CR), and average relative deviation (RD) [41,44], which are defined as follows:

$$B = \frac{1}{n} \sum_{i=1}^{n} \left(Q_{s,upper}^{i} - Q_{s,lower}^{i} \right) \tag{14}$$

$$CR = \frac{n_{Q_{in}}}{n} \times 100\% \tag{15}$$

$$RD = \frac{1}{n} \sum_{i=1}^{n} \frac{\left| \frac{1}{2} \left(Q_{s,upper}^{i} + Q_{s,lower}^{i} \right) - Q_{o}^{i} \right|}{Q_{o}^{i}} \tag{16}$$

Here, $Q_{s,upper}^{i}$ and $Q_{s,lower}^{i}$ are the upper and lower boundary values of the confidence interval; Q_{o}^{i} is the observed flow; and $n_{Q_{in}}$ is the number of observed values in the confidence interval.

4. Results

4.1. Comparison of Different Acceptability Thresholds

By the application of Monte Carlo method, 10,000 parameter sets were generated within the ranges listed in Table 1. The four objective criteria illustrated in Equations (1)–(4) were calculated by running the SWMM with these sampled parameter sets.

We set the threshold of each objective criterion according to the catchment characteristics and time scale, which are list in Table 2. The number of parameter sets which met the threshold requirements are also list in the table.

Table 2. Numbers of behavior parameter sets under different criteria.

Criterion	Number	Criterion	Number	Criterion	Number	Criterion	Number
$NSE \geq 0.7$	2506	$NSE \geq 0.7$ and $VB \leq 0.3$	2226	$NSE \geq 0.7$ and $VB \leq 0.3$ and $PB \leq 0.2$	1598	$NSE \geq 0.7$ and $VB \leq 0.3$ and $PB \leq 0.2$ and $R \geq 0.8$	1598
$VB \leq 0.3$	3958	$NSE \geq 0.7$ and $PB \leq 0.2$	1772	$NSE \geq 0.7$ and $VB \leq 0.3$ and $R \geq 0.8$	2226		
$PB \leq 0.2$	4215	$NSE \geq 0.7$ and $R \geq 0.8$	2506	$NSE \geq 0.7$ and $PB \leq 0.2$ and $R \geq 0.8$	1772		
$R \geq 0.8$	6918	$VB \leq 0.3$ and $PB \leq 0.2$	2557	$VB \leq 0.3$ and $PB \leq 0.2$ and $R \geq 0.8$	2396		
		$VB \leq 0.3$ and $R \geq 0.8$	3413				
		$PB \leq 0.2$ and $R \geq 0.8$	3927				

As shown in Table 2, 2506 parameter sets could satisfy the requirement of NSE threshold, which is the least in the 4 criteria and indicates NSE is the strictest criterion among them. Nevertheless, R threshold is the most flexible one, 6918 parameter sets can satisfy the criterion. Additionally, all parameter sets that satisfied other criteria thresholds can satisfy R threshold simultaneously.

From Table 2 we can also observe that the number of behavioral parameter sets decreased when more criteria were considered. There are 2226 parameter sets that satisfied the NSE and VB thresholds simultaneously, and 1772 parameter sets that satisfied the NSE and PB thresholds. When we considered all criteria thresholds in GLUE-TOPSIS, the number of behavioral parameter sets reduced to 1598 finally.

4.2. Comparison of Posterior Distribution

We can obtain the posterior probability distributions of the SWMM parameters from the behavioral parameter sets listed in Table 2. The posterior probability distributions from the single criterion (NSE) and GLUE-TOPSIS are shown in Figure 3.

Obvious difference can be observed between the posterior probability distributions from single criterion (NSE) and those from GLUE-TOPSIS. In general, we can find the obvious areas with high frequency distributions in the posterior probability distribution curves from GLUE-TOPSIS. However, those from single criterion (NSE) are relative flat.

Overall, the posterior probability distributions of Imperv, N-perv, D-perv, MaxRate, Decay, and N-conduit are similar for the two methods. Nevertheless, N-imperv, D-imperv, MinRate, and N-river exhibit a significantly higher frequency interval under GLUE-TOPSIS conditions, which results in higher sensitivity. Notably, there are obvious high-frequency intervals in the parameters characterizing the impervious area, owing to the large impervious area in the DHM catchments. Additionally, the parameter spatial distribution has more than one high frequency interval, which may reflect the "equifinality" of parameter sets proposed by Beven and Binley [25].

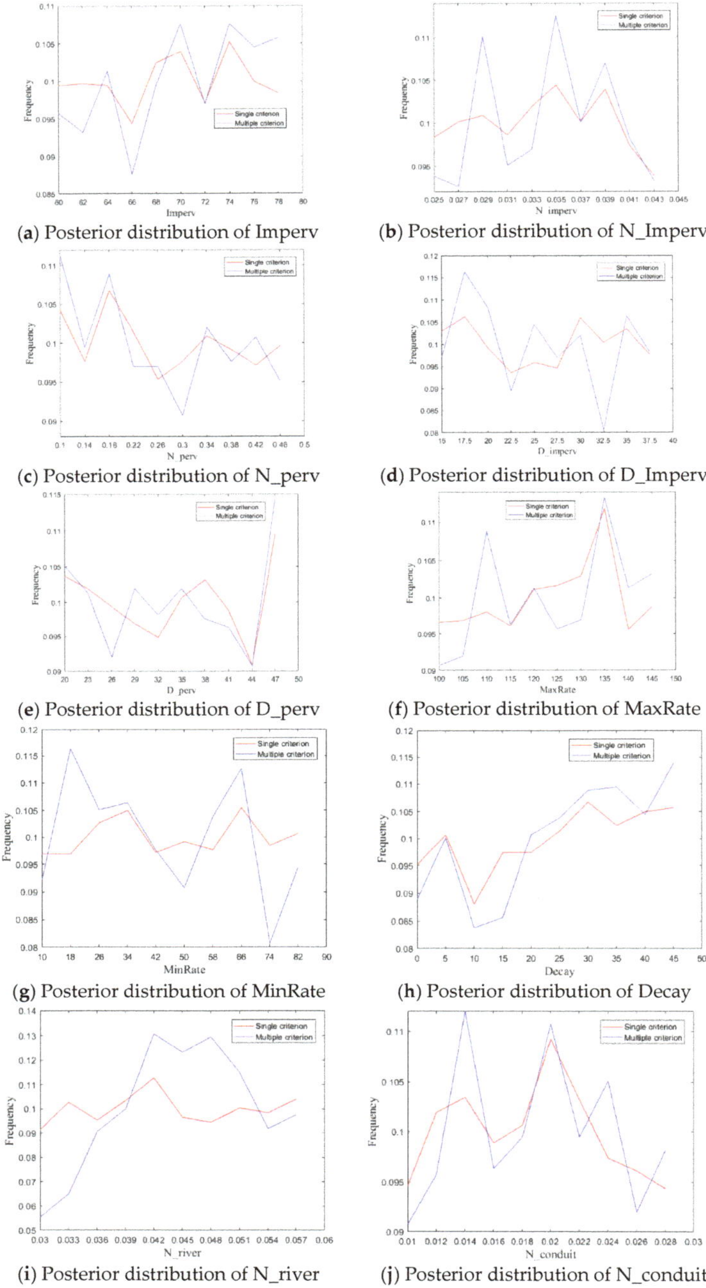

(**a**) Posterior distribution of Imperv

(**b**) Posterior distribution of N_Imperv

(**c**) Posterior distribution of N_perv

(**d**) Posterior distribution of D_Imperv

(**e**) Posterior distribution of D_perv

(**f**) Posterior distribution of MaxRate

(**g**) Posterior distribution of MinRate

(**h**) Posterior distribution of Decay

(**i**) Posterior distribution of N_river

(**j**) Posterior distribution of N_conduit

Figure 3. Parameter posterior distribution for single criterion and multiple criteria.

4.3. Uncertainty Estimation of Discharge Simulation

We simulated the SWMM discharge with parameter sets from the single-criterion and GLUE-TOPSIS method. The normalized simulations were sorted by the likelihood value to determine the 95% and 5% uncertainty bounds (90 confidence interval). Figures 4 and 5 present the plotted simulation results obtained by the single criterion and GLUE-TOPSIS methods for five rainfall events during the calibration and validation periods. To analyze the effectiveness of the uncertainty ranges, the evaluation results of the uncertainty indices under the two methods are listed in Table 3.

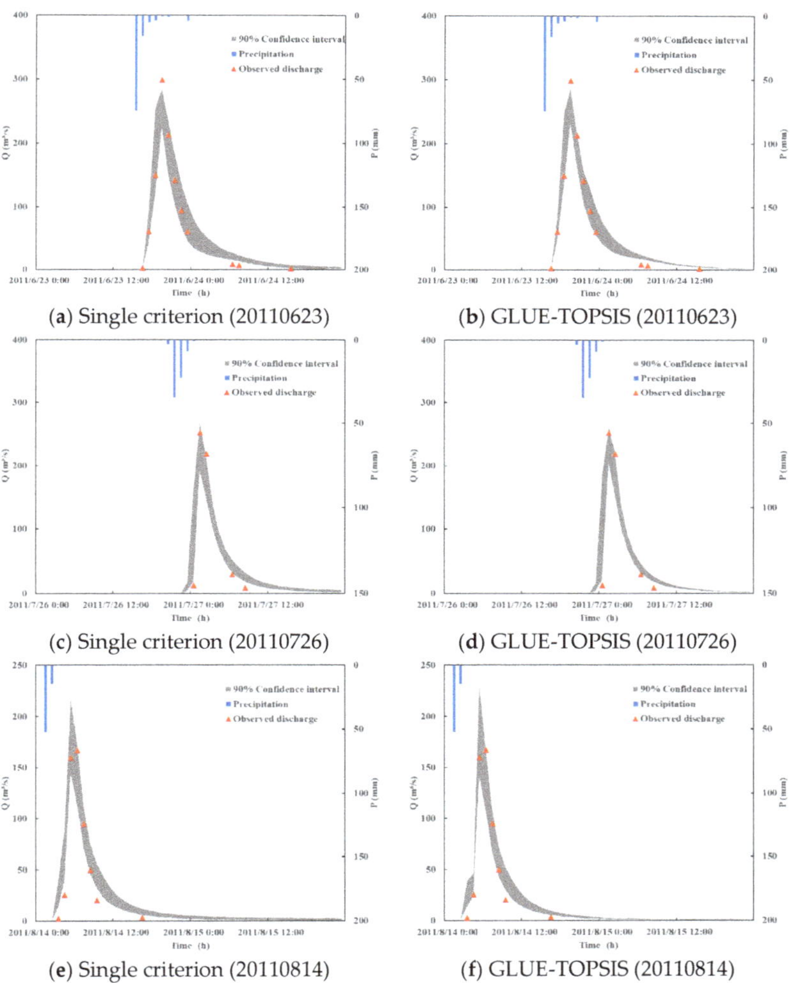

Figure 4. MCDA analysis results for rainstorm events 20110623, 20110726, and 20110814 in calibration period.

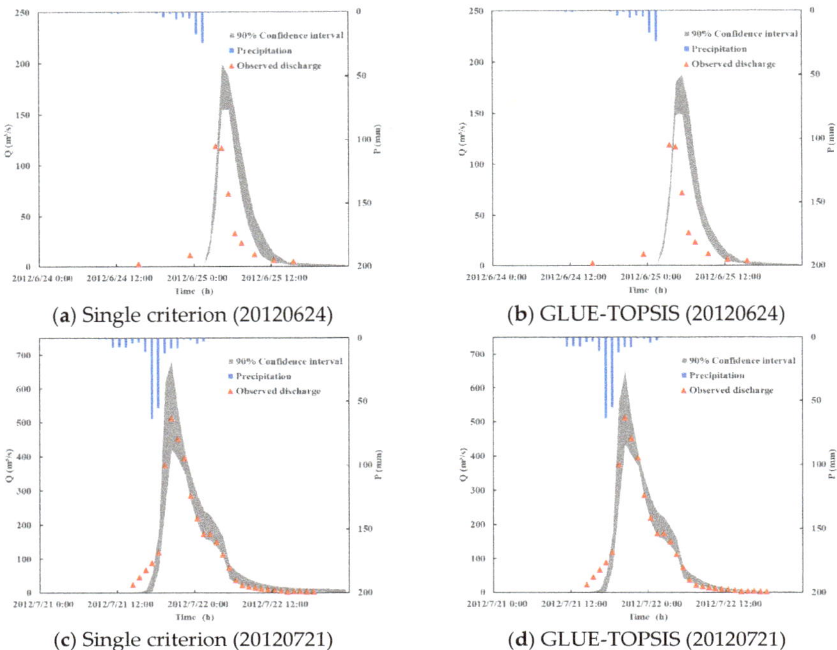

(**a**) Single criterion (20120624)

(**b**) GLUE-TOPSIS (20120624)

(**c**) Single criterion (20120721)

(**d**) GLUE-TOPSIS (20120721)

Figure 5. Multiple-criteria decision analysis (MCDA) analysis results for rainstorm event 20120624, 20120721 in validation period.

Table 3. Uncertainty interval evaluation results.

Method		Criteria	B(m^3/s)	CR (%)	RD (m^3/s)
Calibration period	20110623	Single criteria	48.308	54.5	13.606
		GLUE-TOPSIS	35.17	54.5	12.804
	20110726	Single criteria	50.563	20.0	41.454
		GLUE-TOPSIS	43.723	40.0	40.551
	20110814	Single criteria	39.829	50.0	18.995
		GLUE-TOPSIS	37.908	62.5	16.276
	Average	Single criteria	46.233	41.5	24.685
		GLUE-TOPSIS	38.934	52.3	23.211
Validation period	20120624	Single criteria	17.100	10.0	17.100
		GLUE-TOPSIS	14.100	10.0	14.150
	20120721	Single criteria	53.953	81.8	17.140
		GLUE-TOPSIS	41.839	84.8	13.358
	Average	Single criteria	35.526	45.9	17.120
		GLUE-TOPSIS	27.970	47.4	13.754

As shown in Figures 4 and 5, the widths of the uncertainty bound from GLUE-TOPSIS method obviously narrower than those from single criterion method, particularly for the flood peak area and drop section of the discharge curve. Most of the observations fell within the 90% confidence interval, except rainstorm 20120624, which is the smallest storm and may influenced by the strobe operation. However, SWMM model show high performance in rainstorm 20120721, which is one of the most disastrous rainstorms in history.

As presented in Table 3, the band width (B), coverage (CR), and relative deviation (RD) were 46.233, 41.5%, and 24.685, respectively, for single criterion method in the calibration period, and 38.934, 52.3%, and 23.211, respectively, for GLUE-TOPSIS method. In the verification period, B, CR, and RD were 35.526, 45.9%, and 17.120, respectively, for the single criterion method, and 27.970, 47.4%, and 13.754, respectively, for GLUE-TOPSIS method. The results indicate that the GLUE-TOPSIS method show superiority in uncertainty bounds assessment over the single criterion methods, with lower values in average band width (B), average relative deviation (RD) and higher values in coverage (CR).

From the simulation results of the median GLUE estimates in Table 4, it can be seen that under the GLUE-TOPSIS method, the simulation results of NSE, VB are significantly improved compared with the single criterion method, although the VB of 20110623 has a slight increase. The median estimates improved, with an increase of the NSE by 1.6% in the calibration period, and by 10.0% in the validation period. However, the cost criteria PB increased by 0.010 during the calibration period, from the perspective of single field precipitation, the PB of the 20110814 precipitation and the 20120624 precipitation have increased, indicating that the flood peak flow simulation effect is poor. And the benefit criteria R decreased by 0.078.

Table 4. Simulation results of the median Generalized Likelihood Uncertainty Estimation (GLUE) estimates.

Method		Criteria	NSE	VB	PB	R
Calibration period	20110623	Single criteria	0.863	0.003	0.118	0.994
		GLUE-TOPSIS	0.876	0.02	0.112	0.995
	20110726	Single criteria	0.936	0.086	0.096	0.955
		GLUE-TOPSIS	0.949	0.123	0.077	0.941
	20110814	Single criteria	0.966	0.184	0.085	0.962
		GLUE-TOPSIS	0.982	0.103	0.138	0.991
	Average	Single criteria	0.921	0.091	0.099	0.970
		GLUE-TOPSIS	0.936	0.082	0.109	0.975
Validation period	20120624	Single criteria	0.429	0.402	0.375	0.790
		GLUE-TOPSIS	0.530	0.300	0.428	0.663
	20120721	Single criteria	0.936	0.041	0.077	0.989
		GLUE-TOPSIS	0.974	0.005	0.059	0.962
	Average	Single criteria	0.683	0.222	0.226	0.890
		GLUE-TOPSIS	0.752	0.153	0.244	0.812

5. Discussion

The uncertainties of urban hydrological models have been addressed by several researches [45–48]. Most researches used single objective function such as NSE or VB as evaluation index [45,46]. As shown in the Figure 6a,b, we found that VB and PB still involved large uncertainties when the NSE was higher than 0.8. This means that a high NSE value can still lead to the poor simulation of the flow volume and flood peak.

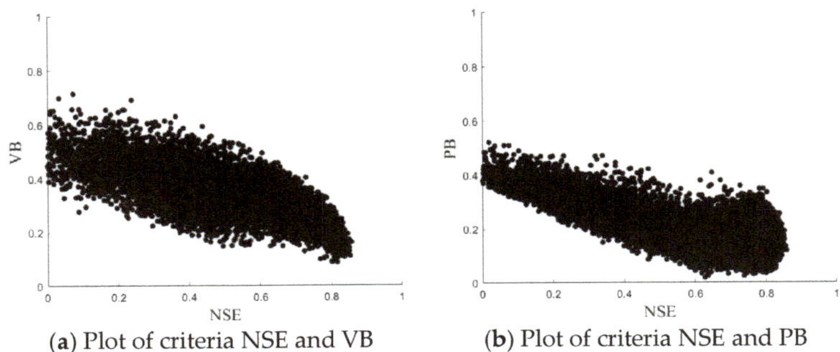

(**a**) Plot of criteria NSE and VB   (**b**) Plot of criteria NSE and PB

Figure 6. Relationship between different criteria.

Some attempts were made to construct one likelihood function [47,49] considering different objectives. This likelihood function will not work when the objective change. Other important objectives such as overflow and flow in sewer system should be considered in uncertainty analysis of urban hydrological model [15]. The proposed method avoids the process of function construction and is feasible to be applied considering different objectives.

Considering the completeness and consistent of data, five heavy storms between 2011 to 2012 were selected for uncertainty analysis. These are representative enough to test the model reliability for heavy storms. However, small storms are easily affected by local Blue-Green Infrastructure in urban areas that make it difficult to predict [50,51]. Further work needs to test the reliability of the propose method for different magnitude of flood.

6. Conclusions

Urban hydrological models are extensively used in flood forecasting, sponge cities design, and pollution management, etc. The uncertainty assessment of an urban hydrological model is important when evaluating the model's reliability. Because of the various applications of urban hydrological models, a single criterion is hard to evaluate their performance. In this research, we proposed a multiple-criteria uncertainty analysis method, namely GLUE-TOPSIS, and applies it to the uncertainty assessment of SWMM model in Dahongmen catchment, Beijing. Five typical rainstorm events that occurred during 2011–2012 were investigated and used to test the performance of the proposed method. The conclusions can be summarized as follows:

1. 10,000 parameter sets generated by the Monte Carlo sampling in GLUE framework revealed that none of the four commonly used objective criteria could fully represent the urban flow process. Notably, the NSE, which is widely used in assessing the performance of hydrological models also cannot describe the flow characteristics alone, which highlights the need for adopting multi-criteria methods.

2. The GLUE-TOPSIS method provided more precise uncertainty bounds and median estimates than traditional GLUE method which used NSE as single criterion. The GLUE-TOPSIS method reduced the bandwidth and deviation of the uncertainty bounds with a higher coverage than these from single criterion. The median estimates of GLUE-TOPSIS are also superior to these from single criterion according to the four objective criteria.

3. The SWMM model performed well in the flood simulation of Dahongmen catchment in Beijing, PRC. Most observed flows fell within the 90% uncertainty interval, which suggests that the parameter uncertainty analysis has a relatively high contribution toward improving the simulation accuracy of flood prediction. The comparison results for the posterior distribution revealed that the parameters characterizing the impervious area had obvious high frequency intervals, owing to the large impervious area of Dahongmen catchment.

The GLUE-TOPSIS method provide a feasible way to assess the uncertainty of urban hydrological models, which have been used various aspects in urban water resources management. The users can select objective criteria flexibly and combine them in the GLUE-TOPSIS framework according to their actual needs. We will proceed our research and apply GLUE-TOPSIS to a wider area in water resource management.

Author Contributions: Conceptualization, B.P., G.Z. and R.S.; methodology, B.P., D.P. and Z.Z.; validation, S.S.; writing—original draft preparation, B.P. and G.Z.; writing—review and editing, S.S. and G.Z.; visualization, S.S.; supervision, B.P. and G.Z.; project administration, B.P.; funding acquisition, B.P. All authors have read and agreed to the published version of the manuscript.

Funding: This research was funded by three research programs: (1) National Natural Science Funds of China (Grant No. 51879008); (2) China Scholarship Council (Grant No. 201906045024); (3) China Scholarship Council-University of Bristol Joint PhD Scholarships Programme (Grant No. 201700260088)

Acknowledgments: We thank Liwen Bianji, Edanz Editing China (www.liwenbianji.cn/ac), for editing the English text of a draft of this manuscript.

References

1. Shukla, S.; Gedam, S. Assessing the impacts of urbanization on hydrological processes in a semi-arid river basin of Maharashtra, India. *Modell. Earth Sys. Environ.* **2018**, *4*, 699–728. [CrossRef]
2. Barbosa, A.E.; Fernandes, J.N.; David, L.M. Key issues for sustainable urban stormwater management. *Water Res.* **2012**, *46*, 6787–6798. [CrossRef]
3. Hapuarachchi, H.A.P.; Wang, Q.J.; Pagano, T.C. A review of advances in flash flood forecasting. *Hydrol. Process.* **2011**, *25*, 2771–2784. [CrossRef]
4. Hallegatte, S.; Green, C.; Nicholls, R.J.; Corfee-Morlot, J. Future flood losses in major coastal cities. *Nat. Clim. Chang.* **2013**, *3*, 802–806. [CrossRef]
5. Suttles, K.M.; Singh, N.K.; Vose, J.M.; Martin, K.L.; Emanuel, R.E.; Coulston, J.W.; Saia, S.M.; Crump, M.T. Assessment of hydrologic vulnerability to urbanization and climate change in a rapidly changing watershed in the Southeast US. *Sci. Total Environ.* **2018**, *645*, 806–816. [CrossRef]
6. Cristiano, E.; Ten Veldhuis, M.; van de Giesen, N. Spatial and temporal variability of rainfall and their effects on hydrological response in urban areas—a review. *Hydrol. Earth Syst. Sc.* **2017**, *21*, 3859–3878. [CrossRef]
7. Cullmann, J.; Krausse, T.; Philipp, A. Enhancing flood forecasting with the help of processed based calibration. *Phys. Chem. Earth* **2008**, *33*, 1111–1116. [CrossRef]
8. Xu, Z. Hydrological Models: Past, present and feature. *J. Beijing Normal Univ. Nat. Sci.* **2010**, *46*, 278–289.
9. Fonseca, A.; Ames, D.P.; Yang, P.; Botelho, C.; Boaventura, R.; Vilar, V. Watershed model parameter estimation and uncertainty in data-limited environments. *Environ. Modell. Softw.* **2014**, *51*, 84–93. [CrossRef]
10. Zahmatkesh, Z.; Karamouz, M.; Nazif, S. Uncertainty based modeling of rainfall-runoff: Combined differential evolution adaptive Metropolis (DREAM) and K-means clustering. *Adv. Water Resour.* **2015**, *83*, 405–420. [CrossRef]
11. Blasone, R.; Vrugt, J.A.; Madsen, H.; Rosbjerg, D.; Robinson, B.A.; Zyvoloski, G.A. Generalized likelihood uncertainty estimation (GLUE) using adaptive Markov chain Monte Carlo sampling. *Adv. Water Resour.* **2008**, *31*, 630–648. [CrossRef]
12. Lindblom, E.; Madsen, H.; Mikkelsen, P.S. Comparative uncertainty analysis of copper loads in stormwater systems using GLUE and grey-box modeling. *Water Sci. Technol.* **2007**, *56*, 11–18. [CrossRef] [PubMed]
13. Liu, Y.R.; Li, Y.P.; Huang, G.H.; Zhang, J.L.; Fan, Y.R. A Bayesian-based multilevel factorial analysis method for analyzing parameter uncertainty of hydrological model. *J. Hydrol.* **2017**, *553*, 750–762. [CrossRef]
14. Beven, K.; Freer, J. Equifinality, data assimilation, and uncertainty estimation in mechanistic modelling of complex environmental systems using the GLUE methodology. *J. Hydrol.* **2001**, *249*, 11–29. [CrossRef]
15. Thorndahl, S.; Beven, K.J.; Jensen, J.B.; Schaarup-Jensen, K. Event based uncertainty assessment in urban drainage modelling, applying the GLUE methodology. *J. Hydrol.* **2008**, *357*, 421–437. [CrossRef]
16. Setegn, S.G.; Srinivasan, R.; Melesse, A.M.; Dargahi, B. SWAT model application and prediction uncertainty analysis in the Lake Tana Basin, Ethiopia. *Hydrol. Process.* **2010**, *24*, 357–367. [CrossRef]

17. Gupta, H.V.; Kling, H.; Yilmaz, K.K.; Martinez, G.F. Decomposition of the mean squared error and NSE performance criteria: Implications for improving hydrological modelling. *J. Hydrol.* **2009**, *377*, 80–91. [CrossRef]

18. Gupta, H.V.; Sorooshian, S.; Yapo, P.O. Toward improved calibration of hydrologic models: Multiple and noncommensurable measures of information. *Water Resour. Res.* **1998**, *34*, 751–763. [CrossRef]

19. Madsen, H. Automatic Calibration of a Conceptual Rainfall-Runoff Model Using Multiple Objectives. *J. Hydrol.* **2000**, *235*, 276–288. [CrossRef]

20. Fenicia, F.; Savenije, H.H.G.; Matgen, P.; Pfister, L. A comparison of alternative multiobjective calibration strategies for hydrological modeling. *Water Resour. Res.* **2007**, *43*, 93–99. [CrossRef]

21. Gill, M.K.; Kaheil, Y.H.; Khalil, A.; Mckee, M.; Bastidas, L. Multiobjective particle swarm optimization for parameter estimation in hydrology. *Water Resour. Res.* **2006**, *42*, 257–271. [CrossRef]

22. Beven, K.; Binley, A. GLUE: 20 years on. *Hydrol. Process.* **2014**, *28*, 5897–5918. [CrossRef]

23. Hwang, C.L.; Lai, Y.J.; Liu, T.Y. A new approach for multiple-objective decision-making. *Comput. Oper. Res.* **1993**, *20*, 889–899. [CrossRef]

24. Hwang, C.; Yoon, K. *Methods for Multiple Attribute Decision Making*, 3rd ed.; Springer: Berlin/Heidelberg, Germany, 1981; pp. 58–191.

25. Beven, K.; Binley, A. The future of distribute models-model calibration and uncertainty prediction. *Hydrol. Process.* **1992**, *6*, 279–298. [CrossRef]

26. Mantovan, P.; Todini, E. Hydrological forecasting uncertainty assessment: Incoherence of the GLUE methodology. *J. Hydrol.* **2006**, *330*, 368–381. [CrossRef]

27. Stedinger, J.R.; Vogel, R.M.; Lee, S.U.; Batchelder, R. Appraisal of the Generalized Likelihood Uncertainty Estimation (GLUE) Method. *Water Resour. Res.* **2009**, *42*. [CrossRef]

28. Clark, M.P.; Kavetski, D.; Fenicia, F. Pursuing the method of multiple working hypotheses for hydrological modeling. *Water Resour. Res.* **2011**, *47*, 178–187. [CrossRef]

29. Freer, J.; Beven, K.; Ambroise, B. Bayesian Estimation of Uncertainty in Runoff Prediction and the Value of Data: An Application of the GLUE Approach. *Water Resour. Res.* **1996**, *32*, 2161–2173. [CrossRef]

30. Beven, K.; Smith, P.; Freer, J. Comment on "Hydrological forecasting uncertainty assessment: Incoherence of the GLUE methodology" by Pietro Mantovan and Ezio Todini. *J. Hydrol.* **2007**, *338*, 315–318. [CrossRef]

31. Loperfido, J.V.; Noe, G.B.; Jarnagin, S.T.; Hogan, D.M. Effects of distributed and centralized stormwater best management practices and land cover on urban stream hydrology at the catchment scale. *J. Hydrol.* **2014**, *519*, 2584–2595. [CrossRef]

32. Nash, J.E.; Sutcliffe, J.V. River flow forecasting through conceptual models part I—A discussion of principles. *J. Hydrol.* **1970**, *10*, 290. [CrossRef]

33. Pang, B.; Guo, S.; Xiong, L.; Li, C. A nonlinear perturbation model based on artificial neural network. *J. Hydrol.* **2007**, *333*, 504–516. [CrossRef]

34. Moriasi, D.N.; Gitau, M.W.; Pai, N.; Daggupati, P. Hydrologic and water quality models: Performance measures and evaluation criteria. *T. Asabe* **2015**, *58*, 1763–1785.

35. Yilmaz, K.K.; Gupta, H.V.; Wagener, T. A multi-criteria penalty function approach for evaluating a priori model parameter estimates. *J. Hydrol.* **2015**, *525*, 165–177. [CrossRef]

36. Cheng, K.; Lien, Y.; Wu, Y.; Su, Y. On the criteria of model performance evaluation for real-time flood forecasting. *Stoch. Env. Res. Risk A.* **2017**, *31*, 1123–1146. [CrossRef]

37. Zanakis, S.H.; Solomon, A.; Wishart, N.; Dublish, S. Multi-attribute decision making: A simulation comparison of select methods. *Eur. J. Oper. Res.* **1998**, *107*, 507–529. [CrossRef]

38. Zoppou, C. Review of urban storm water models. *Environ. Modell. Softw.* **2001**, *16*, 195–231. [CrossRef]

39. Gironas, J.; Roesner, L.A.; Rossman, L.A.; Davis, J. A new applications manual for the Storm Water Management Model (SWMM). *Environ. Modell. Softw.* **2010**, *25*, 813–814. [CrossRef]

40. Rossman, L.A. *Storm Water Management Model User's Manual*, 5th ed.; Environment Protection Agency: Cincinnati, OH, USA, 2005; pp. 125–137.

41. Xu, Z.; Zhao, G. Impact of urbanization on rainfall-runoff processes: Case study in the Liangshui River Basin in Beijing, China. *Proc. Int. Assoc. Hydrol. Sci.* **2016**, *373*, 7–12. [CrossRef]

42. Zhao, G.; Pang, B.; Xu, Z.; Du, L.; Zhong, Y. Simulation of urban storm an Dahongmen drainage area by SWMM. *J. Beijing Normal Univ. Nat. Sci.* **2014**, *50*, 452–455.

43. Shi, R.; Zhao, G.; Pang, B.; Jinag, Q.; Zhen, T. Uncertainty Analysis of SWMM Model Parameters Based on GLUE Method. *J. China Hydrol.* **2016**, *36*, 1–6.
44. Xiong, L.; Wan, M.; Wei, X.; O'Connor, K.M. Indices for assessing the prediction bounds of hydrological models and application by generalised likelihood uncertainty estimation. *Hydrolog. Sci. J.* **2009**, *54*, 852–871. [CrossRef]
45. Zhao, D.; Chen, J.; Wang, H.; Tong, Q. Application of a Sampling Based on the Combined Objectives of Parameter Identification and Uncertainty Analysis of an Urban Rainfall-Runoff Model. *J. Irrig. Drain. Eng.* **2013**, *139*, 66–74. [CrossRef]
46. Wagner, B.; Reyes-Silva, J.D.; Forster, C.; Benisch, J.; Helm, B.; Krebs, P. Automatic Calibration Approach for Multiple Rain Events in SWMM Using Latin Hypercube Sampling. In *Green Energy and Technology*; Mannina, G., Ed.; Springer: Cham, Switzerland, 2019; Volume 23, pp. 435–440.
47. Sun, N.; Hall, M.; Hong, B.; Zhang, L. Impact of SWMM Catchment Discretization: Case Study in Syracuse, New York. *J. Hydrol. Eng.* **2014**, *19*, 223–234. [CrossRef]
48. Zhang, W.; Li, T. The Influence of Objective Function and Acceptability Threshold on Uncertainty Assessment of an Urban Drainage Hydraulic Model with Generalized Likelihood Uncertainty Estimation Methodology. *Water Resour. Manag.* **2015**, *29*, 2059–2072. [CrossRef]
49. Sun, N.; Hong, B.; Hall, M. Assessment of the SWMM model uncertainties within the generalized likelihood uncertainty estimation (GLUE) framework for a high-resolution urban sewershed. *Hydrol. Process.* **2014**, *28*, 3018–3034. [CrossRef]
50. Zhu, Z.; Chen, X. Evaluating the Effects of Low Impact Development Practices on Urban Flooding under Different Rainfall Intensities. *Water* **2017**, *9*, 548. [CrossRef]
51. Elliott, A.H.; Trowsdale, S.A. A review of models for low impact urban stormwater drainage. *Environ. Modell. Softw.* **2007**, *22*, 394–405. [CrossRef]

Article

Parameter Uncertainty of a Snowmelt Runoff Model and Its Impact on Future Projections of Snowmelt Runoff in a Data-Scarce Deglaciating River Basin

Yiheng Xiang [1], Lu Li [2,*], Jie Chen [1,3,*], Chong-Yu Xu [4], Jun Xia [1,3], Hua Chen [1] and Jie Liu [1]

[1] State Key Laboratory of Water Resources and Hydropower Engineering Science, Wuhan University, Wuhan 430072, China; xiangyh@whu.edu.cn (Y.X.); xiajun666@whu.edu.cn (J.X.); chua@whu.edu.cn (H.C.); liujie16@whu.edu.cn (J.L.)
[2] NORCE Norwegian Research Centre, Bjerknes Centre for Climate Research, 5007 Bergen, Norway
[3] Hubei Provincial Key Lab of Water System Science for Sponge City Construction, Wuhan University, Wuhan 430072, China
[4] Department of Geosciences, University of Oslo, P.O. Box 1047, Blindern, 0316 Oslo, Norway; c.y.xu@geo.uio.no
* Correspondence: luli@norceresearch.no (L.L.); jiechen@whu.edu.cn (J.C.)

Received: 16 October 2019; Accepted: 11 November 2019; Published: 18 November 2019

Abstract: The impacts of climate change on water resources in snow- and glacier-dominated basins are of great importance for water resource management. The Snowmelt Runoff Model (SRM) was developed to simulate and predict daily streamflow for high mountain basins where snowmelt runoff is a major contributor. However, there are many sources of uncertainty when using an SRM for hydrological simulations, such as low-quality input data, imperfect model structure and model parameters, and uncertainty from climate scenarios. Among these, the identification of model parameters is considered to be one of the major sources of uncertainty. This study evaluates the parameter uncertainty for SRM simulation based on different calibration strategies, as well as its impact on future hydrological projections in a data-scarce deglaciating river basin. The generalized likelihood uncertainty estimation (GLUE) method implemented by Monte Carlo sampling was used to estimate the model uncertainty arising from parameters calibrated by means of different strategies. Future snowmelt runoff projections under climate change impacts in the middle of the century and their uncertainty were assessed using average annual hydrographs, annual discharge and flow duration curves as the evaluation criteria. The results show that: (1) the strategy with a division of one or two sub-period(s) in a hydrological year is more appropriate for SRM calibration, and is also more rational for hydrological climate change impact assessment; (2) the multi-year calibration strategy is also more stable; and (3) the future runoff projection contains a large amount of uncertainty, among which parameter uncertainty plays a significant role. The projections also indicate that the onset of snowmelt runoff is likely to shift earlier in the year, and the discharge over the snowmelt season is projected to increase. Overall, this study emphasizes the importance of considering the parameter uncertainty of time-varying hydrological processes in hydrological modelling and climate change impact assessment.

Keywords: Snowmelt Runoff Model; parameter uncertainty; data-scarce deglaciating river basin; climate change impacts; generalized likelihood uncertainty estimation

---

1. Introduction

The intergovernmental on Climate Change (IPCC) stated that the temperature will continue to increase in the 21st century [1]. Potential impacts of global warming on water availability, especially in snow- and glacier-dominated regions are of great concern [2]. Due to the sensitivity and vulnerability of snow and glaciers due to climate change, previous studies have paid considerable attention to how climate change impacts snow accumulation [3–5], as well as to the timing and amount of snowmelt [6–8]. Water resources in snow- and glacier-dominated regions will suffer from the probable risk of declining amounts of snow and size of glaciers. In order to assess the climate change impacts on the water resources in snow- and glacier-dominated watersheds, it is common to use data from climate simulations to drive snow/glacier hydrological models [9–11]. In these models, snow- and glacier-related processes are usually simulated by energy budget methods [12–14] and/or temperature index methods [15,16], as well as/or multi-layer snow models with snow depth data assimilation methods [17–19]. The physically-based energy-budget methods and snow models incorporate energy fluxes occurring within the snowpack and interacting with the surface [20], while the empirical temperature-index method assumes a linear relationship between air temperature and the rate of snowmelt [21]. Since the energy budget method requires more meteorological data as input, it is often impractical to conduct at high-altitude snow and glacier basins with sparse observations. Temperature-index methods are more practical for snowmelt simulation in data-scarce mountain regions with insufficient data, as the air temperature is relatively widely available compared to other components [21].

The Snowmelt Runoff Model (SRM) is one of the most commonly used temperature-index models. It is a conceptual model developed to simulate and predict daily streamflow based on the degree-day method in high mountain basins where snowmelt is a major contribution [16]. It has been used to assess the climate change impacts on river discharge in mountainous environments [11,22–26]. For instance, Elias et al., [11] investigated the climate change impacts on streamflow timing and runoff volume utilizing the SRM in the Upper Rio Grande basin, Colorado and New Mexico, USA, and found that the large decrease in runoff volume in summer would pose challenges to water management in this region. Tahir et al., [25] applied the SRM to assess the climate change impacts on snowmelt runoff in the Upper Indus Basin and showed that the SRM is effective in estimating snowmelt runoff in high mountain regions. Xie et al., [26] proposed and applied a progressive segmented optimization algorithm (PSOA) for calibrating time-variant parameters of the SRM on the Manas River basin of Xinjiang, China. Their study showed that the PSOA can effectively calibrate the time-variant model parameters while avoiding the high increase in computational time caused by a significant increase of parameter dimensionality.

However, there are multiple sources of substantial uncertainties when applying the SRM for snowmelt runoff modeling and projection. Snowmelt modeling generally involves three uncertainty sources related to forcing data (i.e., input data and data for calibration), imperfections in the model structure and model parameters [27]. In climate change impact studies, uncertainties arise from different climate scenarios, as well as from parameter instability due to the possible changes in a basins' physical characteristics [28]. Among these uncertainties, the identification of model parameters is an essential step for any hydrological modeling, as there are parameters that cannot be derived directly from watershed characteristics but rather need to be inferred via a trial-and-error process [29,30].

Parameter uncertainty may be especially amplified in future climate change impact studies [31–33]. Therefore, quantifying uncertainties due to parameter instability, identifiability and non-uniqueness should be undertaken routinely, and robust methodologies should be implemented. Much attention has been given to uncertainty in model parameters in the hydrological literature [28,31,34–36]. Among the methods developed to deal with parameter uncertainty, the generalized likelihood uncertainty estimation (GLUE) method [37] has been widely used in recent decades [38–41]. The GLUE method is easy to implement and allows a flexible definition of the so-called likelihood function. The likelihood function, which is used to separate behavioural and non-behavioural solutions is particularly valuable for assessment of integrated, distributed models that operate with multi-variable, multi-site and multi-response criteria [39]. Moreover, a numerical Monte Carlo sampling technique [42] which is used to sample from the prior parameter distribution is usually coupled with GLUE for quantifying parameter uncertainties in hydrological modeling [40,43]. Conditioned on the available observational data and associated uncertainty bounds, this approach produces a well-distributed set of parameter distributions, the acceptability of which are similarly good at producing model simulations.

According to previous studies [44–46], the determination of parameters is essential to utilizing the SRM. Among the SRM parameters, the runoff coefficients (C_s and C_r) expressing the losses as ratios (C_s/C_r corresponds to the ratio of runoff contributed by snowmelt/rainfall to precipitation) vary in time, and so are affected by different climatic conditions and land cover properties in a hydrological year. They are usually calibrated if the agreement between observed and simulated runoffs is not acceptable [47]. The calibration strategy most commonly used for time varying parameters is to calibrate them in several sub-periods during the whole calibration period, and then assume that these parameters are constant in each sub-period. For example, Senzeba et al., [48] used the SRM to simulate the snowmelt runoff in a seasonally snow-covered eastern Himalayan catchment in 2006, 2007 and 2009. Single values of C_S and C_r were applied for each calibration year and the mean value from the three calibration years was used for a validation year (hereafter referred to as 1-period calibration). Zhang et al., [49] divided the hydrological year into two sub-periods (i.e., May–June and July–April) to calibrate the runoff coefficients of the SRM (hereafter referred to as 2-period calibration). In the study of Fuladipanah & Jorabloo [50], more than two sub-periods were divided in a hydrological year based on the variations of climatic conditions (hereafter referred to as multi-period calibration).

In the last few decades, the SRM has been applied successfully to mountainous watersheds all over the world, and it was recommended by Abudu et al., [46] that the SRM has a potential to forecast streamflow and to evaluate the effects of climate change on runoff in the mountainous watersheds of northwestern China, especially in data-scarce watersheds in high-elevation regions. However, the parameter uncertainty of the SRM restricts its application in this respect, as its parameter stability is relatively limited due to the highly dynamic behavior of snowpack and snowmelt.

Accordingly, this study aims to evaluate the parameter uncertainty for SRM simulation by using different calibration strategies, and to assess its impact on future hydrological projections. More specific objectives are to: (1) estimate the uncertainty of SRM simulation based on parameter uncertainty (i.e., the overfitting problem) and different calibration strategies (i.e., yearly calibration and multi-year calibration); and (2) assess the uncertainty of future SRM discharge projections determined by utilizing different calibrated parameter sets in a data-scarce deglaciating river basin.

In order to have a wide review of calibration strategies, we apply yearly calibration (which obtains optimal parameter sets from year to year) and multi-year calibration (which gets one optimal parameter value set for many years) with different sub-periods. The GLUE method is used to assess the flexibility and sensitivity of the parameters and to model the uncertainty arising from calibrated parameters.

2. Study Area and Data

2.1. Study Area

The data-scarce deglaciating Yurungkash watershed was chosen to implement the proposed study (Figure 1). It is one of the tributaries of the snow- and glacier-fed Tarim River Basin, the biggest inland river basin in China. The Yurungkash watershed is located approximately between 80°–82° N latitude and 74°–75° E longitude in the Xinjiang Uygur Autonomous Region and covers a surface area of about 14,600 km^2. It originates from the Akshay Glacier located on the northern slope of Kunlun Mountain, and its elevation ranges from about 1280 m above sea level (asl) up to 6780 m asl. The mean altitude of the watershed is about 4680 m asl and nearly 60% of the area lies above 5000 m asl.

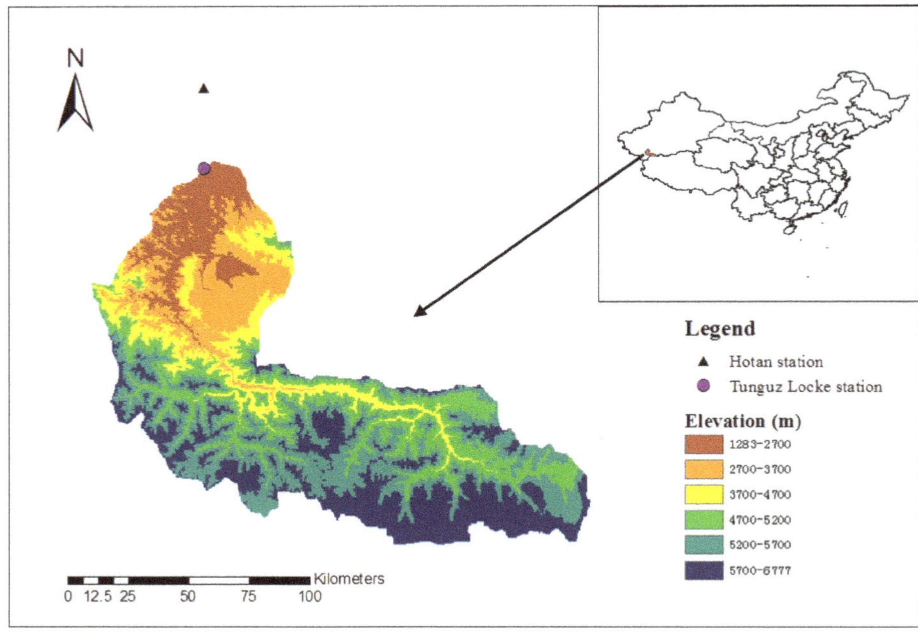

Figure 1. The location and topography of the deglaciating Yurungkash watershed with the meteorological station (Hotan) and the discharge station (Tunguz Locke).

The studied watershed is typically snow- and glacier-characterized, with about 40% of the watershed covered by permanent snow and ice. There is no meteorological station installed within the watershed; Hotan station is the closest one (Figure 1), about 600 km far away from the outlet of the river basin, located at 37.13° N, 79.93° E and 1375 m asl. The mean annual temperature of the study region is about –8 °C, and it receives a mean total annual precipitation of 260 mm with 65% falling between June and September. The Tunguz Locke station is located at the basin outlet (36.81° N, 79.92° E) and records a mean annual flow of 80 m^3/s. The annual runoff presents a large intra-annual variability with about 90% occurring in the snowmelt season (April–September).

2.2. Data

The Shuttle Radar Topography Mission (SRTM), comprised of a modified radar system and offering high-resolution DEM data [51], was used to derive the topography of the study area and delineate the elevation zones. The Yurungkash watershed was divided into six elevation zones and the properties of these zone areas are listed in Table 1. The mean elevation of each elevation zone was calculated by using hypsometric curves, which were also used to derive the temperature of different elevation zones by extrapolation of base station temperature. Daily air temperature observed at the Hotan station and daily precipitation from the China Ground Rainfall Daily Value 0.5° × 0.5° Lattice Dataset (CGRD) [52] for the 2003–2012 period were used for model simulation. Two different datasets were used for temperature and precipitation because precipitation at the Hotan station cannot well represent the whole study watershed, due to the large spatial variability of precipitation in mountain basins. The CGRD was generated by interpolating observed precipitation from more than 2000 meteorological stations in China, utilizing the thin plate spline approach which was developed to interpolate and smooth scattered data in geosciences [53]. The reliability of this dataset has been proven by Zhao et al., [54]. Daily runoff data at the Tunguz Locke station for the 2003–2012 period was used for SRM calibration and validation.

Table 1. Characteristics of elevation zones.

Zone	Elevation Range (m)	Mean Elevation (m)	Area (km^2)	Area (%)
A	1280–2700	2230	1324	9.07
B	2701–3700	3190	2248	15.40
C	3701–4700	4250	2170	14.86
D	4701–5200	4975	2485	17.02
E	5201–5700	5440	3238	22.18
F	5701–6780	6020	3135	21.47
Total	1280–6780	4470	14600	100

The daily snow-covered area (SCA) is another model input, derived from the Moderate Resolution Imaging Spectroradiometer (MODIS) [55] snow cover product. The MODIS 8-day composite snow cover data product (MOD10A2) is generated by compositing observations from the MODIS Snow Cover Daily L3 Global 500 m Grid (MOD10A1) data set. For every eight-day period, the grid is mapped as snow if snow is observed on any single day, and cloud cover is reported only if the grid was cloud-obscured for all eight days. The MOD10A2 can better eliminate the influence of clouds and improves the precision of snow identification over that of the MOD10A1 [56]. The daily SCA derived from the MOD10A2 by using linear interpolating was used for model simulation in this study. The mean monthly watershed-averaged temperature (T), precipitation (P) and snow-covered area (SCA) for the 2003–2012 period is presented in Figure 2. This figure indicates that the T is above 0 °C during May–September. The watershed receives much more precipitation in July and August than in other months. The SCA decreases from April and increases after August, and then decreases again after October. The minimum SCA of the watershed is greater than 40%, which is the area of its permanent snow/ice.

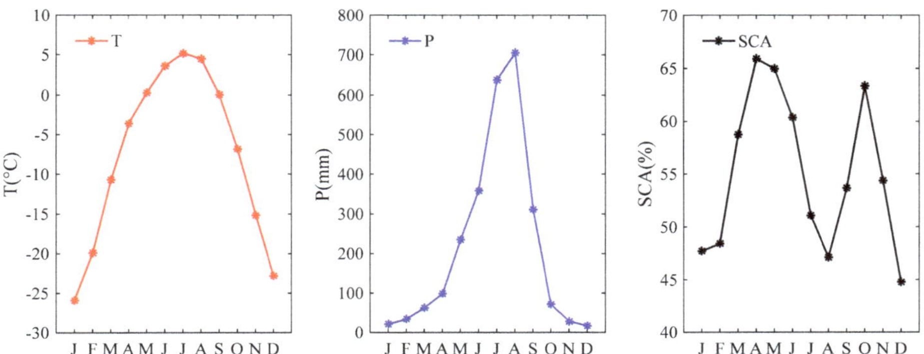

Figure 2. The watershed-averaged temperature (T), precipitation (P) and snow-covered area (SCA) (Period 2003–2012). The T, P and SCA were derived from the Hotan station, China Ground Rainfall Daily Value 0.5° × 0.5° Lattice Dataset and MODIS product, respectively.

For runoff projection, General Circulation Model (GCM) simulations from phase 5 of the Coupled Model Intercomparison Project [57] were used to drive the SRM. Although it is desirable to use as many GCMs as possible, the selection of a representative subset is usually necessary due to the high computational costs of model simulations.

The k-means clustering method [58], which has been used in several previous studies [59–61], was applied to select subsets from 26 GCMs under representative concentration pathway 4.5 (RCP 4.5) and RCP 8.5, respectively. The k-means clustering method divided the ensemble of 26 GCMs into a number of clusters to minimize the within-cluster sum of squared error (SSE). The SSE was characterized by the Euclidean distance based on the standard changes in temperature and precipitation simulated by GCMs between the 2003–2012 and the 2041–2050 periods. A value of four clusters was determined to offer good group separation as well as a manageable number of simulations. Figure 3 presents the selected subsets of GCMs under RCP 4.5 and RCP 8.5. In each cluster, the colour marker which is closest to the centroid is the selected one. Table 2 presents the detailed information of the seven GCMs used in this study.

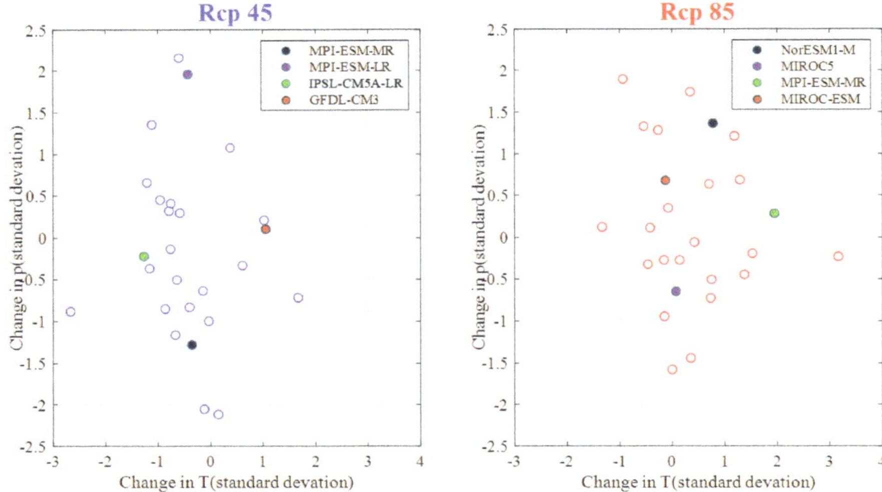

Figure 3. Sub-sets of the four selected General Circulation Model (GCMs) (color-filled markers) under representative concentration pathway (RCP) 4.5 and RCP 8.5 found by the k-means clustering method from a total of 26 GCMs (non-color-filled markers). These are based on the standard changes in temperature and precipitation simulated by GCMs between the 2003–2012 and the 2041–2050 periods.

Table 2. Seven General Circulation Model simulations' details of CMIP5.

Model Name	Institute/Country	Horizontal Resolution (lon×lat)
MPI-ESM-MR	Max Planck Institute for Meteorology, Germany	192 × 96
MPI-ESM-LR	Max Planck Institute for Meteorology, Germany	192 × 96
IPSL-CM5A-LR	Institute Pierre-Simon Laplace, France	96 × 96
GFDL-CM3	USA	144 × 90
NorESM1-M	Norwegian Climate Centre, Norway	144 × 96
MIROC5	MIROC, Japan	256 × 128
MIROC-ESM	MIROC, Japan	128 × 64

3. Methods

3.1. SRM Modeling

The SRM is a conceptual, simple degree-day-based model that was designed to simulate and forecast daily streamflow resulting from snowmelt and rainfall in mountain basins. Since it was developed by Martinec et al., [16], the SRM has been successfully applied to many mountain basins of various sizes ranging from 0.76 to 917,444 km^2, and elevations ranging from 0 to 8848 m asl [62]. In recent years, the SRM has also been applied to investigate the impacts of climate change on runoff [7,11,25].

In this study, the SRM is applied to simulate and predict daily runoff for the Yurungkash watershed. It requires three daily input variables (temperature, precipitation and snow-covered area) and has seven parameters (runoff coefficients Cs and Cr, the degree-day factor, critical temperature, rainfall contribution area, recession coefficient and the time lag). A detailed description of the SRM model can be found in the Appendix A.

3.2. SRM Calibration Methods

Among the SRM parameters, the runoff coefficients (*Cs* and *Cr*) vary with time, and so they are the parameters that are most sensitive to different climatic conditions and land cover properties over a year-long time frame [47]. Therefore, they are usually the parameters chosen for calibration in SRM modelling. In order to have a broad perspective for calibrating runoff coefficients, both the yearly and multi-year calibration strategies with 1, 2, 4, 5, 6, and 7 sub-periods were used for comparison. X-period calibration means that a hydrological year is divided into × sub-periods based on similar characteristics in climatic data for runoff calibration. In total, our approach used 12 calibration strategies. Additional details, including the beginning and ending dates of each sub-period of the different calibration strategies, are presented in Figure 4. For the 2-period calibration, the ending date of the last period is 31st may in the following year. For the other calibration strategies, the ending date of the last period is 31st March in the following year. Apart from C_s and C_r, other model parameters are presented in Table 3, which were determined as described in the appendix. Calibration was carried out in the odd-numbered years and validation was performed in the even-numbered years for the 2003–2012 period.

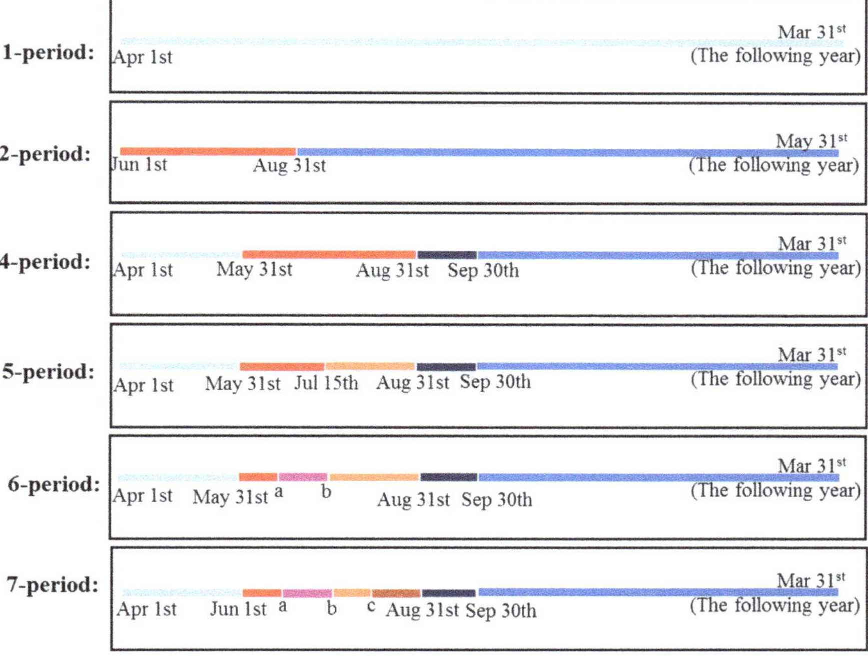

Figure 4. The beginning and ending dates of each period in a hydrological year for different calibration strategies. Each color in each calibration strategy represents a sub-period. For example, there are 2 sub-periods in the 2-period calibration, the 1st period is June 1 August 31 and the 2 period is September 1 May 31. *a* represents June 20, *b* represents July 15 and *c* represents August 5.

Table 3. Parameters determined by hydrological judgment as described in the appendix.

Model Parameters	Values
Degree Day Factor (cm/°C/day)	0.3
Lapse Rate (°C/100 m)	0.65
Threshold Temperature, T_{crit}	1 (June–August); 3 (September–May)
Rainfall Contributing Area, RCA	1 (May–September); 0 (October–April)
Recession Coefficient, K, which is Determined by: $K_{n+1} = \frac{Q_{n+1}}{Q_n} = x \times Q_n^{-y}$	$X_c = 0.9895$; $X_y = 0.026$

The Nash-Sutcliffe coefficient (NSE) [63] and the relative volume error (VE), two classical efficiency criteria in hydrological modelling, were used to evaluate the performance of the SRM. The NSE and the VE are computed as follows:

$$\text{NSE} = 1 - \frac{\sum_{i=1}^{n}\left(Q_i - Q_i'\right)^2}{\sum_{i=1}^{n}\left(Q_i - \overline{Q}\right)^2} \tag{1}$$

$$\text{VE}[\%] = \frac{V_R - V_R'}{V_R} \cdot 100 \tag{2}$$

where Q_i is the measured daily discharge, Q_i' is the simulated daily discharge, $\overline{Q}$ is the average measured discharge of the given time period or snowmelt season, V_R is the measured yearly or seasonal runoff volume, V_R' is the simulated yearly or seasonal runoff volume, and n is the number of daily discharge values.

3.3. Bias Correction of GCM Outputs

_ENREF_12GCM outputs are too coarse and biased to directly drive an SRM in climate change impact studies; therefore, a downscaling or bias correction process is needed [64]. In this study, the watershed-averaged value of the GCM–simulated precipitation was first calculated by the Thiessen Polygon method [65]. The GCM–simulated temperature was interpolated to the Hotan station using the Inverse Distance Weighted method [66] in order to be extrapolated later to the higher elevation zones. The precipitation and temperature were then corrected, respectively based on the watershed-averaged precipitation from CGRD and temperature at the Hotan station, by a distribution-based bias correction method, called the Daily Bias Correction (DBC) method in Chen et al., [67]. The DBC is a hybrid method combining the Local intensity scaling (LOCI) method [68] to correct the precipitation occurrence and the Daily translation (DT) method [69] to correct the frequency distributions of precipitation amounts and temperatures. Here are the two steps of this DBC method:

1. The LOCI method was used to correct the precipitation occurrence, which ensures that the frequency of the precipitation occurrence simulated by GCMs at the reference period equals that of the observed data for a specific month. A threshold for precipitation occurrence determined at the historical period was then applied to the future period.
2. The DT method was used to correct the empirical distribution of GCM-simulated precipitation and temperature magnitudes in terms of 100 quantiles from 0.01 to 1 with an interval of 0.01.

3.4. Future Snow Covered Area Projection

A changed climate would cause a change in the SCA in mountain snow basins. The method for future SRM simulation developed by Rango & Martinec [70] was used to evaluate the impact of future temperature and precipitation on SCA in this study. This method estimates the time shift of the SCA in the present climate over the snowmelt period to produce the SCA in a future climate. The characteristics of the watershed (i.e., the number of elevation zones, the area and the mean elevation of each zone) and the SCA parameters (i.e., the degree-day factor, the temperature lapse rate and the critical temperature) determined in the present climate, as well as the bias corrected temperature and precipitation in the future climate are required for future SCA projection. More detailed procedures for calculating SCA can be found in Rango and Martinec [70].

3.5. GLUE Method for Uncertainty Estimates

The GLUE method [37,71] coupled with Monte Carlo sampling [42] was used in this study to investigate the parameter uncertainty of the SRM. In this method, a large number of model runs were made by using many different parameter sets randomly selected from a priori probability distribution. The acceptability of each run was evaluated against observed values and, the posterior parameter distribution was extracted based on a certain subjective threshold to maintain a good population of behavioral solutions. The parameter set with the corresponding model run considered to be non-behavioral was removed from further analysis. In this study, the GLUE scheme associated with the SRM includes the following steps:

1. 100,000 Monte Carlo sampling points of C_s and C_r were implemented from a feasible parameter space (0–1) with uniform distribution;
2. The likelihood values were calculated for all 100,000 model runs;
3. The likelihood functions (NSE and VE) and the threshold values (0.55 and ±10%, respectively) were specified as behavioral parameter sets; and
4. Posterior parameter sets were extracted depending on the threshold of the likelihood functions.

The sample sizes of the posterior parameter distributions were all over 1000. These posterior parameter sets were then used in validation and projection. It should be mentioned that for the yearly calibration, the posterior parameter sets of each year were used individually in the validation and projection.

The following three indices were used to evaluate the resolution, reliability and efficiency of the uncertainty interval estimates:

1. The Average Relative Interval Length (ARIL) [34] measures the resolution of the predictive distributions, which is defined over the entire simulation time period as:

$$ARIL = \frac{1}{n} \sum \frac{Q_{Upper,t} - Q_{Lower,t}}{Q_{obs,t}} \tag{3}$$

where n is the number of days in the observed record, $Q_{Upper,t}$ and $Q_{Lower,t}$ are the upper and lower simulated discharges of the 95% confidence interval, respectively, and $Q_{obs,t}$ is the observed discharge.

The percentage of observations bracketed by the Confidence Interval (PCI) [72] measures the amount of the observed data within a range of simulated intervals, defined as:

$$PCI = \frac{Q_{obs,in}}{n} \tag{4}$$

where $Q_{obs,in}$ is the number of observed discharges that are contained within the 95% confidence interval. A smaller ARIL means a narrower uncertainty interval and a larger PCI means that the interval has a greater reliability; and

2. The percentage of observations bracketed by the Unit Confidence Interval (PUCI) [35] based on ARIL and PCI, which is defined as:

$$PUCI = \frac{(1 - Abs(PCI - 0.95))}{ARIL} \tag{5}$$

Larger PUCI values represent smaller uncertainty of 95% confidence interval of discharge in general.

For the climate change impacts on runoff projection for the 2041–2050 period, the sub-periods' division in a hydrological year may be different due to the changes in climatic conditions and land cover properties, which may bring further uncertainty to hydrological projections. This study assumed there is no change of sub-period division in the future climate. To better represent the changed climate, the value of C_s, which reflects the decline of the snow coverage and the stage of vegetation growth, was shifted earlier by 30 days in the climate run, as recommended by previous studies [11,16,73]. The climate change impacts on hydrology were assessed at the 95% confidence intervals of average annual hydrographs, annual discharge, and flow duration curves.

4. Results

4.1. Parameter Uncertainty

The SRM was calibrated based on the yearly and multi-year calibration strategies, with different sub-periods (i.e., 1-, 2-, 4-, 5-, 6- and 7-period strategies) in the odd-numbered years, and validation was performed in the even-numbered years during the 2003–2012 period. Calibration performances with NSE > = 0.55 and −10% < VE < 10%, as well as the corresponding validation performances are shown in Figure 5 for different calibration strategies. This figure shows that: (1) for both yearly and multi-year calibration strategies, the maximum NSE values at the calibration period increase from 1– to 5-period approaches, while there is no obvious improvement for the 6- and 7-period strategy; (2) the maximum NSE values are larger at the calibration period for yearly calibration, but smaller at the validation period than those for multi-year calibration; and (3) there are much more samples with NSE values below 0.55 and absolute values of VE above 10% when using more than two sub-periods at the validation period, especially for yearly calibration.

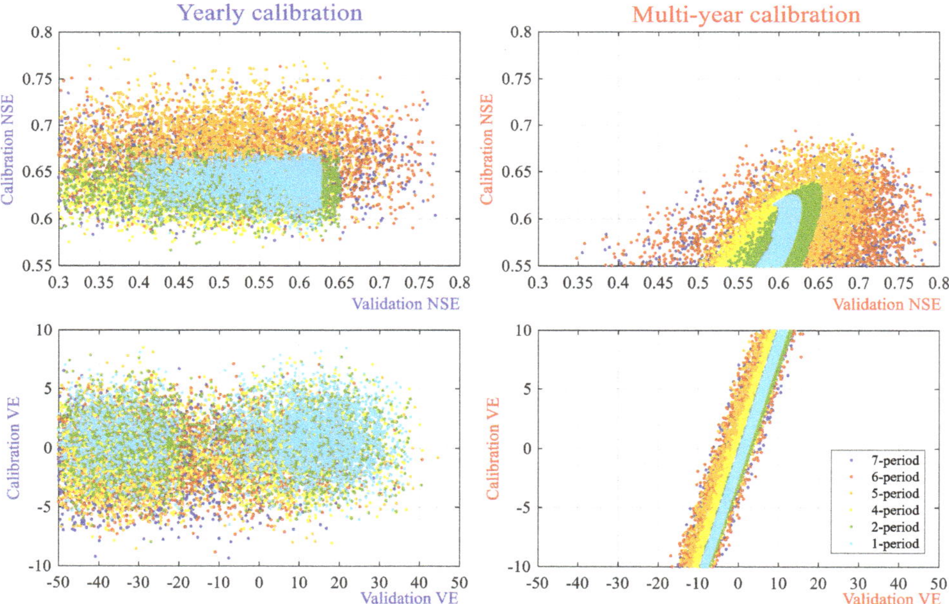

Figure 5. Scatter plots of the Nash-Sutcliffe coefficient (NSE) (upper) and the relative volume error VE (lower) of calibration versus validation for different calibration strategies. The left subplots present the yearly calibration and the right subplots present the multi-Year calibration.

4.2. Confidence Interval of Discharge

Three uncertainty indices for evaluating 95% confidence intervals of discharge derived from different calibration strategies are given in Table 4. The following features can be observed: (1) both

ARIL and PCI increase with the increase of sub-periods for both yearly and multi-year calibration strategies, making it difficult to identify which calibration strategy is better; (2) considering the PUCI, the 1– and 2-period calibrations have a larger value of PUCI than the other sub-periods, which indicates their results are more reliable and less uncertain; and (3) the PUCI values of the multi-year calibrations are generally larger than those of the yearly calibrations, which means the former approach is better.

Table 4. The Average Relative Interval Length (ARIL), the percentage of observations bracketed by the Confidence Interval (PCI) and the percentage of observations bracketed by the Unit Confidence Interval (PUCI) values of the different calibration strategies for the 95% confidence interval of discharge by different calibrations.

Sub-Period(s)	Yearly Calibration			Multi-Year Calibration		
	ARIL	**PCI**	**PUCI**	**ARIL**	**PCI**	**PUCI**
1	0.454	0.204	0.559	0.440	0.178	0.517
2	1.009	0.430	0.475	0.848	0.376	0.501
4	1.287	0.500	0.427	1.192	0.470	0.436
5	1.531	0.586	0.416	1.361	0.538	0.432
6	1.546	0.596	0.418	1.356	0.556	0.447
7	1.548	0.606	0.424	1.396	0.570	0.444

The 95% confidence intervals of the simulated daily discharge are presented in Figure 6. The results from the 5-, 6- and 7-period calibrations are not shown in this figure because they are very similar to those from the 4-period calibration. Figure 6 shows that the confidence intervals of discharge derived by yearly calibration are slightly larger than those derived by multi-year calibration. The confidence intervals also become larger with the increase in sub-periods, which is consistent with the higher PCI values in Table 4. Furthermore, discharges can be observed lying outside the confidence intervals, which indicates that there are some other uncertainty sources related to forcing data and imperfections in the model structure that impact SRM discharge simulation in addition to the parameter uncertainty.

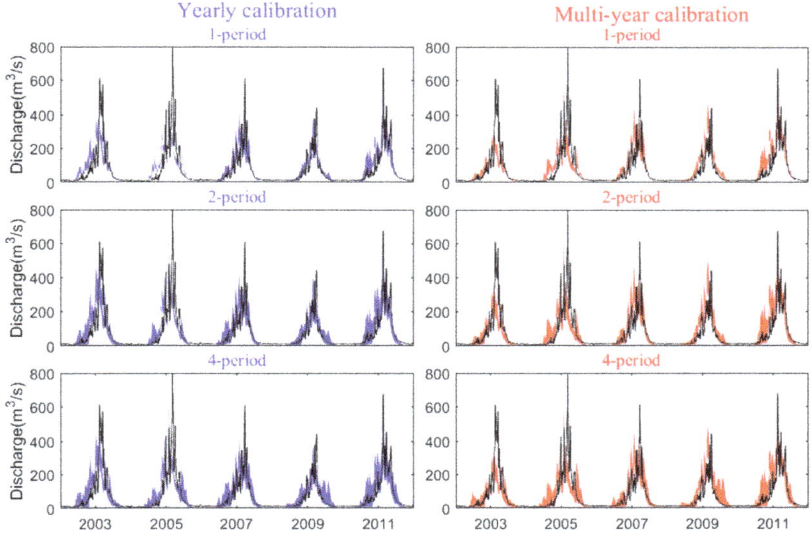

Figure 6. The 95% confidence intervals of daily discharge simulated by posterior parameter sets (color envelope), compared with observed runoff (black lines).

4.3. Projected Future Temperature, Precipitation, and SCA

By the degree-day method, daily temperature in a future climate modifies the SCA in the present climate to the approximate future SCA in the snowmelt season (April–September). Figure 7 depicts the daily watershed-averaged temperature, precipitation and SCA in snowmelt season from four selected GCMs under RCP 4.5 and RCP 8.5 for the 2041–2050 period (hereafter 'seasonal' means the average value from the snowmelt season). Four GCMs suggest an increase in temperature over the catchment between 1.2 to 2.6 °C under RCP4.5, and an increase of between 1.2 to 3.3 °C under RCP8.5. The precipitation over the catchment is also projected to increase by 12.0% to 39.8% under RCP4.5, and by 12.2% to 61.8% under RCP8.5. The SCA is projected to decrease in response to the high temperatures in late June. The figure shows that the predicted SCA decreases rapidly from *a* (June 21st, 58.0%) to *b* (July 31st, 40.1%) under RCP4.5 and from *c* (June 11st, 60.3%) to *d* (July 25th, 42.8%) under RCP8.5. The relative decrease in seasonal SCA ranges from 5.7% to 17.7% under RCP4.5, and from 1.2% to 14.8% under RCP8.5.

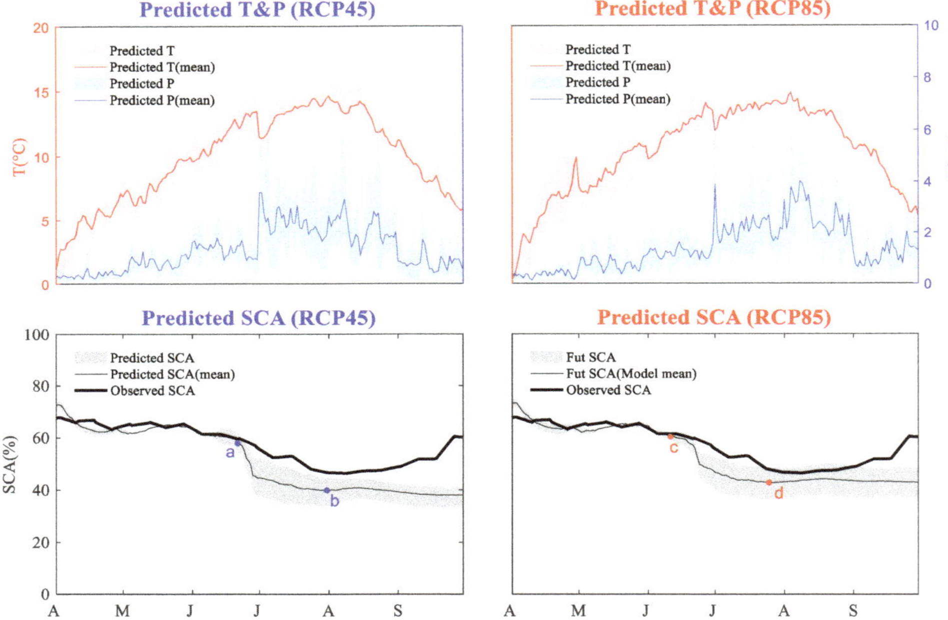

Figure 7. Projected mean monthly temperature (T), precipitation (P) and snow coverage area (SCA) (Period 2041–2050) compared with observations (Period 2003–2012). The pink shading represents the four chosen GCMs, whose mean value is represented by the red line. The same representation is used for the precipitation, indicated in blue. *a* represents (June 21st, 58.0%), *b* represents (July 31st, 40.1%), *c* represents (June 11st, 60.3%), *d* represents (July 25th, 42.8%).

4.4. Uncertainty in the Discharge Projections

The daily discharges are simulated by the SRM using bias corrected temperature and precipitation from the chosen GCMs under RCP4.5 and RCP8.5, as well as the projected SCA, utilizing the posterior parameter sets derived by the yearly and multi-year calibration strategies with 1-, 2- and 4-period segments. Figure 8 presents the 95% confidence interval of the mean daily discharge for the snowmelt season of the 2041–2050 period. The observed mean daily discharge of the 2003–2012 period is also presented for comparison. The confidence intervals represent uncertainties from four different GCMs as well as from the calibrated parameters. The following results can be observed: (1) the median values of the future discharge simulated by the yearly and multi-year calibration strategies are higher than those of the historical discharge, with the onset of snowmelt runoff shifted earlier. (2) The parameter uncertainty of the yearly calibration is larger than that of the multi-year calibration, as indicated by the wider pink intervals. (3) The discharge confidence interval is wider when there are more sub-periods, in particular for the 4-period strategy, which is much wider than the others.

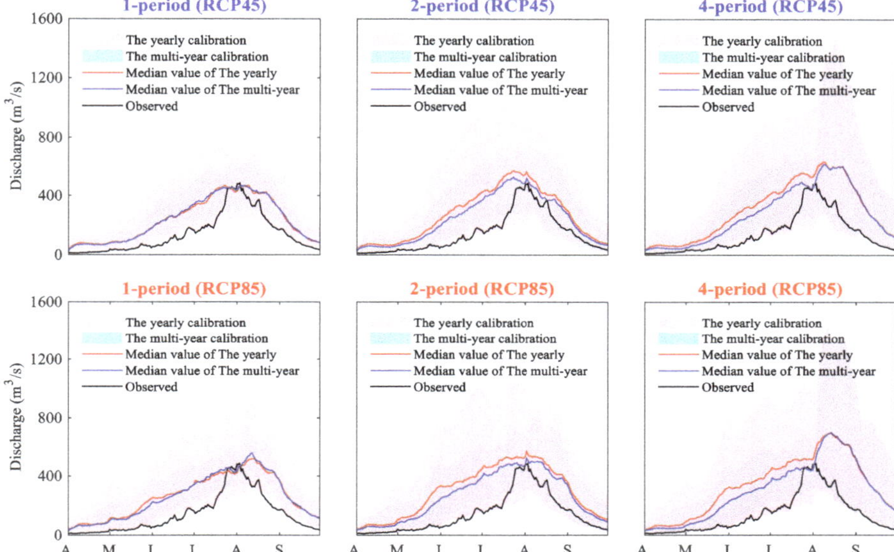

Figure 8. The 95% uncertainty intervals for the GCM–srm predicted daily discharge (Period 2041–2050). The bottom and top of the colored intervals represent the 2.5th and 97.5th percentile values, respectively. In each subplot, the pink and blue shadings represent the yearly and the multi-year calibration strategies, respectively. Median values of the pink and blue envelopes are shown as the red and blue curves, respectively.

Figure 9 shows the 95% confidence intervals of the mean discharge in the snowmelt season from four GCMs under RCP4.5 and RCP8.5 for the 2041–2050 period. The observed mean discharge during the snowmelt season for the 2003–2012 period is also shown. Generally, the projections suggest increases in the mean discharge for the future period, especially under RCP8.5. The confidence intervals of mean discharge from the 1- and 2-period calibrations are smaller than those from the 4-period calibration. In order to compare different sources for uncertainty, these discharge confidence intervals were grouped into five sources (RCPs, GCMs, multi-year/yearly calibration strategies, division of sub-periods, and calibrated parameters). For example, to investigate the uncertainty related to the RCP, discharges were grouped by two RCPs. Each group includes discharges from four GCM, and yearly and multi-year calibration strategies with different sub-periods. The results show that uncertainty ranges related to the choice of RCP, GCM, multi-year/yearly calibration, sub-period and calibrated parameter are [283.11, 308.25], [260.92, 347.93], [267.30, 318.49], [201.86, 229.22] and [165.18, 307.91] with the unit of m^3/s, respectively. In terms of the uncertainty range, the largest uncertainty is found to be related to the SRM parameters.

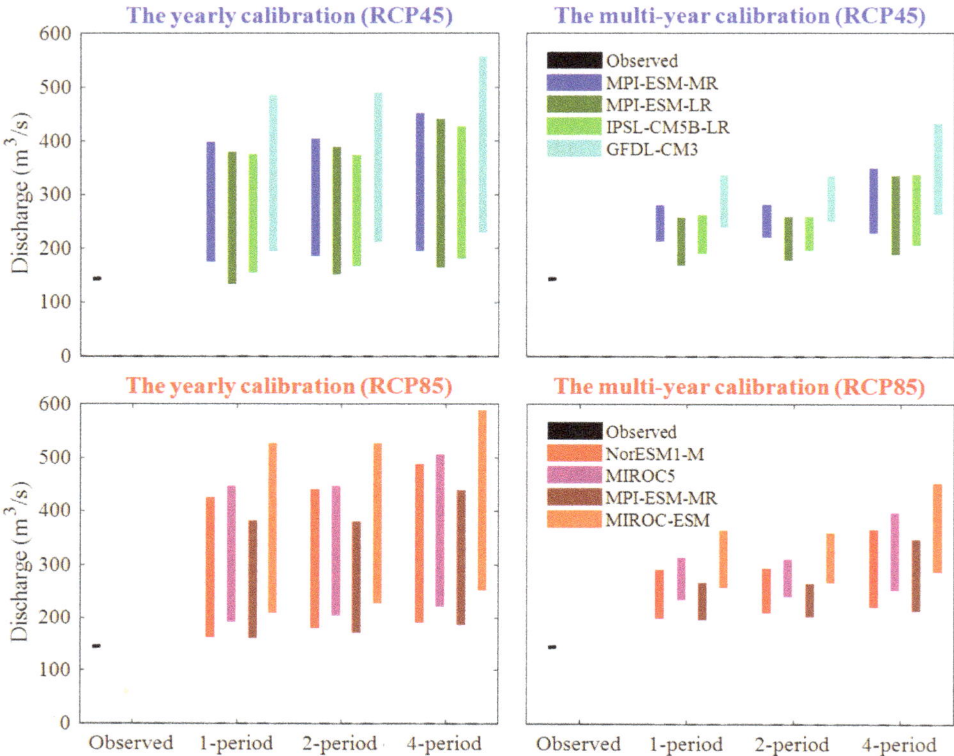

Figure 9. The 95% confidence intervals for the GCM–SRM predicted average discharge of the snowmelt season (Period 2041–2050). The bottom and top of the colored bars represent the 2.5th and 97.5th percentile values, respectively. The black bar represents the average snowmelt season discharge during 2003–2012.

The 95% confidence intervals of the flow duration curve (FDC) for four GCMs under RCP8.5 for the 2041–2050 period are shown in Figure 10. Only the high-representative concentration pathway of RCP8.5 is presented to illustrate the uncertainties of the FDC and discuss the differences between GCMs and calibration strategies. Figure 10 reveals that the low, mid, and high flows are all predicted to increase in the Yurungkash watershed for the 2041–2050 period. The highest flow, which occurs less than 10% of the time, is predicted to increase or decrease for different GCMs. This suggests that the number of days with flows exceedingly above 10% will increase in the future, but there will be no significant change in the number of days with the highest flows.

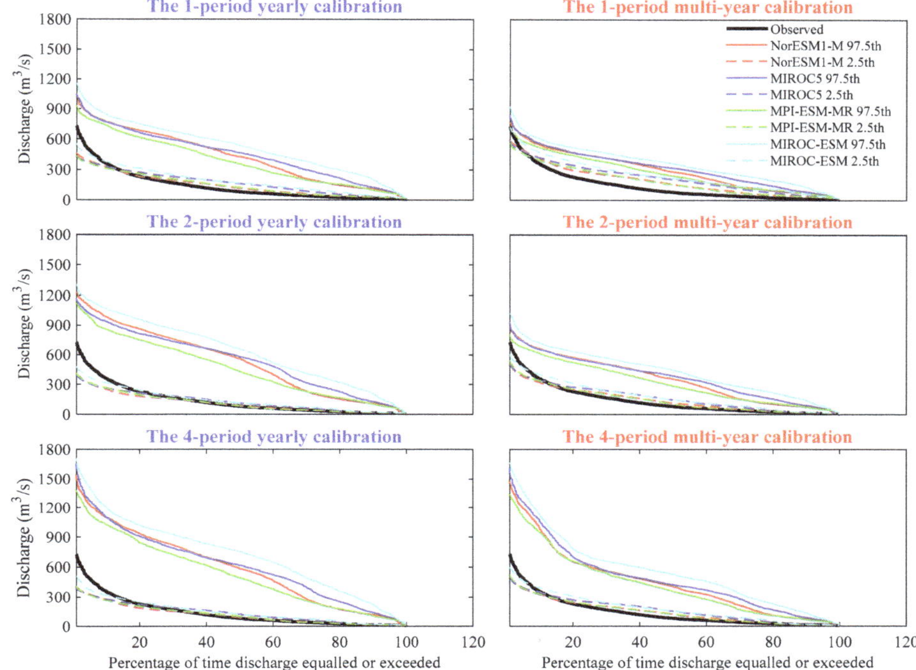

Figure 10. The 95% confidence intervals for the GCM–SRM predicted flow duration curves under RCP8.5 (Period 2041–2050) compared with the observations (Period 2003–2012) shown as black line. The dotted and solid lines represent the 2.5th and 97.5th percentile values, respectively.

5. Discussion

It is widely acknowledged that calibration plays an important role in any hydrological modelling due to parameters that are spatially and temporally heterogeneous or that cannot be measured. Hence, the problem of parameter optimization and the associated uncertainty must be addressed [74]. Our results show that the simulation performance of the SRM is sensitive to calibration strategies with different sub-periods for calibrating the time-varying parameters. In terms of the NSE and the VE, the model's calibration performance becomes better with more sub-periods. However, there are more NSE values below 0.55 when the sub-period is more than 4, indicating higher uncertainty. According to previous studies [32,75], increasing the number of calibrated parameters could yield better calibration performance, while also bringing larger uncertainties at the validation period. An overfitting problem may arise from over-parameterization when the amount of information contained in a hydrograph used for calibration is not enough to estimate a larger number of parameters [76]. An over-fitted model results in a smaller error for calibration but a larger error for validation. Hence, the 1– and 2-period strategies appear to be the most rational choice for calibration when considering the balance between simulation performances and the overfitting problem. In addition, the multi-year calibration would be more stable than the yearly calibration for SRM modeling, as the former performs better at the validation period.

In addition, much attention has been paid to parameter uncertainty assessment in hydroclimatic modeling recently [33,77]. In this study, both the yearly and multi-year calibration strategies with 1-, 2- and 4-periods were used for hydrological climate change impact assessment. Although the runoff projection contains large uncertainty, it shows that the onset of snowmelt runoff shifts earlier and the mean discharge of the snowmelt season increases. This result is similar to that of Wang et al., [78], in which the response of snowmelt runoff to climate change was investigated in a basin located in northwest China.

Particularly, it is noticeable that the predicted hydrographs derived from the 4-period calibration do not shift earlier for peak river runoff. This is not consistent with the common view that the effects of a warming climate will lead to a peak flow shifting from summer and/or autumn to winter and/or early spring [2]. Elias et al., [11] applied a forward shift of 30 days for C_s based on a 2-period calibration in the climate change impact assessment and predicted higher streamflow in March and April, which is similar to the results derived from the 1- and 2-period calibrations in our study. The result of the 4-period calibration is different from our expectations because it shifts a large value of C_s is shifted to August, which leads to the peak flow occurring in August under climate change. This result indicates that the shift in runoff timing predicted by more sub-periods contains greater uncertainty. In addition, the predicted hydrograph and mean discharge show much wider intervals with the 4-period calibration than with the 1- and 2-period calibrations, which also indicates larger uncertainties. Hence, the 1- and 2-periods are recommended for SRM hydrological climate change impact assessment. This work also suggests that a multi-year calibration strategy would be more stable than a yearly calibration strategy in climate change impact studies.

It is necessary to point out some of the limitations of this study. This study adopted the GLUE method to estimate parameter uncertainty, one drawback of which is that the derived parameter distributions and uncertainty intervals are sensitive to some subjective factors, e.g., the threshold values of the estimation criteria [40,43]. Lower threshold values could result in a wider uncertainty interval of the posterior distribution [34] which may explain why the parameter uncertainty of SRM has the largest impact on the future hydrological projection compared to other uncertainties in this study. This result is similar to that of Tian et al. [79], in which parameter was found to be the major source of uncertainty for simulating future high flows over a southern river basin in China. However, this result is different from that of previous studies [64,80], which showed that hydrological model parameter uncertainty was relatively small compared to other uncertainty sources (RCP and GCM). In previous studies, hydrological model parameters were calibrated by a certain optimization algorithm, which is not recommended for the SRM [47]. However, the use of thresholds in this study cannot guarantee that all parameters are optimized. Nevertheless, the results indicate that the uncertainty associated with a parameter should be highlighted in SRM modeling. It is acknowledged that the parameter distribution and uncertainty intervals derived by GLUE have no clear statistical meaning [81]. Comparison of the results by different threshold values is needed to test the sensitivity of the GLUE, as recommended by Jin et al., [34].

Another limitation of this study is that only different calibration strategies were taken into account in the parameter estimation. It would be more comprehensive to calibrate parameters over different periods, such as wet and dry climate periods [82], as well as by a multi-objective calibration approach [83]. In addition, Seiller et al., [77] showed that the diagnosis of the impacts of climate change on water resources is very much affected by the objective functions used for hydrological model calibration. Investigation of the parameters' nonstationarity in a changed climate could be another avenue for future studies [84].

Also, it is necessary to mention that although temperature-index models are more practical, they are often less accurate than full energy balance models [85]. In addition, the skill of temperature-index models is likely to decrease for future climate change impact studies due to the changes in the energy balance which is not a direct result of temperature (e.g., changes in albedo). The degree day factor has been shown to change over time scales of a decade or even over a single melt season [85–87].

6. Conclusions

This study evaluated the parameter uncertainty for SRM simulation by different calibration strategies and its impact on future hydrological projections in a data-scarce deglaciating river basin.

Different sub-periods were compared and the uncertainty of future discharge derived by the calibrated parameter sets was assessed. The following conclusions can be drawn:

1. The strategy with a division of 1 or 2 sub-period(s) in a hydrological year is appropriate for SRM modeling when considering the balance between simulation performance and the overfitting problem/uncertainty. In addition, the multi-year calibration approach is more stable than the yearly calibration for SRM hydrological simulation and projection, as the latter presents a lower validation performance combined with higher future projection uncertainty.

2. The future runoff projection contains large uncertainties, among which parameter uncertainty plays a significant role. The projection results indicate that the onset of snowmelt runoff is likely to shift earlier, and the discharge of the snowmelt season is projected to increase for the 2041–2050 period.

Overall, this research highlights the parameter uncertainty of SRM and its impact on hydrological climate change assessment in a data-scarce deglaciating river basin. The results imply that for hydrological models, like the SRM, time-variant parameters could result in over-parameterization issues and bring uncertainties in the hydrological simulation. Both of these have a significant effect on climate change impact assessment, and therefore should be routinely considered when using these models.

Author Contributions: Data curation, Y.X.; Funding acquisition, L.L. and J.C.; Investigation, Y.X.; Methodology, Y.X. and L.L.; Writing – original draft, Y.X.; Writing – review & editing, L.L., J.C., C.-Y.X., J.X., H.C. and J.L.

Funding: This research was funded by [the State Key Laboratory of Water Resources and Hydropower Engineering Science funding] grant number [2017SWG02], [the National Natural Science Foundation of China] grant number [51779176, 51525902, 51539009], and [the Thousand Youth Talents Plan from the Organization Department of CCP Central Committee (Wuhan University, China)]. The APC was funded by [the State Key Laboratory of Water Resources and Hydropower Engineering Science funding] grant number [2017SWG02].

Acknowledgments: This work was partially supported by the State Key Laboratory of Water Resources and Hydropower Engineering Science funding (No. 2017SWG02), the National Natural Science Foundation of China (Grant No. 51779176, 51525902, 51539009) and the Thousand Youth Talents Plan from the Organization Department of CCP Central Committee (Wuhan University, China). The authors would like to acknowledge the contribution of the World Climate Research Program Working Group on Coupled Modeling, and to thank climate modeling groups for making available their respective climate model outputs. The authors wish to thank the China Meteorological Data Sharing Service System and the Xinjiang Tarim River Management Bureau for providing the dataset for the Yurungkash river basin.

Conflicts of Interest: The authors declare no conflicts of interest.

Appendix A

Appendix A.1 SRM Modelling

The following equation shows the daily discharge estimated by superimposing snowmelt and rainfall on the calculated recession flow:

$$Q_{n+1} = [C_{sn} \times a_n(T_n + \Delta T_n)S_n + C_{rn} \times P_n] \times A \times \alpha \times (1 - K_{n+1}) + Q_n K_{n+1} \tag{A1}$$

where:

Q = the average daily discharge (m^3/s)

C_s/C_r = the runoff coefficient expressing the losses as a ration of runoff to precipitation, with C_s referring to snowmelt and C_r referring to rainfall

a = the degree-day factor (cm/°C/day)

T = the number of degree-days (°C day)

ΔT = the adjustment by temperature lapse rate when extrapolating the temperature from the station to the mean elevation of the zone (°C day)

S = the ratio of the snow covered area to the total area (%)

P = the precipitation contributing to runoff (cm), which is determined by a preselected threshold temperature, T_c, to be rainfall or snowfall. The contribution of rainfall is immediate while snowfall will be kept on storage until melting conditions occur.

A = the area elevation zone (km^2)

α = 10,000/86,400, the coefficient converting data from a runoff depth (cm×km^2/day) to discharge (m^3/s)

K_{n+1} = the recession coefficient, $K_{n+1} = \frac{Q_{n+1}}{Q_n}$

n = the day number

It is recommended to subdivide the basin into several elevation zones if the elevation range exceeds 500 m. The simulated discharge of each zone is summed to estimate the runoff of the basin. The study area was divided into 6 elevation zones, characters of which are present in Table 1. Model structure are shown above, the determination of input variables and parameters are declared below:

Appendix A.1.1 Input Variables

T, P, S are three basic inputs of SRM.

The daily snowmelt depth is calculated by the number of degree-days, which is determined by the daily temperature. In order to run SRM for each elevation zone, daily temperature from Hotan station was extrapolated to the mean elevation of each zone based on the global lapse rate value of 6.5 °C/km.

The daily watershed-averaged precipitation was derived from the China Ground Rainfall Daily Value 0.5° × 0.5° Lattice Dataset by using the Thiessen polygon (TP) method, which was used for each elevation zone.

The daily SCA for each elevation zone was generated from the SCA of 8-day composite MOD10A2 satellite images by using linear interpolating.

Appendix A.1.2 Parameters

Cs, Cr, a, T_c, RCA, K, and time lag are the seven parameters to set up SRM, whose value are presented in Table 3.

The runoff coefficients C_s and C_r, account for the losses between the runoff contributed by snowmelt and rainfall to precipitation. C_s and C_r are time-varying parameters, affected by lots of controlling factors such as climatic conditions and land cover properties. They can vary over seasonal, monthly, or even daily scale, which are the primary candidates for tuning if a runoff simulation is not immediately successful [47].

The degree-day factor a (cm/°C/day), converts the degree-days T (°C day) into the daily snowmelt depth M (cm) by Equation (A2). a is related to snow properties. According to the research for degree-day factors in western China conducted by Zhang et al. [88], 0.3 is adopted in this study.

$$M = a \times T \tag{A2}$$

The critical temperature T_c is used to determine whether the precipitation is rainfall ($T > T_c$) or snowfall ($T < T_c$). The T_c values for June–August and September–May are set to 3 °C and 0 °C, respectively, As recommended by Martinec [47].

The rainfall contributing area RCA represents the way that how to treat the rainfall determined by T_c and it needs to be determined according to the basin characteristics beforehand. When RCA is set to be 0, it is assumed that the rain falling on the snowpack is retained by the snow which is dry and deep. In this situation, rainfall depth is reduced by the ratio snow-free area/zone area. RCA is set to be 1 when the snowpack is ripe, which means that rainfall from the entire area is added to snowmelt. In this study, the snowpack is assumed to be ripe during May–September when the temperature is above 0 °C.

The recession coefficient K indicates the decline of discharge in a period without snowmelt or rainfall, and $(1 - K)$ presents the proportion of the meltwater production immediately appearing in the runoff. K is determined by Equation (A3) [47], in which parameters x and y can be obtained through linear regression analysis using the historical discharge data.

$$K_{n+1} = \frac{Q_{n+1}}{Q_n} = x \times Q_n^{-y} \tag{A3}$$

There is a time lag (L) between the temperature cycle and the resulting discharge cycle in Equation (A1). In this study, a lag time of 18 h is adapted depending on the relation between L and basin size [16]. In the case of 18 h, temperature measured on the nth day corresponds to the discharge on the $n + 1$ day.

References

1. Stocker, T.F.; Qin, D.; Plattner, G.-K.; Tignor, M.; Allen, S.K.; Boschung, J.; Nauels, A.; Xia, Y.; Bex, V.; Midgley, P.M. (Eds.) *IPCC. Climate Change 2013: The Physical Science Basis. Contribution of Working Group I to the Fifth Assessment Report of the Intergovernmental Panel on Climate Change*; Cambridge University Press: Cambridge, UK, 2013; p. 1535.
2. Barnett, T.P.; Adam, J.C.; Lettenmaier, D.P. Potential impacts of a warming climate on water availability in snow-dominated regions. *Nature.* **2005**, *438*, 303. [CrossRef] [PubMed]
3. Stewart, I.T. Changes in snowpack and snowmelt runoff for key mountain regions. *Hydrol. Process.* **2009**, *23*, 78–94. [CrossRef]
4. Klein, I.M.; Rousseau, A.N.; Frigon, A.; Freudiger, D.; Gagnon, P. Evaluation of probable maximum snow accumulation: Development of a methodology for climate change studies. *J. Hydrol.* **2016**, *537*, 74–85. [CrossRef]
5. Kudo, R.; Yoshida, T.; Masumoto, T. Uncertainty analysis of impacts of climate change on snow processes: Case study of interactions of GCM uncertainty and an impact model. *J. Hydrol.* **2017**, *548*, 196–207. [CrossRef]
6. Hamlet, A.F.; Mote, P.W.; Clark, M.P.; Lettenmaier, D.P. Effects of Temperature and Precipitation Variability on Snowpack Trends in the Western United States*. *J. Clim.* **2005**, *18*, 4545–4561. [CrossRef]
7. Khadka, D.; Babel, M.S.; Shrestha, S.; Tripathi, N.K. Climate change impact on glacier and snow melt and runoff in Tamakoshi basin in the Hindu Kush Himalayan (HKH) region. *J. Hydrol.* **2014**, *511*, 49–60. [CrossRef]
8. Mukhopadhyay, B.; Khan, A. A reevaluation of the snowmelt and glacial melt in river flows within Upper Indus Basin and its significance in a changing climate. *J. Hydrol.* **2015**, *527*, 119–132. [CrossRef]
9. Adam, J.C.; Hamlet, A.F.; Lettenmaier, D.P. Implications of global climate change for snowmelt hydrology in the twenty-first century. *Hydrol. Process.* **2009**, *23*, 962–972. [CrossRef]
10. Vicuña, S.; Garreaud, R.D.; McPhee, J. Climate change impacts on the hydrology of a snowmelt driven basin in semiarid Chile. *Clim. Chang.* **2011**, *105*, 469–488. [CrossRef]
11. Elias, E.H.; Rango, A.; Steele, C.M.; Mejia, J.F.; Smith, R. Assessing climate change impacts on water availability of snowmelt-dominated basins of the Upper Rio Grande basin. *J. Hydrol. Reg. Stud.* **2015**, *3*, 525–546. [CrossRef]
12. Quick, M.; Pipes, A. Daily and seasonal runoff forecasting with a water budget model. In *Role of Snow and Ice in Hydrology Proceedings of the UNESCO/WMO/IAHS Symposium*; World Meteorological Organization: Banff, AB, Canada, 1972; pp. 1017–1034.
13. Leavesley, G.H.; Lichty, R.W.; Troutman, B.M.; Saindon, L.G. Precipitation-runoff modeling system: User's manual. *Geol. Surv. Water Ivestig.* **1983**, 83–4238.
14. Jordan, R. *A One-Dimensional Temperature Model for a Snow Cover: Technical Documentation for SNTHERM. 89*; Cold Regions Research and Engineering Lab Hanover NH: Washington, DC, USA, 1991.
15. Bergstrom, S.; Forsman, A. Development of a conceptual deterministic rainfall-runoff model. *Nord. Hydrol.* **1973**, *4*, 147–170. [CrossRef]
16. Martinec, J. Snowmelt-Runoff Model for Stream Flow Forecasts. *Nord. Hydrol.* **1975**, *6*, 145–154. [CrossRef]
17. Charrois, L.; Cosme, E.; Dumont, M.; Lafaysse, M.; Picard, G. On the assimilation of optical reflectances and snow depth observations into a detailed snowpack model. *Cryosphere* **2016**, *10*, 1021–1038. [CrossRef]

18. Magnusson, J.; Gustafsson, D.; Hüsler, F.; Jonas, T. Assimilation of point swe data into a distributed snow cover model comparing two contrasting methods. *Water Resour. Res.* **2014**, *50*, 7816–7835. [CrossRef]

19. Magnusson, J.; Winstral, A.; Stordal, A.S.; Essery, R.; Jonas, T. Improving physically based snow simulations by assimilating snow depths using the particle filter. *Water Resour. Res.* **2017**, *53*, 1125–1143. [CrossRef]

20. USDA-NRCS. *National Engineering Handbook: Part 630—Hydrology*; USDA Soil Conservation Service: Washington, DC, USA, 2004.

21. Hock, R. Temperature index melt modeling in mountain areas. *J. Hydrol.* **2003**, *282*, 104–115. [CrossRef]

22. Seidel, K.; Martinec, J.; Baumgartner, M.F. Modeling runoff and impact of climate change in large himalayan basins. In Proceedings of the International Conference on Integrated Water Resources Management (ICIWRM), Roorke, India, 19–21 December 2000.

23. Nazari, M.A.; Saleh, F.N.; Chavoshian, S.A. Flood forecasting and river flow modeling in mountainous basin with significant contribution of snowmelt runoff. Presented at the International Conference on Flood Management, Tsukuba, Japan, 25 September 2011.

24. Ye, L.; Zhou, J.; Zeng, X.; Guo, J.; Zhang, X. Multi-objective optimization for construction of prediction interval of hydrological models based on ensemble simulations. *J. Hydrol.* **2014**, *519*, 925–933. [CrossRef]

25. Tahir, A.A.; Hakeem, S.A.; Hu, T.; Hayat, H.; Yasir, M. Simulation of snowmelt-runoff under climate change scenarios in a data-scarce mountain environment. *Int. J. Digit. Earth* **2017**, 1–21. [CrossRef]

26. Xie, S.; Du, J.; Zhou, X.; Zhang, X.; Feng, X.; Zheng, W.; Xu, C.-Y. A progressive segmented optimization algorithm for calibrating time-variant parameters of the snowmelt runoff model (SRM). *J. Hydrol.* **2018**, *566*, 470–483. [CrossRef]

27. Refsgaard, J.C.; Storm, B. Construction, Calibration and Validation of Hydrological Models. *Distrib. Hydrol. Model.* **1990**, *22*, 41–54. [CrossRef]

28. Wilby, R.L. Uncertainty in water resource model parameters used for climate change impact assessment. *Hydrol. Process.* **2005**, *19*, 3201–3219. [CrossRef]

29. Vrugt, J.A.; Gupta, H.V.; Bouten, W.; Sorooshian, S. A Shuffled Complex Evolution Metropolis algorithm for optimization and uncertainty assessment of hydrologic model parameters. *Water Resour. Res.* **2003**, *39*. [CrossRef]

30. Finger, D.; Vis, M.; Huss, M.; Seibert, J. The value of multiple data set calibration versus model complexity for improving the performance of hydrological models in mountain catchments. *Water Resour. Res.* **2015**, *51*, 1939–1958. [CrossRef]

31. Bastola, S.; Murphy, C.; Sweeney, J. The role of hydrological modeling uncertainties in climate change impact assessments of Irish river catchments. *Adv. Water Resour.* **2011**, *34*, 562–576. [CrossRef]

32. Brigode, P.; Oudin, L.; Perrin, C. Hydrological model parameter instability: A source of additional uncertainty in estimating the hydrological impacts of climate change? *J. Hydrol.* **2013**, *476*, 410–425. [CrossRef]

33. Joseph, J.; Ghosh, S.; Pathak, A.; Sahai, A.K. Hydrologic impacts of climate change: Comparisons between hydrological parameter uncertainty and climate model uncertainty. *J. Hydrol.* **2018**, *566*, 1–22. [CrossRef]

34. Jin, X.; Xu, C.-Y.; Zhang, Q.; Singh, V.P. Parameter and modeling uncertainty simulated by GLUE and a formal Bayesian method for a conceptual hydrological model. *J. Hydrol.* **2010**, *383*, 147–155. [CrossRef]

35. Li, L.; Xu, C.-Y.; Xia, J.; Engeland, K.; Reggiani, P. Uncertainty estimates by Bayesian method with likelihood of AR (1) plus Normal model and AR (1) plus Multi-Normal model in different time-scales hydrological models. *J. Hydrol.* **2011**, *406*, 54–65. [CrossRef]

36. Raje, D.; Krishnan, R. Bayesian parameter uncertainty modeling in a macroscale hydrologic model and its impact on Indian river basin hydrology under climate change. *Water Resour. Res.* **2012**, *48*. [CrossRef]

37. Beven, K.; Binley, A. The future of distributed models: Model calibration and uncertainty prediction. *Hydrol. Process.* **1992**, *6*, 279–298. [CrossRef]

38. Saltelli, A.; Annoni, P. *Sensitivity Analysis*; Wiley: Hoboken, NJ, USA, 2000.

39. Blasone, R.S.; Madsen, H.; Rosbjerg, D. Uncertainty assessment of integrated distributed hydrological models using glue with markov chain monte carlo sampling. *J. Hydrol.* **2008**, *353*, 18–32. [CrossRef]

40. Li, L.; Xia, J.; Xu, C.-Y.; Singh, V.P. Evaluation of the subjective factors of the GLUE method and comparison with the formal Bayesian method in uncertainty assessment of hydrological models. *J. Hydrol.* **2010**, *390*, 210–221. [CrossRef]

41. Fuentes-Andino, D.; Beven, K.; Halldin, S.; Xu, C.-Y.; Baldassarre, G.D. Reproducing an extreme flood with uncertain post–event information. *J. Hydrol. Earth Syst. Sci.* **2017**, *21*, 3597–3618. [CrossRef]

42. Metropolis, N.; Ulam, S. The Monte Carlo Method. *J. Am. Stat. Assoc.* **1949**, *44*, 335–341. [CrossRef]
43. Blasone, R.-S.; Vrugt, J.A.; Madsen, H.; Rosbjerg, D.; Robinson, B.A.; Zyvoloski, G.A. Generalized likelihood uncertainty estimation (GLUE) using adaptive Markov Chain Monte Carlo sampling. *Adv. Water Resour.* **2008**, *31*, 630–648. [CrossRef]
44. Prasad, V.H.; Roy, P.S. Estimation of Snowmelt Runoff in Beas Basin, India. *Geocarto Int.* **2005**, *20*, 41–47. [CrossRef]
45. Li, X.; Williams, M.W. Snowmelt runoff modeling in an arid mountain watershed, Tarim Basin, China. *Hydrol. Process.* **2008**, *22*, 3931–3940. [CrossRef]
46. Abudu, S.; Cui, C.L.; Saydi, M.; King, J.P. Application of snowmelt runoff model (SRM) in mountainous watersheds: A review. *Water Sci. Eng.* **2012**, *5*, 123–136.
47. Martinec, J.; Rango, A.; Major, E. *The Snowmelt-Runoff Model (S.R.M.) User's Manual*; NASA Reference Publication 1100: Washington, DC, USA, 1983.
48. Senzeba, K.T.; Rajkumari, S.; Bhadra, A.; Bandyopadhyay, A. Response of streamflow to projected climate change scenarios in an eastern Himalayan catchment of India. *J. Earth Syst. Sci.* **2016**, *125*, 443–457. [CrossRef]
49. Zhang, G.; Xie, H.; Yao, T.; Li, H.; Duan, S. Quantitative water resources assessment of Qinghai Lake basin using Snowmelt Runoff Model (SRM). *J. Hydrol.* **2014**, *519*, 976–987. [CrossRef]
50. Fuladipanah, M.; Jorabloo, M. The estimation of snowmelt runoff using SRM case study (Gharasoo basin, Iran). *World Appl. Sci. J.* **2012**, *17*, 433–438.
51. Farr, T.G.; Rosen, P.A.; Caro, E.; Crippen, R.; Duren, R.; Hensley, S.; Alsdorf, D. The Shuttle Radar Topography Mission. *Rev. Geophys.* **2007**, *45*. [CrossRef]
52. Available online: http://data.cma.cn (accessed on 10 November 2019).
53. Andrew, T.; Roddy, H.; Richard, T.; Zheng, X.G. Thin plate smoothing spline interpolation of daily rainfall for New Zealand using a climatological rainfall surface. *Int. J. Climatol.* **2006**, *26*, 2097–2115.
54. Zhao, Y.F.; Zhu, J. Assessing quality of grid daily precipitation datasets in china in recent 50 years. *Plateau Meteorol.* **2015**, *34*, 50–58.
55. Available online: http://nsidc.org/data (accessed on 10 November 2019).
56. Huang, X.; Zhang, X.; Li, X.; Liang, T. Accuracy analysis for MODIS snow products of MOD10A1 and MOD10A2 in northern Xinjiang area. *J. Glaciol. Geocryol.* **2007**, *29*, 722–729.
57. Taylor, K.E.; Stouffer, R.J.; Meehl, G.A. An overview of CMIP5 and the experiment design. *Bull. Am. Meteorol. Soc.* **2012**, *93*, 485–498. [CrossRef]
58. Hartigan, J.A.; Wong, M.A. Algorithm AS 136: A k-means clustering algorithm. *J. R. Stat. Soc. Ser. C (Appl. Stat.)* **1979**, *28*, 100–108. [CrossRef]
59. Cannon, A.J. Selecting GCM Scenarios that Span the Range of Changes in a Multimodel Ensemble: Application to CMIP5 Climate Extremes Indices*. *J. Clim.* **2015**, *28*, 1260–1267. [CrossRef]
60. Chen, J.; Brissette, F.P.; Lucas-Picher, P. Transferability of optimally-selected climate models in the quantification of climate change impacts on hydrology. *Clim. Dyn.* **2016**, *47*, 3359–3372. [CrossRef]
61. Wang, H.M.; Chen, J.; Cannon, A.J.; Xu, C.Y.; Chen, H. Transferability of climate simulation uncertainty to hydrological impacts. *Hydrol. Earth Syst. Sci.* **2018**, *22*, 3739–3759. [CrossRef]
62. Martinec, J.; Rango, A.; Roberts, R.T. *Snowmelt Runoff Model (SRM) User's Manual*; New Mexico State University Press: New Mexico, NM, USA, 2008; pp. 19–39.
63. Nash, J.E.; Sutcliffe, J.V. River flow forecasting through conceptual models part I—A discussion of principles. *J. Hydrol.* **1970**, *10*, 282–290. [CrossRef]
64. Chen, J.; Brissette, F.P.; Poulin, A.; Leconte, R. Overall uncertainty study of the hydrological impacts of climate change for a Canadian watershed. *Water Resour. Res.* **2011**, *47*. [CrossRef]
65. Liu, J.; Zhu, A.-X.; Duan, Z. Evaluation of trmm 3b42 precipitation product using rain gauge data in Meichuan watershed, Poyang Lake Basin, China. *J. Resour. Ecol.* **2012**, *3*, 359–366.
66. Bartier, P.M.; Keller, C.P. Multivariate interpolation to incorporate thematic surface data using inverse distance weighting (idw). *Comput. Geosci.* **1996**, *22*, 795–799. [CrossRef]
67. Chen, J.; Brissette, F.P.; Chaumont, D.; Braun, M. Performance and uncertainty evaluation of empirical downscaling methods in quantifying the climate change impacts on hydrology over two North American river basins. *J. Hydrol.* **2013**, *479*, 200–214. [CrossRef]
68. Schmidli, J.; Frei, C.; Vidale, P.L. Downscaling from GCM precipitation: A benchmark for dynamical and statistical downscaling methods. *Int. J. Climatol.* **2006**, *26*, 679–689. [CrossRef]

69. Mpelasoka, F.S.; Chiew, F.H.S. Influence of Rainfall Scenario Construction Methods on Runoff Projections. *J. Hydrometeorol.* **2009**, *10*, 1168–1183. [CrossRef]

70. Rango, A.; Martinec, J. Areal extent of seasonal snow cover in a changed climate. *Hydrol. Res.* **1994**, *25*, 233–246. [CrossRef]

71. Ratto, M.; Tarantola, S.; Saltelli, A. Sensitivity analysis in model calibration: GSA-GLUE approach. *Comput. Phys. Commun.* **2001**, *136*, 212–224. [CrossRef]

72. Li, L.; Xia, J.; Xu, C.Y.; Chu, J.J.; Wang, R.; Cluckie, I.D.; Mynett, A. Analyse the sources of equifinality in hydrological model using GLUE methodology. Paper presented at the Hydroinformatics in Hydrology, Hydrogeology and Water Resources. In Proceedings of the Symposium JS.4 at the Joint IAHS IAH Convention, Hyderabad, India, 6–12 September 2009.

73. Katwijk, V.F.; Rango, A.; Childress, A.E. Effect of Simulated Climate Change on Snowmelt Runoff Modeling in Selected Basins. *J. Am. Water Resour. Assoc.* **1993**, *29*, 755–766. [CrossRef]

74. Matott, L.S.; Babendreier, J.E.; Purucker, S.T. Evaluating uncertainty in integrated environmental models: A review of concepts and tools. *Water Resour. Res.* **2009**, *45*. [CrossRef]

75. Butts, M.B.; Payne, J.T.; Kristensen, M.; Madsen, H. An evaluation of the impact of model structure on hydrological modeling uncertainty for streamflow simulation. *J. Hydrol.* **2004**, *298*, 242–266. [CrossRef]

76. Jakeman, A.J.; Hornberger, G.M. How much complexity is warranted in a rainfall-runoff model? *Water Resour. Res.* **1993**, *29*, 2637–2649. [CrossRef]

77. Seiller, G.; Roy, R.; Anctil, F. Influence of three common calibration metrics on the diagnosis of climate change impacts on water resources. *J. Hydrol.* **2017**, *547*, 280–295. [CrossRef]

78. Wang, J.; Li, H.; Hao, X. Responses of snowmelt runoff to climatic change in an inland river basin, Northwestern China, over the past 50 years. *Hydrol. Earth Syst. Sci.* **2010**, *14*, 1979–1987. [CrossRef]

79. Tian, Y.; Xu, Y.P.; Booij, M.J.; Wang, G. Uncertainty in future high flows in Qiantang river basin, China. *J. Hydrometeorol.* **2015**, *16*, 363–380. [CrossRef]

80. Wilby, R.L.; Harris, I. A framework for assessing uncertainties in climate change impacts: Low-flow scenarios for the River Thames, UK. *Water Resour. Res.* **2006**, *42*. [CrossRef]

81. Stedinger, J.R.; Vogel, R.M.; Lee, S.U.; Batchelder, R. Appraisal of the generalized likelihood uncertainty estimation (GLUE) method. *Water Resour. Res.* **2008**, *44*. [CrossRef]

82. Li, Z.; Shao, Q.; Xu, Z.; Cai, X. Analysis of parameter uncertainty in semi-distributed hydrological models using bootstrap method: A case study of SWAT model applied to Yingluoxia watershed in northwest China. *J. Hydrol.* **2010**, *385*, 76–83. [CrossRef]

83. Ruelland, D.; Hublart, P.; Tramblay, Y. Assessing uncertainties in climate change impacts on runoff in Western Mediterranean basins. *Proc. Int. Assoc. Hydrol. Sci.* **2015**, *371*, 75–81. [CrossRef]

84. Vaze, J.; Post, D.A.; Chiew, F.H.S.; Perraud, J.M.; Viney, N.R.; Teng, J. Climate non-stationarity—Validity of calibrated rainfall–runoff models for use in climate change studies. *J. Hydrol.* **2010**, *394*, 447–457. [CrossRef]

85. Van den Broeke, M.R.; Smeets, C.J.P.P.; van de Wal, R.S.W. The seasonal cycle and interannual variability of surface energybalance and melt in the ablation zone of the west Greenland ice sheet. *Cryosphere* **2011**, *5*, 377–390. [CrossRef]

86. Bougamont, M.; Hunke, E.; Tulaczyk, S. Sensitivity of ocean circulation and sea-ice conditions to loss of west antarctic ice shelves and ice sheet. *J. Glaciol.* **2007**, *53*, 490–498. [CrossRef]

87. Huss, M.; Farinotti, D.; Bauder, A.; Funk, M. Moddelling runoff from highly glacierized alpine drainage basins in a changing climate. *Hydrol. Process.* **2008**, *22*, 3888–3902. [CrossRef]

88. Zhang, Y.; Liu, S.; Ding, Y. Observed degree-day factors and their spatial variation on glaciers in western China. *Ann. Glaciol.* **2006**, *43*, 301–306. [CrossRef]

Article

Application of Empirical Mode Decomposition Method to Synthesize Flow Data: A Case Study of Hushan Reservoir in Taiwan

Tai-Yi Chu and Wen-Cheng Huang *

Department of Harbor and River Engineering, National Taiwan Ocean University, Keelung 20224, Taiwan; 20352006@email.ntou.edu.tw
* Correspondence: b0137@mail.ntou.edu.tw; Tel.: +886-2-2462-2192

Received: 7 March 2020; Accepted: 19 March 2020; Published: 25 March 2020

Abstract: Although empirical mode decomposition (EMD) was developed to analyze nonlinear and non-stationary data in the beginning, the purpose of this study is to propose a new method—based on EMD—to synthesize and generate data which be interfered with the non-stationary problems. While using EMD to decompose flow record, the intrinsic mode functions and residue of a given record can be re-arranged and re-combined to generate synthetic time series with the same period. Next, the new synthetic and historical flow data will be used to simulate the water supply system of Hushan reservoir, and explore the difference between the newly synthetic and historical flow data for each goal in the water supply system of Hushan reservoir. Compared the historical flow with the synthetic data generated by EMD, the synthetic data is similar to the historical flow distribution overall. The flow during dry season changes in significantly (± 0.78 m^3/s); however, the flow distribution during wet season varies significantly (± 0.63 m^3/s). There are two analytic scenarios for demand. For Scenario I, without supporting industrial demand, the simulation results of the generation data of Method I and II show that both are more severe than the current condition, the shortage index of each method is between 0.67–1.96 but are acceptable. For Scenario II, no matter in which way the synthesis flow is simulated, supporting industrial demand will seriously affect the equity of domestic demand, the shortage index of each method is between 1.203 and 2.12.

Keywords: empirical mode decomposition; Hushan reservoir; data synthesis

---

1. Introduction

Water resources are necessary for human survival, and are an important part of maintaining socio-economic development and the sustainability of the ecological environment as well. Taiwan is located at the border of tropical and subtropical areas. The main source of water resources is rainfall. The annual average rainfall in Taiwan is about 2500 mm, which is 2.5 times of the world's average. In recent years, statistics show that rainfall characteristics have become more extreme in Taiwan. Not only the raining days gradually decreasing, but the intensity gradually is increasing. On the whole, although Taiwan seems to be a country with abundant rainfall, the overall water resources environment is not easy to manageme because of climate change.

The impact of climate change tends to cause non-stationary conditions in hydrological data. Traditionally, hydrological analysis and data generation only deal with stationary data. Empirical mode decomposition (EMD) can be applied to any type of signal decomposition. Unlike singular spectrum analysis, Fourier transform, and Wavelet transform, EMD does not require any pre-determined basis functions and can extract intrinsic mode function (IMF) components from the original signal in a self-adaptive way [1]. Through the characteristics of EMD, non-stationary and nonlinear signals can be processed, which effectively overcomes the limitations of traditional methods.

EMD has been proposed [2] as an adaptive time-frequency data analysis method. In this study, the EMD method and the decomposed IMFs were used for data analysis and synthesis. Many studies use EMD to analyze different time series and decompose the implied fluctuation periods [3–6]. In recent years, there are many carefully studies of nonlinear and non-stationary for the hydrological field: Lee and Ouarda [7,8] proposed model reproduces Nonstationary oscillation (NSO) processes by utilizing EMD and nonparametric simulation techniques; Molla et al. [9,10] show that the EMD successively extracts the IMFs with the highest local frequencies in a recursive way, which yields effectively a set low-pass filters based entirely on the properties exhibited by the data. In addition, there are many studies based on the EMD method to predict and estimate streamflow data: Huang et al. [11] and Meng et al. [12] developed a modified EMD-based support vector machine for non-stationary streamflow prediction; Zhang et al. [13] improved the efficiency of a prediction approach which can be used as a general technique for non-stationary time series forecasting.

However, the mode mixing problem may occur during the EMD. To avoid mode mixing, Wu and Huang [14] proposed a new noise-assisted data analysis method, the Ensemble EMD (EEMD), which defines the true IMF components as the mean of an ensemble of trials, each consisting of the signal plus a white noise of finite amplitude. Zhao and Chen [15] and Zhang et al. [16] used the EEMD and other methods to build a hybrid model for annual runoff time series forecasting. Many of EEMD applications in hydrology focus on the diagnostic analysis of historical data, such as features of temperature variation trends [17], forward prediction of runoff data in data-scarce basins [18], and forecast monthly reservoir inflow [19].

The analysis methods of water resources allocation system are mainly divided into two categories: simulation method and optimization method. Yeh [20] and Wurbs [21] introduced the theory development of the simulation model and optimization model in reservoir system analysis. Chang et al. [22] present the usefulness of the neural network in deriving general operating policies for a multi-reservoir system. An optimization-based approach is typically required for the optimal operation of a reservoir system, in order to obtain the optimal solutions to support the decision-making process [23]. In this study, we use the simulation method for the analysis of water resources systems. Lee and Huang [24] evaluated the impact of climate change on the water supply of the reservoir in central Taiwan.

The Hushan reservoir is a new water supply resource system in central Taiwan (Figure 1). Since some problems exist in the water supply system of the Zhuoshui river, such as its high concentration of sediment, and low efficiency for the Hushan reservoir while it considers a joint operation with Jiji weir [25]. Therefore, Chu and Huang [26] recommends the Hushan reservoir should operate independently. Under the assumption that simulation results show that the Hushan reservoir is capable of supplying more than 100,000 m^3 per day (up to 380,000 m^3/day) and the reservoir utilization rate can reach 2.41 from 1.18 while Shortage Index (SI) = 1.

This study proposes a novel procedure, which is based on EEMD and its derivation (IMF and residue) as an implement for synthesizing non-stationary data, which will be applied in reservoir inflow and impact assessment for water supply. This study will continue the research results of Huang et al. [27]. Huang et al. [27] proposed a surface temperature and rainfall synthesis method base on EMD. The method utilizes the recombination of the IMF of the segmented data, as well as the characteristics of the residuals to generate the data. There is a plan to use the IMFs and residues obtained, after EMD decomposition to permutation and combination, to generate new flow data. There is also a plan to try to apply the newly synthetic flow data to simulate the water supply system of the Hushan reservoir (as shown in Figure 1). Finally, several evaluation indicators are used to evaluate water supply system simulation results.

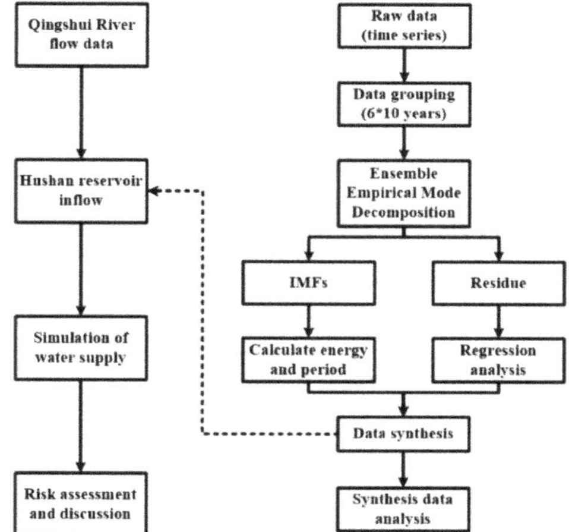

Figure 1. Research flowchart. EMD, empirical mode decomposition components.

2. Materials and Method

2.1. Study Area

2.1.1. Hushan Reservoir

Hushan reservoir is an off-channel reservoir. It is located between Douliu City and Gukeng Township, Yunlin County, 10 km southeast away from the Douliu City center (as shown in Figure 2). The reservoir was set up in 2016 and first reached its full water level in June 2019. The main purpose of the Hushan Reservoir is to reduce pumping groundwater and to slow down the problem of land subsidence in Yunlin county. The elevation-area-capacity relationship of Hushan Reservoir is shown in Table 1. The elevation of the remaining water level is 165 m, the full water level is 211.5 m, and the effective storage capacity of the reservoir is 53.47×106 m^3. The elevations of the water outlets are 165 m and 180 m respectively. The designed flow of the domestic water channel and the permanent river outlet are 12.27 m^3/s and 1.00 m^3/s, respectively. The watershed of Hushan reservoir is only 6.58 km^2 and leads to limited water resources. Therefore, the Tongtou weir, the intake of Hushan reservoir, on the Qingshui river needs to be guided through the diversion tunnel.

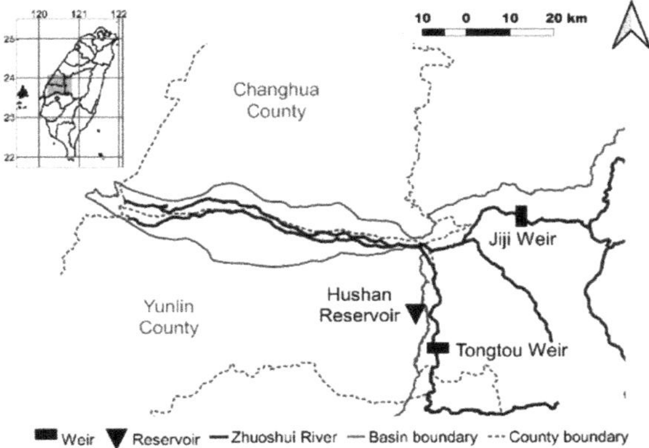

Figure 2. Location of study area.

Table 1. Elevation-area-capacity relationship of Hushan reservoir.

Water Level (m)	Area (ha)	Volume (10^4 m^3)
Dead water level: 165	12	36
170	32	147
175	76	418
180	99	857
185	107	1371
190	132	1969
195	142	2653
200	170	3432
205	181	4310
210	210	5288
Full water level: 211.5	214	5612

2.1.2. Tongtou Weir

Tongtou weir is the intake of Hushan reservoir. In order to ensure the river base flow and the downstream water requirements—such as agriculture and domestic—of Tongtou weir will be satisfied. The weir diversion operation will give priority to the base flow and water right; next, the intake of Hushan reservoir from Tongtou weir on Qingshui river be considered. The plan of the Tongtou weir diversion water accounts for about 10% of the Qingshui river during the wet season (May to October), and about 13% during the dry season. The Tongtou weir is located 70 m downstream of the Tongtou suspension bridge in the Qingshui river. There is a sedimentation basin before the water be taken from Tongtou weir to Hushan reservoir. The intake channel is 2.8 km long and designed intake capacity is 20 m^3/s, with a gravity type.

2.2. Research Method

2.2.1. Empirical Mode Decomposition

EMD is proposed for nonlinear and nonstationary time series analysis [2]. For the hydrological time series, EMD divides the sequence into two parts: IMFs and residue. The IMFs represent the implied period sequence of the data, and the residue represents the non-period sequence of the data. Although the methods used in traditional signal analysis fields such as fast Fourier transform, harmonic analysis, and wavelet analysis can also express the original sequence as a superposition of multiple period functions; these Fourier basic function-based analyses usually only deal with stationary time series. Contrary to the previous methods, EMD is intuitive, direct, a posteriori, and adaptive, with the basis of the decomposition based on, and derived from, the data [2]. EMD is more suitable for analyzing non-stationary hydrological time series.

EMD decomposes the original signal into different time scale of oscillation components called IMF. Unlike singular spectrum analysis, Fourier transform, and Wavelet transform, EMD does not require any pre-determined basis functions and can extract IMF components from the original signal in a self-adaptive way [1]. Each IMF component should satisfy the following two conditions:

1. In the whole data series, the number of extrema and the number of zero-crossings must either equal or differ at most by 1;
2. At any point, the mean value of the envelope defined by the local maxima and the envelope defined by the local minima is zero.

An IMF represents a simple oscillatory mode as a counterpart to the simple harmonic function, but it is much more general. For time series data x(t) (t = 1, 2, ... , n), the procedure of EMD can be described as follows [16]:

1. Identify all the local maxima and minima of the original time series x(t);
2. Using the three-spline interpolation function to create the upper envelopes $e_{up}(t)$ and the lower envelopes $e_{low}(t)$ of the time series;
3. Calculate mean value m(t) of the upper and lower envelopes $\left(m(t) = \left[e_{up}(t) + e_{low}(t)\right]/2\right)$;
4. Calculate the difference value d(t) between time series x(t) and mean value m(t), $(d(t) = x(t) - m(t))$;
5. Check the difference value d(t): (a) if d(t) satisfies the two IMF conditions, then d(t) is defined as the ith IMF, the residue r(t) = x(t) − d(t) replace the x(t). The *i*th IMF is denoted as $c_i(t)$; (b) if d(t) is not an IMF, then d(t) replace the x(t).
6. Repeat (1)–(5) until the residue item r(t) becomes a monotone function or the number of extrema is less than or equal to 1, so that the IMF component cannot be decomposed again.

Finally, original time series x(t) can be denoted as sum of IMFs $c_i(t)$ and residue r(t).

$$x(t) = \sum_{i=1}^{n} c_i(t) + r(t) \tag{1}$$

where n, $c_i(t)$ and r(t) represent the number of IMF, the ith IMF and the residue, respectively. The residue r(t) also represents the overall trend or the mean value of the original time series data. Figure 3 shows the results of the EMD method decomposition.

The procedure is illustrated in Huang et al. [2] and Zhang et al. [16].

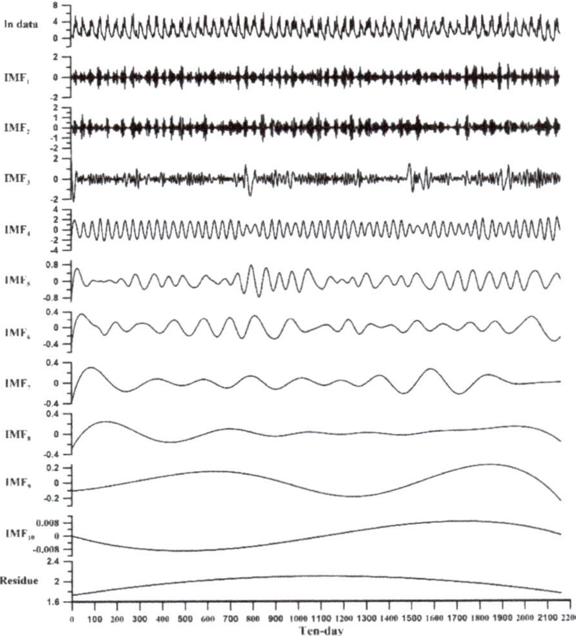

Figure 3. Results of EMD (empirical mode decomposition) method decomposition.

2.2.2. Ensemble Empirical Mode Decomposition

EEMD, an improvement of EMD, was proposed by Wu and Huang [14]. Although EMD has obvious advantages in signal analysis, there are unavoidable defects such as the boundary-effect and mode-mixing. Mode mixing is a problem often encountered when using EMD to decompose sequences. Mode mixing means that a certain IMF contains waves with very different periodicity, often caused by intermittent signal interference. In particular, the mode-mixing will not only cause the mixing of various scale vibration modes but can even lose the physical meaning of individual IMF. To overcome this problem of the EMD method, a new noise-assisted data analysis method is proposed—the EEMD [14]—which defines the true IMF components as the mean of an ensemble of trials, each consisting of the signal plus a white noise of finite amplitude.

The main step of EEMD is described as follows [16]:

1. Add white noise w(t) to the original time series x(t). The new time series can be defined as:

$$X(t) = x(t) + w(t) \tag{2}$$

2. Decompose the new time series into IMFs using EMD method;
3. Repeat steps (1) and (2) with different white noises series each time;
4. Obtain the mean of the ensemble corresponding IMFs of the decompositions as the final result.

Differently from EMD, the EEMD method first adds white noise to the original data before decomposition and uses the statistics feature of white noise, the expectation is equal to zero, to perform the screening process. After adding white noise many times, the effect of adding white noise will be offset by the mean of the ensemble IMF. Therefore, the decomposition using EEMD not only keeps the inherent feature of the original signal, but also overcomes the mode-mixing.

The effect of the added white noise should decrease using the equation [14]

$$\varepsilon_n = \frac{\varepsilon}{\sqrt{N}} \tag{3}$$

where N is the number of ensemble members, ε is the amplitude of the added noise and ε_n is the final standard deviation of error, which is defined as the difference between the input signal and the corresponding IMF(s). According to Equation (3), when the average number of times increases, the error will reduce, and when n reaches a certain number of times, its error value can be ignored.

2.2.3. Zero Up-Cross Method

Statistical analysis of a IMF, a wave set, requires establishment of the definition of individual wave heights and periods. The standard is the employment of the zero-crossing definition. The zero line or the line of the mean wave level is set for a record of wave registration, and each wave is defined with the reference point where the instantaneous water level crosses the zero line. When the upward crossing point is employed as the beginning of an individual wave, the method is called the zero up-crossing method [28]. The definition of upward crossing point is [29]

$$\eta_i \times \eta_{i+1} < 0; \eta_{i+1} > 0 \tag{4}$$

where η_i is the wave height in given time.

The wave period, T, of an individual wave will be defined as the time between two successive upward crossing point, as Figure 4 showing. For any individual wave, the peak and valley are expressed as

$$\eta_{i-1} < \eta_i \cap \eta_i > \eta_{i+1} \text{ (highest point)} \tag{5}$$

$$\eta_{i-1} > \eta_i \cap \eta_i < \eta_{i+1} \text{ (lowest point)} \tag{6}$$

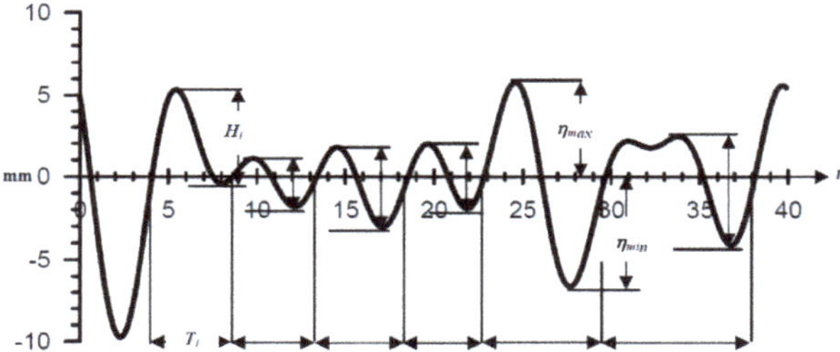

Figure 4. Show of the zero up-cross method [30]. H, The wave high; n_{max}: The average value of the highest point (peak).

The average value of η_{max} the highest point (peak) and the lowest point (valley) of η_{min} is used as the amplitude of each wave, as shown in Equation (7)

$$\text{amplitude} : (\eta_{max} + \eta_{min})/2 \tag{7}$$

In this study, we use the IMF decomposed by the EEMD to perform a zero up cross analysis method, analyze the periods of each IMF, and discuss it.

After finding the period in the fluctuation using the zero up cross analysis method, the sum of squares of each IMFs is regarded as its energy, and its value can correspond to the period relationship in the fluctuation. The total energy value of each group is added up to form the IMF percentage of each group compared with the energy of each group. The relational expressions of the energy percentages are

$$E_i = \sum_{i=1}^{n} x_i^{2} \tag{8}$$

$$W_i = \frac{E_i}{\sum_{i=1}^{n} E_i} \times 100\% \tag{9}$$

where x_j^2 is the square value at each time point of each group of IMF; E_i is the total energy; W_i is weight of energy.

2.2.4. Data Synthesis

The impact of climate change tends to cause non-stationary conditions in hydrological data. Therefore, we use the EEMD [14] in this study, and proposes a new way to synthesize and generate data that can solve the non-stationary data problems. For a given flow time series, the IMFs and residue, the EEMD decomposed result, would be replace by corresponding component in another period; next, the new combination of IMFs and residue are summed up according to the completeness of EMD [2]; then, the synthetic new flow time series will be produced. Thus, if there are many EEMD decomposed results with the same length but different period, we could synthesize a flow data set through permutation and combination of the corresponding IMFs and residues.

In this study, the historical 60-year (1956–2015) flow data of Tongtou streamflow gauge are segmented in groups of 10 years. Through the EEMD method divides the sequence into two parts: IMFs represent the inherent period sequence of the data, and the residue represents the non-period sequence of the data. We can go through the permutations and combinations of the IMFs to get n^i sequences. Where n is the number of groups of data grouping; i is the number of IMFs by the group.

After completing the permutations and combinations of IMFs as shown in Figure 5, we can get the new synthesis data of equal length streamflow. In this study, two permutations and combinations methods are used to synthesize new 10-year flow series data. The method is described as follows:

1. Regression analysis is performed on the different residue of each group to fit a new residue to represent the residue value. By adding the results of the IMFs permutation and combination with the representative values of the residue, we can get 10-year new synthetic flow data for the n^i groups.

2. Permutation and combination of the different remainders of each group with IMFs, we can get 10-year new synthetic flow data for the n^{i+1} groups.

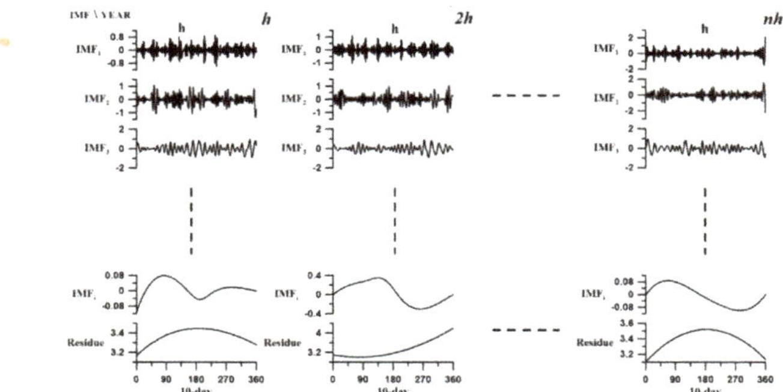

Figure 5. Graphs of the data synthesis process. h: segment.

2.2.5. Water Supply System Simulation of Hushan Reservoir

The simulation method is commonly used in the analysis of water resources systems. It conceptualizes the actual operation of the system to explore the use of water resources. In this study, we use the simulation method to simulate the water supply system and explore the difference between the newly synthesized flow data for each goal in the water supply system of Hushan reservoir.

First of all, the historical (1956–2015) flow data of Qingshui river were used to simulate accordance with the Hushan reservoir operation rules. Because the Hushan reservoir takes water from the Qingshui river through a water diversion tunnel, and there is some agricultural water right downstream of the Tongtou weir (as shown in Figure 6). Therefore, before taking water to the Hushan reservoir, Tongtou weir has to consider and satisfy the agricultural water downstream. The remaining flow is the amount of water that can be taken to the Hushan reservoir. The main thing of simulation is that the water taken into the Hushan reservoir should not affect the downstream agricultural water right.

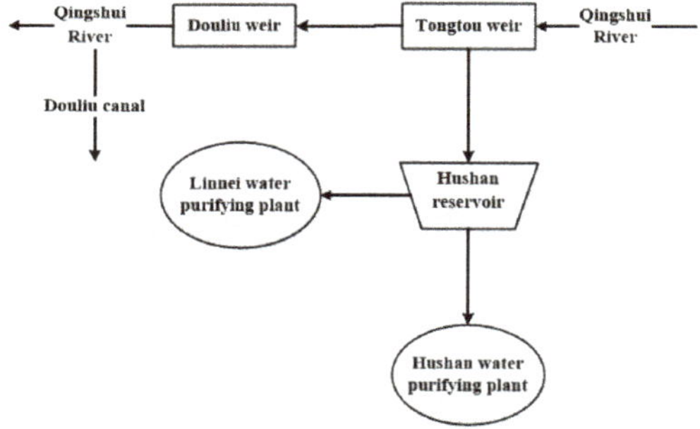

Figure 6. Water supply system of Hushan reservoir.

This study will assume different scenarios to simulate the water supply system of Hushan reservoir. The main difference of these two scenarios is the object of water supply (as shown in Table 2). The local water supply system is shown in Figure 6, and the simulation program calculation process is shown in Figure 7.

Table 2. Scenario description.

Scenario	I	II
Water supply range	Supply the domestic water	Support the industrial water
The operational regulations	The upper and lower rule was set at the water level of 190 and 185 m in Hushan reservoir, respectively.	

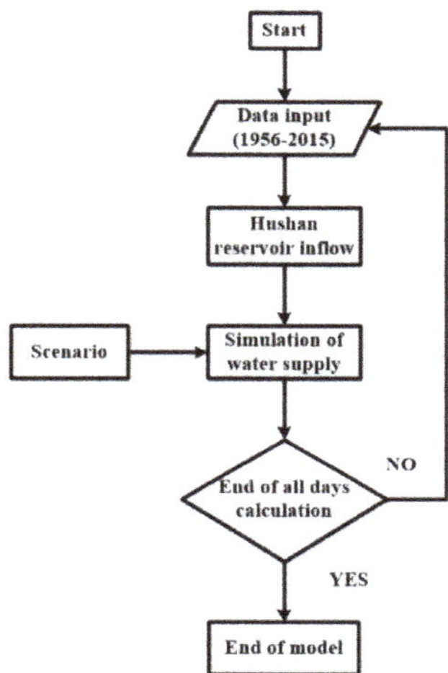

Figure 7. Program simulation flowchart.

In the past, the Linnei water purifying plant only supplied 1.2×10^5 m^3 water per day, and the deficit of demand was filled by pumping groundwater. After the completion of the Hushan reservoir, the situation of over pumping will be improved. The Hushan reservoir was originally intended to be used in conjunction with the Jiji weir to supply water for domestic demand in the Yunlin area. However, the conjunction operation of Hushan reservoir and Jiji weir will result in a low reservoir utilization rate of only 1.18 [25]. Therefore, we assume the Hushan reservoir operates independently in this study. The domestic demand in the Yunlin area is 2.8×10^5/day, and the support demand for industrial is 3.0×10^5/day. Based on the previous study, the operational regulations of reservoir were added [26], and the upper and lower rule curve is set at the water level of 190 m and 185 m for Hushan reservoir, respectively. In order to explore the situation of water resources utilization in the Yunlin area of the Hushan reservoir under different demand scenarios, the operation of reservoir should follow these principles:

1. Supply of the Hushan reservoir must not be less than the projected demand if the water surface elevation exceeds the upper curve.
2. Supply of the Hushan reservoir should satisfy 100% of the projected domestic demand, 90% of the projected industrial demand, and 50% of the projected irrigation demand if the water surface elevation is between the upper and lower curve.
3. Supply of the Hushan reservoir should satisfy 80% of the planning domestic demand and 0% of the projected irrigation and industrial demand if the water surface elevation is under the lower curve.

2.2.6. Indices for Impact and Risk Assessment

To assess the impact and the risk under different scenarios, the shortage index (SI), satisfaction and reliability of the water supply were determined. Equations (10)–(16) define these indices.

1. Satisfaction:

$$\text{Satisfaction} = \frac{Q_{sup}}{Q_d} \times 100\%$$

(10)

2. Reliability:

$$\text{Reliability} = \frac{N_{sat}}{N} \times 100\%$$

(11)

3. Shortage Index:

$$SI = \frac{100}{N} \sum_{t=1}^{N} \left(\frac{|Q_d - Q_{sup}|}{Q_d} \right)^2$$

(12)

where Q_d and Q_{sup} are the quantities of water demand and supply, respectively; N is the given time scale of the simulation, and N_{sat} is the number of days for which the demand has been satisfied.

4. Reservoir Efficiency:

$$RE = \frac{\text{Annual actual water supply}}{\text{reservoir effective capacity}}$$

(13)

5. The average duration of water shortage events:

$$\bar{l}_n = \frac{1}{M} \sum_{j=1}^{M} l_j$$

(14)

where M is the number of occurrences of water shortage events, l_j is the duration of each water shortage event, $\bar{l}_n$ is the average duration of water shortage events.

6. The average water shortage in the water shortage event (10^6 m^3):

$$\bar{d}_n = \frac{1}{M} \sum_{j=1}^{M} d_j$$

(15)

where d_j is the total water shortage in each event, and $\bar{d}_n$ is the average water shortage in the water shortage event.

7. The daily average water shortage in the water shortage event (10^6 m^3/day):

$$\frac{\bar{d}_n}{\bar{l}_n}$$

(16)

3. Results and Discussion

3.1. 60-Year Flow Data Analysis with EEMD

We analyze the historical flow data from 1956 to 2015 in this study. Because the skewness of flow data will affect the decomposition process, the historical flow data is taken nature log transformation before decomposing by the EEMD. The 60-year historical data is decomposed by the EEMD method, and the decomposed IMFs are shown in Figure 8.

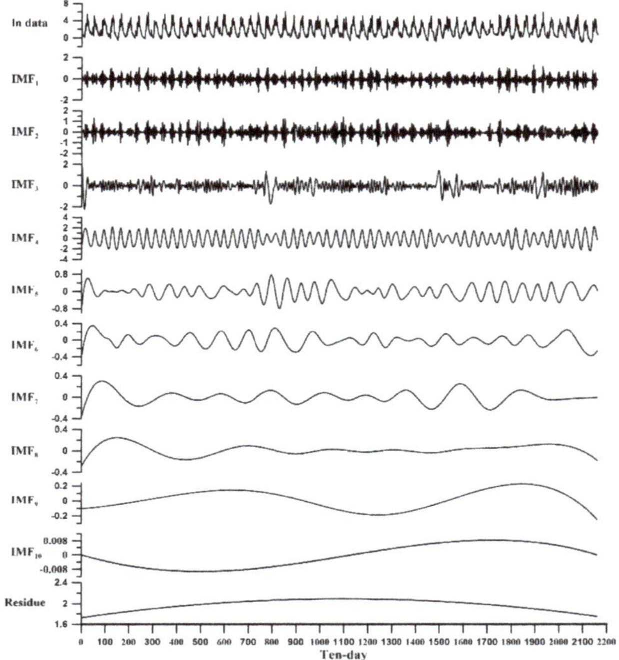

Figure 8. Decomposition of 60-year flow data by EMD.

Figure 8 shows that there are 10 IMFs and 1 residue. The termination condition of the EEMD is according to following literatures: Wu and Huang [14] show that the number of IMFs approaching $\log_2 n$ can be obtained by EEMD decomposition of the original data; in addition, Huang et al. [27] also explained that the IMF obtained after the data is decomposed will be between $\lceil \log_2 n \rceil - 1$ and $\log_2 n$. Therefore, the result that deals with 60-year daily flow by EEMD will introduce 10 IMFs and 1 residue. Among them, the oscillation of IMF_1–IMF_{10} gradually decrease from ± 3 to ± 0.008, but the range of the residue gradually increased from 1.73 to 2.09 and then decreased to 1.75, which is a monotonic function. From the decomposition, we can find that the residue part plays an important role, which often affects the trend of whole time series.

Historical flow data is regarded as a signal for a certain period of time. After decomposing the flow data through EEMD, the physical meaning of IMFs will be diagnosed. The period of each IMF is estimated according to zero up-cross analysis. However, the true energy value cannot be calculated, the square of the IMF value only be regarded as direct proportion with the true energy. In other words, the energy representation, the summation of square of each IMF, should be regarded as the 'weight' of oscillation for IMFs.

The values of the IMFs decomposed by the EEMD are squared and the sum is taken as the representative value of energy, as shown in Figure 9 and Table 3. After the zero up cross analysis method is used to average the IMF period of each group, the average period represented by the IMF can be obtained. IMF_{10} is NAN because the period of the sequence cannot get after the zero up cross analysis method.

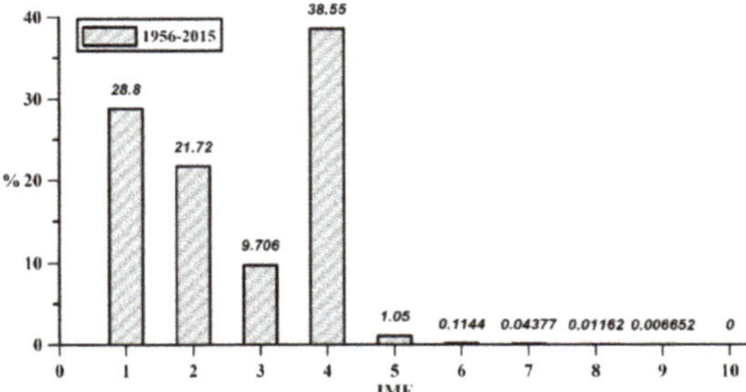

Figure 9. Weight of each IMF for a 60-year flow.

Table 3. Results of 60-year data by EMD with energy and period.

IMF	1	2	3	4	5	6	7	8	9	10
Energy (%)	29%	22%	10%	39%	1%	0%	0%	0%	0%	0%
Period (ten-day)	3.06	6.10	12.51	36.15	68.15	122.69	219.52	506.72	1262.74	N/A

As can be seen from Figure 9, the energy value is mostly concentrated in IMF_4, reaching 39% and the energy of IMF_4 is higher than that of other IMFs. Corresponding to IMF_4 in Figure 10, it can be found that the corresponding period is about 36 ten-day, which represents that the flow data has a strong annual (36 ten-day) period characteristic. In addition, the IMF_1 and IMF_2 still have a certain proportion of the energy percentage, which are 29% and 22% respectively, and their periods correspond to about 3 and 6 ten-day, respectively. Therefore, it can be seen that seasonal characteristics exists in the flow data.

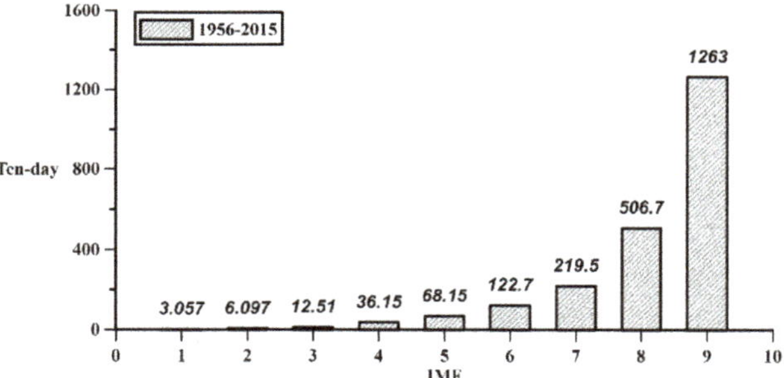

Figure 10. Average period of each IMF for a 60-year flow.

3.2. Analysis of 6 Groups of 10-Year Flow Data

The historical flow data (1956–2015) is divided into 6 decades, and the EEMD screening is implement to get the IMFs and the residue. The relationship between the period, energy, and flow data of different segment IMFs will be discussed. The period and energy percentage distribution of IMFs for each decade are shown in Table 4 and Figures 11 and 12.

Table 4. Results of decadal data by EMD with energy and period.

Energy and Period of Each IMFs Title		1956–1965	1966–1975	1976–1985	1986–1995	1996–2005	2006–2015
IMF_1	Energy (%)	29.21	29.80	30.01	32.27	36.61	28.63
	Period (ten-day)	3.34	3.39	3.04	2.96	3.05	3.18
IMF_2	Energy (%)	17.89	21.43	23.75	27.34	24.22	16.89
	Period (ten-day)	5.89	6.38	6.79	6.29	6.17	5.72
IMF_3	Energy (%)	7.94	7.57	11.98	8.09	17.82	9.12
	Period (ten-day)	11.80	11.48	14.16	12.28	15.97	11.87
IMF_4	Energy (%)	44.32	40.85	30.22	31.93	20.54	44.46
	Period (ten-day)	37.03	35.97	33.00	36.98	36.78	36.90
IMF_5	Energy (%)	0.60	0.31	3.47	0.28	0.72	0.59
	Period (ten-day)	68.73	83.47	63.53	61.08	75.53	78.63
IMF_6	Energy (%)	0.04	0.04	0.56	0.10	0.09	0.31
	Period (ten-day)	93.06	155.07	95.69	114.80	129.14	173.41
IMF_7	Energy (%)	0.00	0.00	0.00	0.00	0.00	0.00
	Period (ten-day)	-	-	-	-	-	-

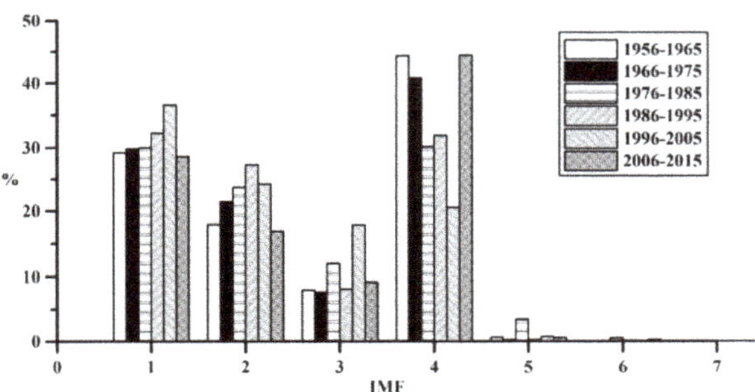

Figure 11. Weight of each IMF for a 10-year flow.

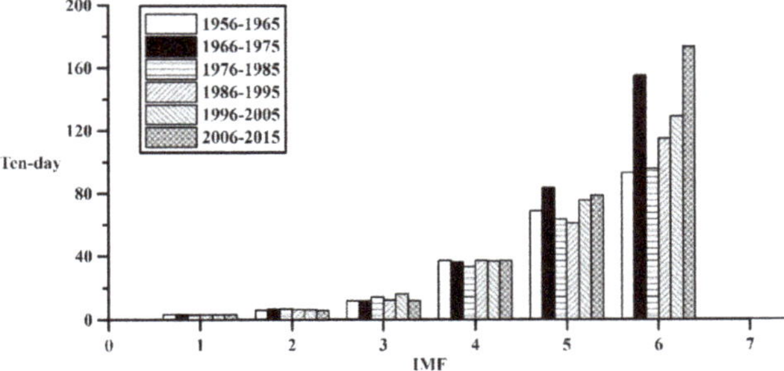

Figure 12. Average period of each IMF for a 10-year flow.

The corresponding relationship between the energy percentage of each IMFs and the average period is shown in Figures 11 and 12. We can find the energy distribution concentrates in IMF_1–IMF_4, which correspond to an average period from 2.96 to 37.03 ten-days. In addition, it can be found that the energy percentage of IMF_4 for the 6 groups has the most highest weight, with an average of about 20.5–44.5% compared with total weight of all IMFs, and it corresponds to an average period around

36 ten-days. Moreover, the minor IMFs, IMF_1 and IMF_2, are have around 28–36% and 18–27% energy of the total respectively, and their corresponding periods are about 3 and 6 ten-days. These results are similar to the 60-year flow analysis, as shown in Table 4.

3.3. Data Synthesis

This section explains the synthesis of flow data. The historical flow data (1956–2015) is divided into six groups of 10-year segments in order. After decomposing through EEMD, multiple IMFs and residue corresponding to them can be obtained. The decadal residues are shown in Figure 13. Regression analysis of different residue in each group is fitted to a new residue representing the value.

This study used two methods to synthesize new decadal flow data as the basis for simulation research. Method (I): first permutations and combinations all IMFs to get n^i ($6^7 = 279,936$) new IMF ($IMF_1 + IMF_2 + \ldots + IMF_7$) set. Where n is the number of groups and i is the number of IMFs each group. After permutations and combinations of the IMFs, add a representative value of the residue, which is the regression of six residues. The representative value of the residue is shown in Figure 13, the bold dashed line. Regression analysis fits the representative value of the residue, and then adds this residue to the IMF of the 279,936 group to obtain the synthesis flow data of the 279,936 group for 10 years. Method (II): all the IMFs and the residue in each group are permutations and combinations to obtain a total of n^{i+1} ($6^{7+1} = 1,679,616$) new synthetic flow data. In the following, the new flow data synthesized from the two methods and historical flow data are compared with each other, and the differences are explored. In addition, this study will apply new flow data synthesized from two different methods to simulate Hushan reservoir water supply systems, and sequentially discuss the simulation results of different synthetic flow data in different water supply scenarios.

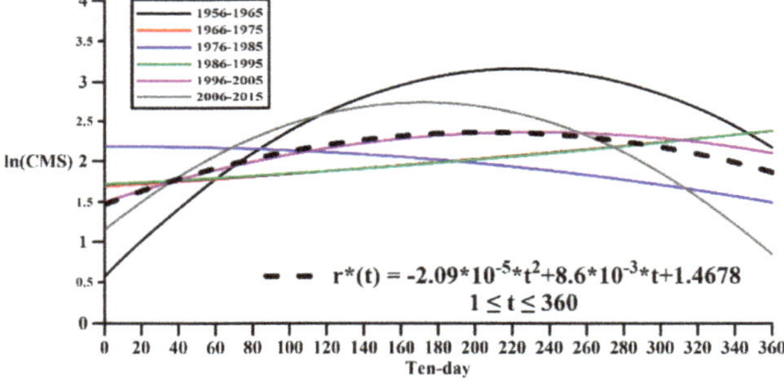

Figure 13. Flow data residues and their representative equation. CMS, m^3/s; r*(t): representative values of the residue.

Result of comparing the synthetic data with historical data divided by 10 years is shown in Table 5 and Figure 14. For the historical data, the flow in Qingshui river is significantly different between the wet and dry season; the flow from wet season, May to October, is significantly higher than November to April. The ratio of wet/dry is about 9:1. The highest flow occurs in August (62.09 m^3/s) and the lowest flow occurs in January (2.05 m^3/s). Compared the historical with the synthetic flow, the monthly distribution is similar between both Method I and Method II. The synthetic flow is concentrated in May–October, and the highest flow always occurs in August, but the lowest flow occurs in the dry season in January and February. In terms of the overall flow during the wet season, the average synthesized by Method I is significantly less than the current situation; in contrast, the average synthesized by Method II is much higher. In the dry season, the average synthesized by Method I and Method II is not much different from the current.

Table 5. Comparison of historical and synthesis flow data.

Flow (cms) \ Month	1	2	3	4	5	6	7	8	9	10	11	12	Average
Observation	2.05	2.56	3.39	5.74	16.85	45.82	41.02	62.09	36.43	10.67	3.88	2.55	19.24
Synthesis (I)	1.91	1.98	2.28	3.44	12.17	33.89	27.69	41.37	25.33	8.59	3.54	2.33	13.71
Synthesis (II)	3.29	3.10	3.70	5.67	20.70	57.92	55.11	80.93	40.40	12.68	5.08	3.54	24.34

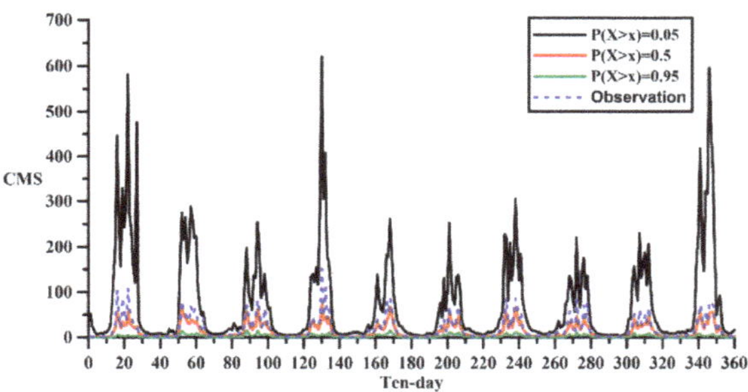

Figure 14. Comparisons of 10-year flow data.

The distribution of the probability of exceedance is shown in the Figures 15 and 16. For the 10-year synthesis, the overall time distribution is consistent with the historical one; furthermore, the synthesis keeps the distribution of wet and dry seasons as well. Through the probability of exceedance of 0.05, the extreme flow of Method II is greater than Method I. In contrast, while the probability of exceedance is 0.95, the flow data of Method II is much less than Method I. This will also lead to the opposite result of the average flow in the simulation of the water supply system.

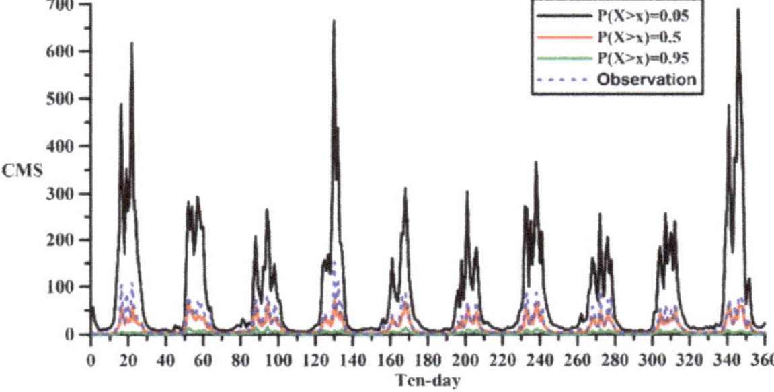

Figure 15. Comparisons of 10-year flow data with Method I. P(X > x): probability of exceedance.

Figure 16. Comparisons of 10-year flow data with Method II.

Although the preceding says that, there are not much variance of the distribution trend between the 10-year flow data synthesized by these two different methods. However, the annual average between two synthesis methods is more significant. The flow data synthesized by Method II has an annual average flow of 24.34 m³/s, which is more than the annual average of Method I, 13.71 m³/s. The water supply system simulation will be based on these three flow sequences, historical data, Method I and Method II synthetic data, to simulate the water supply system of Hushan reservoir.

3.4. Application of Synthesis Data

Next, we apply the synthesized flow data to simulate the water supply system according to the given two demand scenarios. Using the historical flow data, Method I and the Method II synthetic flow data are used for simulation, and the simulation results are shown in the Table 6 and Figure 17.

Table 6. Indices of the water supply for present.

	Simulation Results		
	Unit: Quantity of Water in 10^6 m³/yr		
		Present	
Project	Scenario	No Support Industrial Water	Support Industrial Water (30×10^4 CMD)	Support Industrial Water (10×10^4 CMD)
Reservoir supply		101.93	165.14	128.30
Domestic shortage		0.27	9.74	2.18
Reservoir inflow		101.90	162.80	128.05
Domestic SI		0.002	0.309	0.10
Reservoir efficiency		1.91	3.09	2.40
Public ACDD		9.00	135.00	45.00
ADWS during water shortage events		0.03	0.06	0.05

Note: CMD, m³/day; ACDD, average consecutive dry days; ADWS, average daily water shortage.

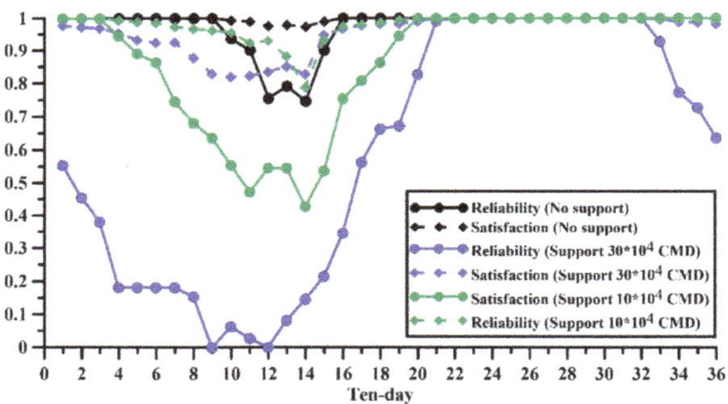

Figure 17. Ten-day reliability and satisfaction of the water supply in present each scenario. CMD, m³/day.

First, we apply the historical flow to simulate the water supply system. For the case of no supporting industrial water in Scenario I, the annual water supply and shortage are 101.93×10^6 m³ and 0.27×10^6 m³, respectively. The water shortage index (SI) is only 0.002. While the water shortage occurring, the average daily water shortage (ADWS) is only 0.03×10^6 m³. The preceding indices show that if the Hushan reservoir only supplies water for domestic demand, the water supply situation will be very stable and shortage will not be easy to occur. In the case of supporting industrial water in Scenario II. If the daily support for industrial water is 300,000 m³, the annual water supply will reach 165.14×10^6 m^3, and the annual water shortage will also increase significantly to 9.74×10^6 m^3. Although the SI only increases to 0.309, and the ADWS only increases to 0.06×10^6 m^3 while the water shortage occurred. The reliability of the water supply reaches 0 already at the 9th and 11th ten-day, as depicted in Figure 17. The result show that the water supply risk in demand Scenario II is relatively severe than Scenario I.

Because the simulation of demand Scenario II indicates severe impact on domestic, we set the domestic SI = 0.1, and try to find the capability of industrial supporting for Hushan reservoir. We found that Hushan reservoir can support 100,000 m^3/day of industrial water, and the annual water shortage has decreased to 2.18×10^6 m^3. The number of average consecutive dry days (ACDD) for domestic has been greatly reduced to 45 days, and the reliability of the water supply is also raised to 0.4–0.5. In the following analyses, we recommend the daily supporting amount for industry from the Hushan reservoir should be 100,000 m^3 as basis.

We simulate the water supply system with the historical flow as applied to method I (279,936 groups) and method II (1,679,616 groups) for newly synthesized flow data. The results are compared based on the ensemble average (as shown in Table 7). Scenario I without considers supporting industrial water, the current annual water supply is 101.93×10^6 m^3, the annual water shortage is only 0.27×10^6 m^3, and the SI is only 0.002. While the water shortage occurring, the ADWS is only 0.03×10^6 m^3.

According to the water supply simulation results of 279,936 sets of flow data in Method I, the annual water supply slightly decreased to 98.25×10^6 m^3, the annual water shortage increased to 3.59×10^6 m^3, and the SI increased significantly to 0.668. While the water shortage occurring, the ADWS is increased slightly to 0.05×10^6 m^3. However, Figure 18 shows that whether the reliability is lower than the current simulation result, they are still above 0.5, the lowest satisfaction is 0.87, which is generally acceptable.

According to the water supply simulation results of 1,679,616 sets of flow data in Method II. The annual water supply decreases to 95.52×10^6 m^3, the annual water shortage has increased significantly to 6.68×10^6 m^3. The SI increased significantly to 1.96 and the ADWS increased to 0.07×10^6 m^3 as well. Figure 18 shows that the reliability is similar to the method I in dry season. However, Method II's reliability is lower than method I value in the wet season. The satisfaction is generally lower than the current situation and the method I, but the minimum is still 0.84. The water supply situation of Method II is still acceptable. There is something odd: why the average annual flow of Method II is significantly more than the current situation and the average flow of Method I, but the simulation results are worse than the other cases? The main reason is that the extreme flow of the Method II often occurs; excessive flow cannot be used efficiently and there are many low flows during wet season. Therefore, the simulation result of Method II is the worst.

Table 7. Comparison of historical and synthesis flow data. SI, Shortage Index.

Simulation Results			
Unit: Quantity of Water in 10^6 m^3/yr			
	No Support Industrial Water		
Project　　　　　　**Scenario**	**Present**	**Method I**	**Method II**
Reservoir supply	101.93	98.25	95.52
Domestic shortage	0.27	3.95	6.68
Reservoir inflow	101.90	94.72	92.07
Domestic SI	0.002	0.668	1.96
Reservoir efficiency	1.91	1.84	1.79
Domestic ACDD	9.00	51.2	59.28
ADWS during water shortage events	0.03	0.05	0.07

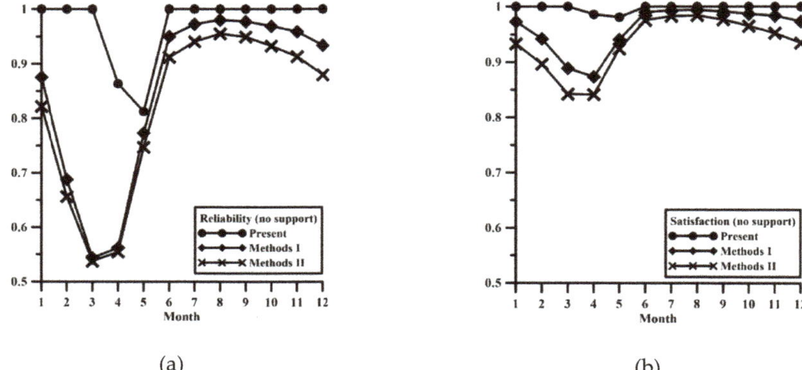

Figure 18. Monthly (**a**) reliability and (**b**) satisfaction of the water supply for Scenario I.

Finally, we simulate the water supply system using the historical flow with Method I and Method II's synthesis flow data. The results are compared to the ensemble average (as shown in Table 8). Under the condition of supporting 100,000 m^3 of industrial demand daily, the current annual water supply is 128.30 × 10^6 m^3, the annual water shortage is 2.18 × 10^6 m^3 and the SI is 0.1. While water shortage occurs, the average daily water shortage is 0.05 × 10^6 m^3.

According to the simulation result of water supply in Method I, the annual water supply decreased slightly to 116.82 × 10^6 m^3, the annual water shortage increased to 11 × 10^6 m^3, and the SI increased significantly to 1.203. While the water shortage occurring, the ADWS is still maintained at 0.05 × 10^6 m^3. Figure 19 shows that the reliability decrease more earlier than the current situation, the lowest monthly reliability reaches 0.24 in March, and recoveries in May. The lowest fulfillment of demand still occurs in March (0.74). In terms of dry season, the risk of water supply shortage is much worse than the current situation.

According to the simulation result of water supply in Method II, the annual water supply decreased to 113.82 × 10^6 m^3, the annual water shortage increased sharply to 13.46 × 10^6 m^3, and the SI increased significantly to 2.12. While occurs the water shortage event, the ADWS also increased to 0.07 × 10^6 m^3. Figure 19 depicts that the reliability is similar to but lower than Method I in the dry season. Compared with the current situation, the water shortage occurs earlier, but it is lower than the Method I in the wet season. The satisfaction is generally lower than the current situation and the Method I, but similar to Method I. The satisfaction rises rapidly in May, which is higher than the current simulation results. However, the satisfaction is decreasing from August to October in wet season, it seems worse than the others.

In summary, no matter what kind of water supply scenario is applied, the current results are better than Method I and Method II. Although the simulation result of Method I is similar to Method II, the ensemble result of Method I is slightly better than the Method II. However, Method II synthesis annual average flow is significantly more than the current flow and the Method I flow, but the simulation results are worse than other cases. The main reason is that the extreme flow of the method II often occurs. Excessive flow cannot be used efficiently and there are many low flows in wet season. Therefore, the simulation result of Method II is the worst.

Table 8. Comparison of historical and synthesis flow data.

Simulation Results			
Unit: Quantity of Water in 10^6 m^3/yr;			
	Support Industrial Water (10×10^4 m^3/s)		
Project \ Scenario	Present	Method I	Method II
Reservoir supply	128.30	116.92	113.82
Public shortage	2.18	11.00	13.46
Reservoir inflow	128.05	112.37	109.43
Public SI	0.10	1.203	2.12
Reservoir efficiency	2.40	2.19	2.13
Public ACDD	45.0	111.0	113.82
ADWS during water shortage events	0.05	0.05	0.07

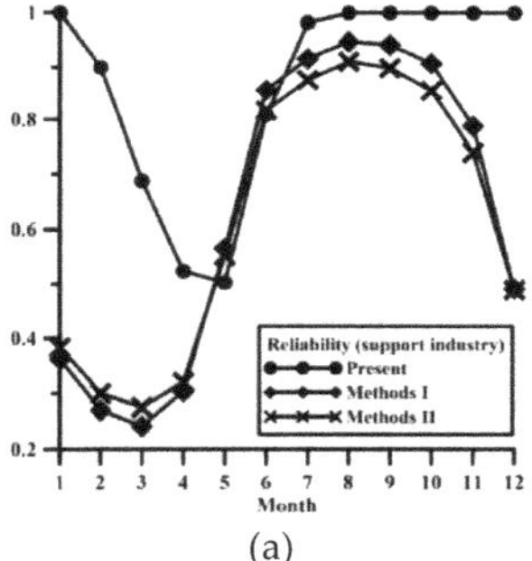

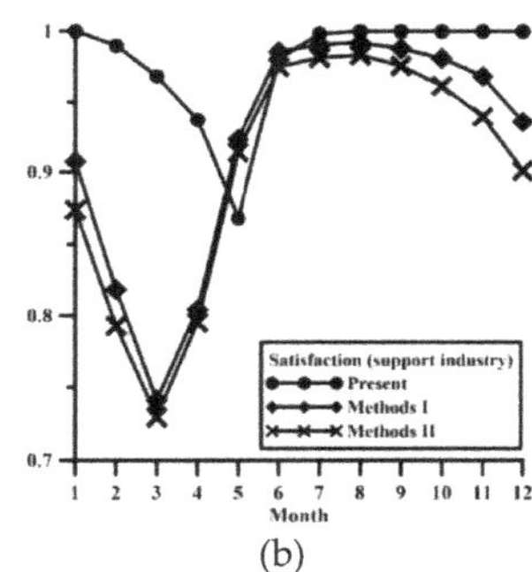

Figure 19. Monthly (**a**) reliability and (**b**) satisfaction of the water supply for Scenario II.

4. Conclusions

The IMFs and residue of the Tongtou weir flow data can be obtained by EEMD decomposition. This method can effectively solve the stationary or nonstationary impacts of hydrological data due to climate change. In this study, two methods are used to synthesize new 10-year flow series data. Method I uses the IMFs segmented data to reorganize the periodic characteristics and sums them up with the residue representation; in contrast, Method II uses the permutation and combination of all IMFs and residues. The new synthetic flow data will provide the input of simulation for the Hushan Reservoir water supply system.

In the energy distribution of IMFs, most of the energy is concentrated between IMF_1 and IMF_4, which IMF_4 has the largest weight (39%), and the corresponding period is about 36 ten-day periods (1 year). However, IMF_1 still has a certain proportion of energy percentage, the corresponding period is about 3 ten-day periods (1 month). This reflects that the flow of Tongtou weir is not only affected by a 1-year periodic sequence, but influenced by the monthly rainfall event. Comparing the current situation with the new data generated by EEMD, the generated data is similar to the current flow distribution, but the flow distribution during the wet season is significantly different (± 0.63 m^3/s), and the flow difference during the dry season is insignificant (± 0.78 m^3/s).

For the Scenario I without considers supporting industrial water, the simulation results of the generation data of Method I show that compared with the current situation, the SI has increased significantly from 0.002 to 0.668, and the number of ACDD of water shortage has also increased significantly from 9 to 51.2 days. The simulation results of Method II show that the SI is more severe than the current situation and Method I, it reaches 1.96. The number of ACDD increased slightly to

59.28 days. In addition, it can be found from the reliability and satisfaction Figure that the simulation results of the synthesis flow in this study will occur water shortages earlier than the current situation.

Scenario II considers supporting industrial water usage, the simulation results of Method I show that the SI has increased significantly to 1.023, and the number of ACDD of water shortage has also increased significantly to 111 days. The simulation results of Method II show that the SI is more severe to 2.12. The number of ACDD increased slightly to 113.82 days. In addition, it can be found from the reliability and satisfaction figure that no matter in which way the synthesis flow is simulated, supporting industrial water will seriously affect the equity of domestic water.

In this study, the multiplication data obtained through permutation and combination after EEMD can show the flow distribution characteristics of the Tongtou weir. In the future, the hydrological factors, such as temperature and rainfall in the same area, can be integrated to explore the correlation between the overall hydrological factors.

Author Contributions: T.-Y.C. analyzed the data and drafted the manuscript; W.-C.H. revised the manuscript. All authors have read and agreed to the published version of the manuscript.

Funding: This research received no external funding.

Acknowledgments: This research was sponsored by the Ministry of Science and Technology in Taiwan (MOST 106-2625-M-019-001).

Conflicts of Interest: The authors declare no conflict of interest.

References

1. Huang, N.E.; Shen, Z.; Long, S.R. A new view of nonlinear water waves: The hilbert spectrum. *Annu. Rev. Fluid Mech.* **1999**, *31*, 417–457. [CrossRef]
2. Huang, N.E.; Shen, Z.; Long, S.R.; Wu, M.C.; Shih, H.H.; Zheng, Q.; Yen, N.C.; Tung, C.C.; Liu, H.H. The empirical mode decomposition and the Hilbert spectrum for nonlinear and nonstationary time series analysis. *Proc. R. Soc. Math. Phys.* **1998**, *454*, 903–995. [CrossRef]
3. Yang, Y.Z. *The Application of Empirical Mode Decomposition on the SASW*; National Cheng Kung University: Tainan, Taiwan, 2008.
4. Huang, Y.; Schmitt, F.G.; Lu, Z.; Liu, Y. Analysis of daily river flow fluctuations using empirical mode decomposition and arbitrary order Hilbert spectral analysis. *J. Hydrol.* **2009**, *373*, 103–111. [CrossRef]
5. Wu, Z.H.; Huang, N.E. On the filtering properties of the empirical mode decomposition. *Adv. Adapt. Data Anal.* **2010**, *2*, 397–414. [CrossRef]
6. Lu, K.Y. *Use EMD to Study the ISO in the Background Circulation of the Typhoon Morakot*; National Central University: Taoyuan, Taiwan, 2011.
7. Lee, T.; Ouarda, T. Prediction of climate nonstationary oscillation processes with empirical mode decomposition. *J. Geophys. Res. Atmos.* **2011**, *116*, D06107. [CrossRef]
8. Lee, T.; Uarda, T. Stochastic simulation of nonstationary oscillation hydroclimatic processes using empirical mode decomposition. *Water Resour. Res.* **2012**, *48*, W02514. [CrossRef]
9. Molla, M.K.I.; Rahman, M.S.; Sumi, A.; Banik, P. Empirical Mode Decomposition Analysis of climate changes with Special Reference to Rainfall Data. *Discret. Dyn. Nat. Soc.* **2006**, *2006*, 1–17. [CrossRef]
10. Molla, M.K.I.; Sumi, A.; Rahman, M.S. Analysis of Temperature Change under Global Warming Impact using Empirical Mode Decomposition. *Int. J. Inf. Tecnol.* **2007**, *3*, 131–139.
11. Huang, S.Z.; Chang, J.X.; Huang, Q.; Chen, Y.T. Monthly streamflow prediction using modified EMD-based support vector machine. *J. Hydrol.* **2014**, *511*, 764–775. [CrossRef]
12. Meng, E.H.; Huang, S.Z.; Huang, Q.; Fang, W.; Wu, L.Z.; Wang, L. A robust method for non-stationary streamflow prediction based on improved EMD-SVM model. *J. Hydrol.* **2019**, *568*, 462–478. [CrossRef]
13. Zhang, H.B.; Wang, B.; Lan, T.; Chen, K. A modified method for non-stationary hydrological time series forecasting based on empirical mode decomposition. *J. Hydroelectr. Eng.* **2015**, *34*, 42–53.
14. Wu, Z.H.; Huang, N.E. Ensemble empirical mode decomposition: A noise-assisted data analysis method. *Adv. Adapt. Data Anal.* **2005**, *1*, 1–41. [CrossRef]

15. Zhao, X.H.; Chen, X. Auto Regressive and Ensemble Empirical Mode Decomposition Hybrid Model for Annual Runoff Forecasting. *Water Resour. Manag.* **2015**, *29*, 2913–2926. [CrossRef]
16. Zhang, X.; Zhang, Q.W.; Zhang, G.; Nie, Z.P.; Gui, Z.F. A Hybrid Model for Annual Runoff Time Series Forecasting Using Elman Neural Network with Ensemble Empirical Mode Decomposition. *Water* **2018**, *10*, 416. [CrossRef]
17. Bai, L.; Xu, J.H.; Chen, Z.S.; Li, W.H.; Liu, Z.H.; Zhao, B.F.; Wang, Z.J. The regional features of temperature variation trends over Xinjiang in China by the ensemble empirical mode decomposition method. *Int. J. Climatol.* **2015**, *35*, 3229–3237. [CrossRef]
18. Yu, Y.H.; Zhang, H.B.; Singh, V.P. Forward Prediction of Runoff Data in Data-Scarce Basins with an Improved Ensemble Empirical Mode Decomposition (EEMD) Model. *Water* **2018**, *10*, 388. [CrossRef]
19. Bai, Y.; Wang, P.; Xie, J.J.; Li, J.T.; Li, C.A. Additive Model for Monthly Reservoir Inflow Forecast. *J. Hydrol. Eng.* **2015**, *20*, 04014079. [CrossRef]
20. Yeh, W.W.-G. Reservoir management and operations models: A state-of-the-art review. *Water Resour. Res.* **1985**, *21*, 1797–1818. [CrossRef]
21. Wurbs, R.A. Reservoir-System Simulation and Optimization Model. *J. Water Resour. Plan. Manag.* **1993**, *119*, 455–472. [CrossRef]
22. Chang, J.X.; Wang, Y.M.; Huang, Q. Reservoir Systems Operation Model Using Simulation and Neural Network. In Proceedings of the Second IFIP Conference on Artificial Intelligence Applications and Innovations, Beijing, China, 7–9 September 2005.
23. Lin, N.M.; Rutten, M. Optimal Operation of a Network of Multi-purpose Reservoir: A Review. *Procedia Eng.* **2016**, *154*, 1376–1384. [CrossRef]
24. Lee, J.L.; Huang, W.C. Climate change impact assessment on Zhoshui River water supply in Taiwan. *Terr. Atmos. Ocean. Sci.* **2017**, *28*, 463–478. [CrossRef]
25. Lee, J.L.; Huang, W.C. Impact on Water Utilization of Zhoshui River Basin by Wushe Reservoir Sediment and Hushan Reservoir Operation. *Taiwan Water Conserv.* **2012**, *60*, 85–97.
26. Chu, T.Y.; Huang, W.C. Water-Resource Utilization of Hushan Reservoir in Taiwan. *Taiwan Water Conserv.* **2016**, *64*, 1–11.
27. Huang, W.C.; Chu, T.Y.; Jang, Y.S.; Lee, J.L. Data Synthesis Based on Empirical Mode Decomposition. *J. Hydrol. Eng.* **2020**. accepted.
28. Goda, Y. Effect of Wave Tilting on Zero-Crossing Wave Heights and Periods. *Coast. Eng. Jpn.* **1986**, *29*, 79–90. [CrossRef]
29. Goda, Y. *Random Seas and Design of Maritime Structures*, 1st ed.; Yokohama National University: Yokohama, Japan, 1985.
30. Tong, C.H. *A Suitability Study on 2-D Irregular Wave Tests*; National Taiwan Ocean University: Keelung, Taiwan, 2013.

Article

Improving Parameter Transferability of GR4J Model under Changing Environments Considering Nonstationarity

Ling Zeng, Lihua Xiong *, Dedi Liu, Jie Chen and Jong-Suk Kim

State Key Laboratory of Water Resources and Hydropower Engineering Science, Wuhan University, Wuhan 430072, China; zengling@whu.edu.cn (L.Z.); dediliu@whu.edu.cn (D.L.); jiechen@whu.edu.cn (J.C.); jongsuk@whu.edu.cn (J.-S.K.)
* Correspondence: xionglh@whu.edu.cn; Tel.: +86-13871078660; Fax: +86-27-68773568

Received: 13 August 2019; Accepted: 25 September 2019; Published: 28 September 2019

Abstract: Hydrological nonstationarity has brought great challenges to the reliable application of conceptual hydrological models with time-invariant parameters. To cope with this, approaches have been proposed to consider time-varying model parameters, which can evolve in accordance with climate and watershed conditions. However, the temporal transferability of the time-varying parameter was rarely investigated. This paper aims to investigate the predictive ability and robustness of a hydrological model with time-varying parameter under changing environments. The conceptual hydrological model GR4J (Génie Rural à 4 paramètres Journalier) with only four parameters was chosen and the sensitive parameters were treated as functions of several external covariates that represent the variation of climate and watershed conditions. The investigation was carried out in Weihe Basin and Tuojiang Basin of Western China in the period from 1981 to 2010. Several sub-periods with different climate and watershed conditions were set up to test the temporal parameter transferability of the original GR4J model and the GR4J model with time-varying parameters. The results showed that the performance of streamflow simulation was improved when applying the time-varying parameters. Furthermore, in a series of split-sample tests, the GR4J model with time-varying parameters outperformed the original GR4J model by improving the model robustness. Further studies focus on more diversified model structures and watersheds conditions are necessary to verify the superiority of applying time-varying parameters.

Keywords: time-varying parameter; GR4J model; changing environments; temporal transferability; western China

---

1. Introduction

The conceptual hydrological models are simplified representations of the complex rainfall-runoff processes in the real world. Hence, great emphases are placed on the calibration of parameters to improve the performance of hydrological models. The optimal constant parameter set for a catchment can be inferred using various approaches [1,2] under the assumption that both the climate and the catchment characteristics are stationary. However, the ability to produce reliable predictions with the constant parameter set has been challenged under changing climate and watershed conditions [3–8].

The transferability of the model parameters between periods with different climate and watershed conditions is essential for reliable application of the hydrological models. Plenty of previous studies have revealed the fact that model parameters are strongly dependent on the data period with which the hydrological models are calibrated [9–11]. In general, loss of robustness and variation in model performance can be witnessed when the model parameters are inappropriately transferred [12]. For example, a model calibrated under wet conditions is usually unlikely to perform well under dry conditions and vice versa. In addition, the optimal parameter set may become no longer suitable once the watershed conditions change significantly as a result of human activities (e.g., the construction of dams, afforestation and deforestation). The impact of such changes on hydrologic processes such as streamflow has been well documented [13,14]. In such cases, hydrological models with time-invariant parameters may show a poor predictive performance due to their inability to explicitly represent the changes within the catchment.

One approach to make the model perform robustly under changing environments is to identify the optimal parameter set with time consistent performance [15–22]. Such approach is based on a series of calibration procedures on different sub-periods, and careful identification of the parameter sets with consistent performance for all these sub-periods. All parameter sets are independently evaluated in every sub-period, and only the parameter set that performs well in all cases are finally selected. Nevertheless, this is achieved at the cost of a possible decline in model performance for some individual sub-periods.

Another approach to improve the predictive performance of hydrological models under changing environments is to allow model parameters to evolve over time [23–33]. For example, Wallner and Haberlandt [25] linked the time-varying model parameters of a modified version of the HBV (Hydrologiska Byråns Vattenbalansavdelning) model with the climate indices. The results showed a significant improvement in model performance for streamflow simulation when using the time-varying parameters, especially for low flows. Pathiraja et al. [26] demonstrated the potential for the data assimilation method to detect the temporal variation in all parameters of the Probability Distributed Model and to improve model performance for streamflow simulation. Westra et al. [27] reported a significantly improved streamflow simulation performance of the GR4J model when the parameter represents the production storage capacity was made dependent on covariates describing seasonality, annual variability, and longer-term trends. Deng et al. [28] assumed the parameters of the two-parameter monthly water balance model to be time-varying as functions of catchment properties. Their case study in southern China suggests that the incorporation of time-varying parameters can enhance the capability of hydrological models to produce reliable predictions under changing environments. These studies suggest that the time-varying parameter may serve as an effective compensation for the possible deficiency of model structure, which failed to represent changes in hydrological dynamics within the watershed.

However, few recent studies focus on the temporal transferability of the time-varying parameters (i.e., robustness under extrapolation). Some previous studies [25,27,28] have tested the temporal transferability of the time-varying parameter from only one calibration period to only one validation period. They reported that the streamflow simulation using a model with the time-varying parameter is better than the original model during the validation procedure. However, these studies failed to fully account for the cases where time-varying parameter is transferred between periods with contrasting climate and watershed conditions (e.g., from driest period to wettest period), which remains an interesting topic that deserves further researches.

This study aims to investigate whether a hydrological model with time-varying parameter would achieve a better streamflow simulation performance and parameter transferability under changing environments than its original form. The GR4J model was chosen for its simplicity in model structure and parsimony in model parameters [34]. The relatively sensitive parameters of the GR4J model were treated as time-varying and assumed to be a function of several external covariates. Several sub-periods with different climate and watershed conditions were set up for a series of split-sample test for the GR4J model with constant parameter and time-varying parameter. The investigation was carried out for Weihe Basin and Tuojiang Basin in western China.

2. Methodology

2.1. The Original GR4J Model and GR4J Model with Time-Varying Parameter

The original GR4J model has four parameters, namely the production storage capacity x_1, the groundwater exchange coefficient x_2, the one day ahead maximum capacity of the routing storage x_3 and the time base of unit hydrograph x_4, as presented in Table 1. More details about the GR4J model can be found in the work of Perrin et al. [34].

Table 1. Description of the GR4J model parameters.

Parameter	Description	Unit	Feasible Range	
			Lower Bound	Upper Bound
x_1	production storage capacity	mm	20	1200
x_2	groundwater exchange coefficient	mm	−5	3
x_3	one day ahead maximum capacity of the routing store	mm	20	500
x_4	time base of unit hydrograph	days	1	5

A comprehensive sensitivity analysis of all four parameters is needed to determinate which parameter is relatively more sensitive and should be treated as time-varying. To this end, the Sobol' sensitivity analysis method [35] was applied in this study. Considering that this study focuses on changing environments where stationary rainfall-runoff relationship is usually absent, the Sobol' sensitivity indices were computed at the monthly timescale. This temporally discretized sensitivity analysis was expected to provide significant information about model behavior and capture the diversity of parameter sensitivity during different periods [36].

The GR4J model with time-varying parameters is referred to as GR4J-T model hereafter. The selected time-varying parameters are treated as functions of several external covariates related to the alternation of climate or watershed conditions. The potential covariates for time-varying parameters are denoted as $V_j (j = 1, 2, \ldots, M)$ and their long-term mean values as $\overline{V}_j (j = 1, 2, \ldots, M)$, where M is the total number of the external covariates. When the parameter $x_i (i = 1, 2, 3, 4)$ of the GR4J model is considered to be time-varying, it is denoted as $x_{i,t}$, where t indicates the time. The general relationship of $x_{i,t}$ with the corresponding external covariates is assumed to be,

$$\frac{x_{i,t} - x_{i,c}}{x_{i,c}} = \sum_{j=1}^{M} \beta_{i,j} f_j \left(\frac{V_{j,t} - \overline{V}_j}{\overline{V}_j} \right) \tag{1}$$

which can be rewritten as,

$$x_{i,t} = x_{i,c} + \sum_{j=1}^{M} \lambda_{i,j} f_j \left(\frac{V_{j,t}}{\overline{V}_j} - 1 \right) \tag{2}$$

where $x_{i,c}$ stands for the constant value of parameter x_i over the calibration period, while $\beta_{i,j}$ and $\lambda_{i,j}$ ($\lambda_{i,j} = \beta_{i,j} \cdot x_{i,c}$), $j = 1, \ldots, M$, are the corresponding regression parameters. The $f_j(\cdot)$ is the link function between the time-varying parameter $x_{i,t}$ and the external covariate V_j, which can take multiple forms such as $f(z) = z$ (linear), $f(z) = e^z$ (exponential), or $f(z) = \ln(z + 1)$ (logarithmic), so as to account for possible linear or non-linear relationship of the time-varying parameter with the external covariates.

Choosing suitable external covariates is a critical issue for the utilization of Equation (1) and (2), but it is usually limited by the availability of dataset. Considering that in practice the daily changes in any external covariates could only cause negligible changes in the time-varying parameters, so only impacts of the monthly changes in any physical covariates on the parameter were investigated in this study, i.e., the time-varying parameters were updated monthly in every simulation run. Inspired by recent work [27,28], 8 external covariates related to variation of climate or watershed underlying conditions, including 1-month, 2-month and 3-month antecedent monthly precipitation and potential evapotranspiration (P_1, P_2, P_3, PET_1, PET_2 and PET_3) and $NDVI$ (Normalized Difference Vegetation Index) of current month ($NDVI_0$), one-month shifted $NDVI$ ($NDVI_1$) (as shown in Table 2), were considered as candidate covariates for the time-varying parameter.

Table 2. Description of external covariates for time-varying parameters.

Covariates	Unit	Description
P_1	mm	One-month antecedent monthly total precipitation
P_2	mm	Two-month antecedent monthly total precipitation
P_3	mm	Three-month antecedent monthly total precipitation
PET_1	mm	One-month antecedent monthly total potential evapotranspiration
PET_2	mm	Two-month antecedent monthly total potential evapotranspiration
PET_3	mm	Three-month antecedent monthly total potential evapotranspiration
$NDVI_0$	-	$NDVI$ value for the current month
$NDVI_1$	-	$NDVI$ value for the previous month

2.2. Model Calibration and Evaluation

The differential split-sample test (DSST) is a common way to evaluate hydrological models between sub-periods with contrasting climate and watershed underlying conditions [37,38]. In practice, usually only two or three contrasting sub-periods can be identified due to the availability of the hydro-meteorological data. This may limit the potential of DSST for accessing the transferability of model parameters and drawing general conclusions. To overcome this limitation, the generalized split-sample test (GSST) proposed by Coron et al. [39,40] was chosen to test the temporal transferability of parameter for the GR4J and GR4J-T model in this study. The main objective of GSST procedure is to evaluate the model performance under as many and as varied climatic and watershed conditions as possible. To this end, a sliding window with a specific length (e.g., five years) is applied to define the sub-periods (as illustrated in Figure 1). In Figure 1, the blue bars indicate the sub-periods while the grey bars represent the remaining part of the time series. Between two adjacent sub-periods, the sliding window is moved by one year.

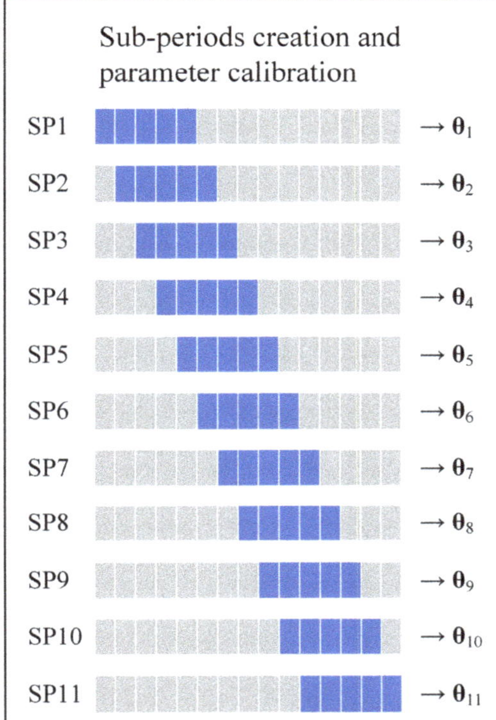

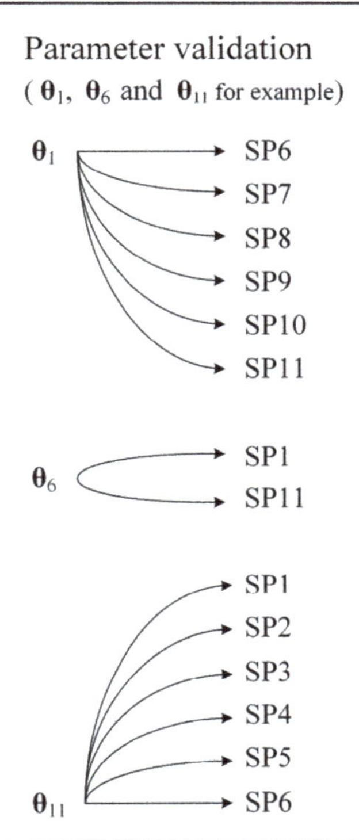

Figure 1. Illustration of the generalized split-sample test (GSST) procedure (example with 15 years available and 5-year subperiods). (Adapted from the work of Coron et al. [40]).

When the sub-periods creation completed, both the GR4J and GR4J-T model were calibrated in each sub-period. Then the parameter sets obtained was used to perform validation test on independent sub-periods, i.e., the sub-periods that do not overlap with the calibration sub-period. For example, the parameter calibrated during SP1 won't be validated in SP2, SP3, SP4 and SP5, as illustrated in Figure 1. For more detail about GSST, the readers can refer to the work of Coron et al. [39,40].

The original GR4J model and the GR4J-T model have different numbers of calibrated parameters. For the original GR4J model, the number of calibrated parameters is 4, i.e., $\theta = \{x_1, x_2, x_3, x_4\}$. In the case of the GR4J-T model, when the parameter $x_i (i = 1, 2, 3, 4)$ is treated to be time-varying, its constant value $x_{i,c}$ and the corresponding regression parameters $\lambda_{i,j} (j = 1, \ldots, M)$ are calibrated together with other time-invariant parameters. For example, when only the parameter x_1 is regarded as the time-varying parameter, parameters that require calibration are $\theta = \{x_{1,c}, \lambda_{1,1}, \ldots, \lambda_{1,M}, x_2, x_3, x_4\}$. Similarly, when multiple parameters are assumed to be time-varying, e.g., x_1 and x_2, the calibrated parameters become $\theta = \{x_{1,c}, \lambda_{1,1}, \ldots, \lambda_{1,M}, x_{2,c}, \lambda_{2,1}, \ldots, \lambda_{2,M}, x_3, x_4\}$.

The widely used Kling-Gupta Efficiency (*KGE*) [41] was applied to assess the overall performance of the GR4J and GR4J-T model. The *KGE* is calculated as follows,

$$KGE = 1 - \sqrt{(r(Q_{act}, Q_{sim}) - 1)^2 + \left(\frac{\mu(Q_{act})}{\mu(Q_{sim})} - 1\right)^2 + \left(\frac{\sigma(Q_{act})}{\sigma(Q_{sim})} - 1\right)^2} \tag{3}$$

where Q_{act} and Q_{sim} are the observed and model-simulated streamflow series, respectively, $r(Q_{act}, Q_{sim})$ is the Pearson's correlation coefficient between the observed and simulated streamflow series, $\mu(Q_{act})$

and $\mu(Q_{sim})$ are the mean values of the observed and simulated streamflow series, respectively, $\sigma(Q_{act})$ and $\sigma(Q_{sim})$ are the standard deviation of the observed and simulated streamflow series, respectively. *KGE* value ranges from $-\infty$ to 1, with a value closer to 1 indicating a better simulation performance.

The bias between the observed and simulated streamflow (*BIAS*) was also applied, which is calculated as follows,

$$BIAS = \frac{\sum_{t=1}^{T} Q_{sim}(t)}{\sum_{t=1}^{T} Q_{act}(t)} - 1 \tag{4}$$

where $Q_{act}(t)$ and $Q_{sim}(t)$ are the observed and model-simulated streamflow at time t, respectively. *BIAS* with a value of 0 indicates no bias, and a value above 0 means an overestimation of the total streamflow volume.

Both the GR4J and GR4J-T models were calibrated using the SCEM (Shuffled Complex Evolution Metropolis) optimization algorithm [2,42], which has been widely used for practical assessment of parameter uncertainty in hydrological modeling. The SCEM algorithm, partially inspired by the SCE-UA (Shuffled Complex Evolution - University of Arizona) algorithm [1], merges the strengths of Metropolis-Hastings sampling [43,44], controlled random search, competitive evolution and complex shuffling to evolve a population of sampled points to an approximation of the posterior distribution of the parameters. Besides, the SCEM method can identify the most likely parameter set and meanwhile its underlying posterior probability distribution in every single run [45]. The likelihood of a parameter set was calculated as the corresponding *KGE* value. Considering that the likelihood value must be nonnegative and monotonically increasing with improved performance, the parameter set that leads to a *KGE* value below 0 was rejected during the evolution procedure within SCEM. The convergence of the algorithm was determined using the Gelman-Rubin statistic [46], which is calculated on the posterior densities to check whether convergence to a stationary target distribution has been reached. For more detail about the SCEM optimization algorithm, the reader can refer to the work of Vrugt et al. [2].

After the convergence of the SCEM optimization algorithm, the final posterior distribution of parameter can be obtained. In total 5000 parameter sets were sampled from the posterior distribution to account for parameter uncertainty in this study. It should be noted that the 5000 parameter sets can lead to very similar streamflow simulation performance in terms of the *KGE* criteria, and the one corresponding to the highest *KGE* value (i.e., the *KGE*-best) was chosen to represent the estimate of the global optimum and used for the model evaluation and selection.

2.3. Temporal Transferability Test of Model Parameters

When a parameter set θ that calibrated from the sub-period D (denoted as "donor period" hereafter) is transferred to the sub-period R (i.e., validation, denoted as "receiver period"), the Kling-Gupta Efficiency (*KGE*) can be rewritten as,

$$KGE_{D \to R} = 1 - \sqrt{(r(Q_{act,R}, Q_{sim,R}[\theta_D]) - 1)^2 + \left(\frac{\mu(Q_{act,R})}{\mu(Q_{sim,R}[\theta_D])} - 1\right)^2 + \left(\frac{\sigma(Q_{act,R})}{\sigma(Q_{sim,R}[\theta_D])} - 1\right)^2} \tag{5}$$

where $Q_{act,R}$ is the simulated streamflow series of sub-period R, $Q_{sim,R}[\theta_D]$ is the simulated streamflow series of sub-period R using parameter set obtained from sub-period D. In addition, $KGE_{R \to R}$ is defined to access the streamflow simulation performance during sub-period R using parameter sets obtained from sub-period R.

Note that the temporal transferability test on one single parameter set could be less persuasive. To fully evaluate the temporal parameter transferability of both the GR4J and GR4J-T model, the 5000 parameter sets from each sub-period were then validated at other independent sub-periods, and the average performance of both models were compared. To this end, a parameter transferability criteria (*PTC*) is defined as:

$$PTC = \frac{\sum_{n=1}^{N} KGE_{D \to R}^{t}(n) - \sum_{n=1}^{N} KGE_{D \to R}^{c}(n)}{N} \tag{6}$$

where $KGE_{D \to R}^{c}(n)$ and $KGE_{D \to R}^{t}(n)$ represent the $KGE_{D \to R}$ value in the case of the GR4J and the GR4J-T model using the nth parameter set, respectively. Here, N takes the value of 5000. Note that PTC with a value above 0 indicates a better parameter transferability for the GR4J-T model over the original GR4J model when transferred from sub-period D to sub-period R.

3. Study Data and Area

Weihe Basin and Tuojiang Basin in Western China were chosen as the study areas. Weihe Basin is located between the coordinates 33°40′–37°26′ N and 103°57′–110°27′ E, with a drainage area of 148,000 km^2 above Huaxian hydrological station, as shown in Figure 2. The elevation within Weihe Basin ranges from 3671 m in the western upstream region to 318 m in the eastern downstream region. Dominated by the semi-arid continental monsoon climate, most precipitation and flood events in Weihe Basin occur in late summer and early autumn. For Weihe Basin, climate variability is significant and human activities have been proved to be the main cause of the alternation of flow regimes. Previous studies have shown that the annual streamflow of the gauge Huaxian has been declining over the past decades [47–49], mainly due to the increasing human activities including the agricultural irrigation, the construction of large water control projects and the implementation of the water-soil conservation projects [50,51]. In addition, the variations of the annual precipitation have also contributed to the reduced annual streamflow in Weihe Basin [52].

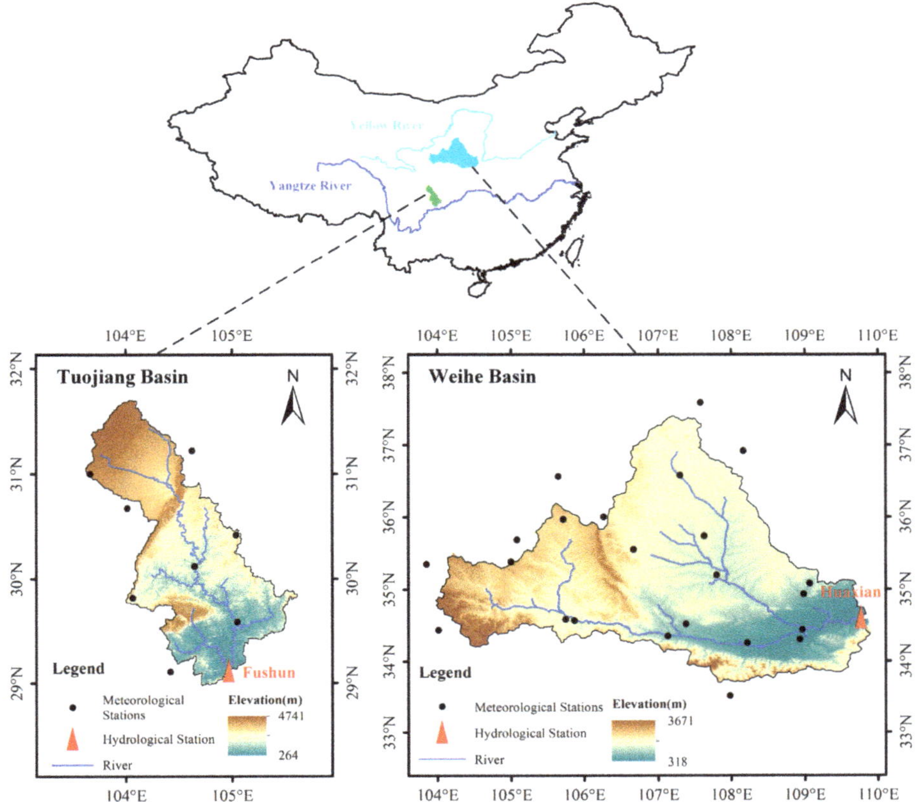

Figure 2. Location of Weihe Basin and Tuojiang Basin in China and the meteorological and hydrological gauges.

Tuojiang Basin (28°88′–30°29′ N, 105°44′–108°20′ E) is an upstream tributary of the Yangtze River, with a total length of 712 km. Tuojiang Basin covers 19,613 km^2 with Fushun hydrological station as the catchment outlet. The elevation within Tuojiang Basin ranges from 264 to 4741 m and decreases from northwest to southeast. Tuojiang Basin is dominated by the subtropical monsoon climate, with most precipitation and flood events occurring in summer and early autumn. The annual streamflow of Tuojiang Basin was also in a downtrend. However, few recent studies focus on Tuojiang Basin and the reason for the declined annual streamflow remains unclear.

The input data for the GR4J model include precipitation (P) and potential evapotranspiration (PET). The precipitation and air temperature data were provided by National Meteorological Information Center of China [53] on a daily basis for the period from 1981 to 2010 and has been widely validated for its quality in previous studies [47–49]. Limited by availability of the meteorological dataset, the potential evapotranspiration (PET) was calculated using the Blaney-Criddle method [54] from the daily average air temperature and the daily sunshine duration. To calibrate and validate the GR4J model, the observed streamflow (Q) series from the gauge Huaxian and Fushun were obtained for the period 1981–2010. The $NDVI$ dataset for the period from 1981–2010 was obtained from the GIMMS (Global Inventory Modelling and Mapping Studies) $NDVI$–3g product [55], which has been widely validated [55,56]. Since its temporal resolution is 15-day, the monthly $NDVI$ is represented by the mean value of the two $NDVI$ values in each month.

4. Results and Discussion

4.1. Diagnostics of Hydrological Nonstationarity

Considering the available hydro-meteorological dataset for Weihe Basin and Tuojiang Basin are for the period 1981–2010, in total 26 sub-periods of 5-year in length are set up following the method described in Figure 1. The reasons for the selection of 5-year as the sub-period length are: (1) 5-year is sufficient for continuous hydrological simulation, (2) a large number of sub-periods (26 in this study) can be set up given that only 30-year records were available. To identify the long-term variation of the hydro-meteorological data series of both basins, the non-parametric Mann-Kendall method [57,58] was applied to examine the trends in the annual P, PET, Q and annual runoff ratios (RR) series of the period 1981–2010, and the result is shown in Table 3. The results show a significant downtrend for the annual Q and RR series, a slight downtrend for annual P series and uptrend for the PET series in Weihe Basin. In the case of Tuojiang Basin, a slight downtrend for annual P series and significant downtrend for annual PET, Q, and RR series are presented. It can be found that both basins witnessed a clear nonstationary in rainfall-runoff relationship. In terms of the annual $NDVI$ series, Weihe Basin shows a clear uptrend while Tuojiang Basin shows no clear changes. This may be attributed to the fact that Weihe Basin has experienced an incessant construction of water-soil conservation projects (e.g., afforestation) since the early 1980s [49].

Variability is also apparent between longer periods, as shown in Figure 3, where the relative mean P, PET, Q, and RR values for all sub-periods (5-year sliding window) are plotted for both basins. Figure 3a–d show how the 5-year hydro-meteorological conditions differ from those of the whole period (1981–2010). For each basin, a line corresponds to the series of the mean values over a 5-year sliding window (sub-period). Each value is expressed relative to the mean value of the whole period (1981–2010) and plotted in the middle year of the corresponding sub-period. The result is in line with the results of the MK-test. For both Weihe Basin and Tuojiang Basin, a significant downtrend of 5-year total streamflow volume and a decreased runoff-ratios are presented. This may question the parameter transferability of the GR4J model, which is detailed in following sections.

Table 3. Mann-Kendal (MK) test results of the annual hydro-meteorological data series, the annual runoff ratios and the annual *NDVI* data series for Weihe Basin and Tuojiang Basin. The critical value of the MK statistic is $Z_{1-\alpha/2} = 1.960$ with $\alpha = 0.05$.

Basin	Data	*Z* statistic	Trend
Weihe Basin	Annual *P*	−1.22	↓
	Annual *PET*	1.56	↑
	Annual *Q*	−2.61	↓**
	Annual *RR*	−2.56	↓**
	Annual *NDVI*	3.96	↑**
Tuojiang Basin	Annual *P*	−1.32	↓
	Annual *PET*	−3.51	↓**
	Annual *Q*	−2.75	↓**
	Annual *RR*	−2.78	↓**
	Annual *NDVI*	0.46	↑

Notes: The symbol "**" indicates a significant uptrend or downtrend.

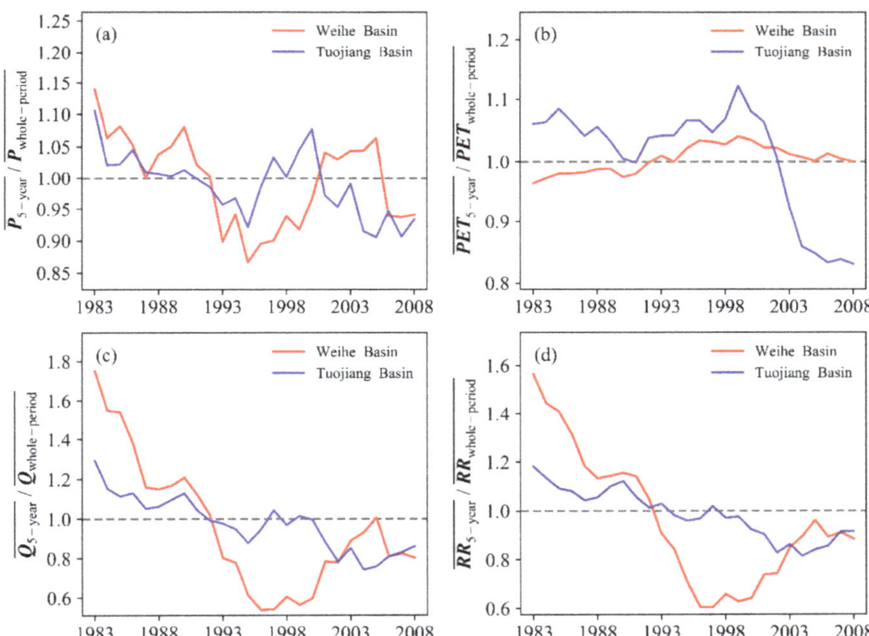

Figure 3. Relative long-term hydro-meteorological variability of (**a**) precipitation (*P*), (**b**) potential evapotranspiration (*PET*), (**c**) runoff (*Q*), and (**d**) runoff ratio (*RR*) over Weihe Basin and Tuojiang Basin.

4.2. Parameter Estimation for the GR4J and GR4J-T Model

To incorporate time-varying sensitivity analysis, the Sobol' sensitivity analysis was repeated at a monthly temporal resolution rather than over the whole period (1981–2010). The monthly sensitivity indices (including the first-order and total-order) of each parameter were therefore obtained with *KGE* as the metric. More detail about the time-varying sensitivity analysis can be found in the work of Herman et al. [36]. Figure 4 provides a simple and direct interpretation of the relationships between the sensitivity of the GR4J's parameters and changing precipitation and streamflow conditions, which was achieved by sorting the monthly sensitivity indices (only the total-order were presented for brevity) for both basins along ascending gradients of monthly precipitation and streamflow.

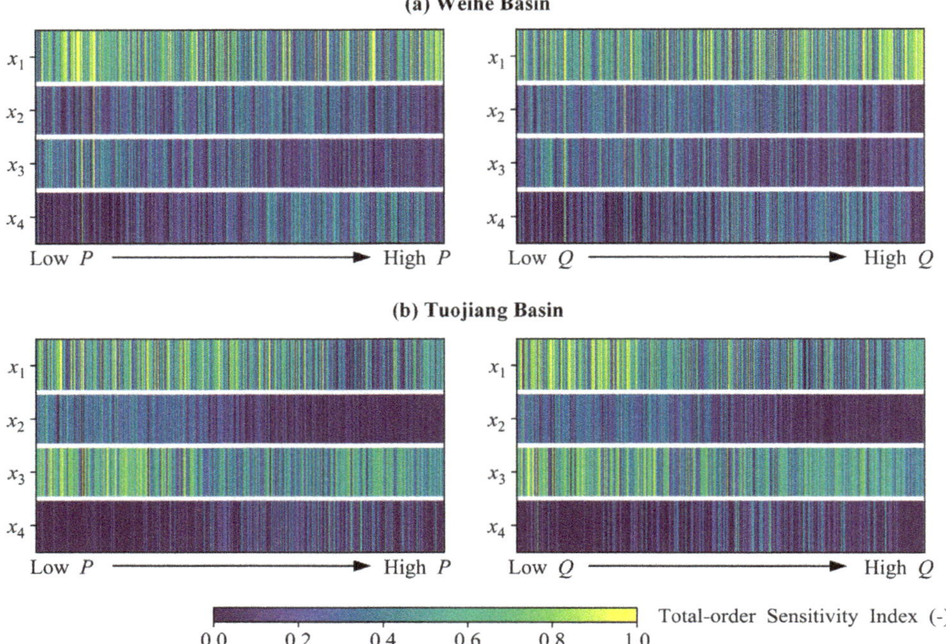

Figure 4. Monthly total-order sensitivity indices of GR4J parameters for (**a**) Weihe Basin and (**b**) Tuojiang Basin. For each basin, the left panel and the right panel share the same sensitivity indices, but they are sorted by monthly total precipitation (left) and monthly streamflow (right), respectively.

For Weihe Basin, the parameter x_1, which controls the production storage capacity, shows a strong sensitivity during both high-flow and low-flow months. In contrast, the parameter x_2, x_3 and x_4 are less sensitive. In the case of Tuojiang Basin, the parameter x_1 and x_3 dominate the months with low-flow and less-precipitation, while the parameter x_2 and x_4 show relatively weaker sensitivity. Thus, the parameter x_1 is assumed to be time-varying for Weihe Basin, and the parameter x_1 and x_3 for Tuojiang Basin.

The correlation between monthly runoff-ratios and monthly external covariate is summarized in Figure 5 in term of Spearman's rank correlation coefficient. Note that the monthly runoff-ratios were calculated as the ratios of monthly runoff depth (mm) and the monthly total precipitation (mm). It can be found that the monthly runoff-ratios has a strong correlation with PET_1 and $NDVI_0$ for Weihe Basin. In the case of Tuojiang Basin, the covariates with high correlation coefficient are P_2 and $NDVI_0$. This correlation analysis was done to justify the reasonability of incorporating the abovementioned external covariates into the time-varying parameters. It may be arbitrary to assert that an external covariate with a weak correlation coefficient is inappropriate and useless. In this study, all 8 external covariates were applied to construct the time-varying parameters for both Weihe Basin and Tuojiang Basin, which is detailed in the following section.

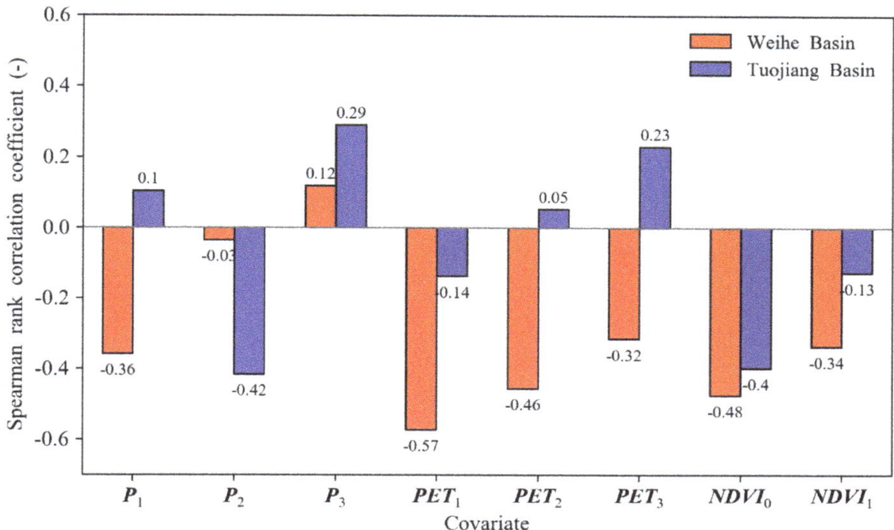

Figure 5. Spearman rank correlation coefficient between monthly covariates and monthly runoff ratio (*RR*) for Weihe Basin and Tuojiang Basin for the whole period (1981–2010).

To reduce the influence of the initial state, a spin-up period of 1-year was used at the parameter calibration procedure for each sub-period, and the simulated streamflow within the spin-up period was excluded in the model assessment procedure. To fully explore the potential for the GR4J-T model in streamflow simulation, all combination of the eight external covariates were investigated and their corresponding performances were compared in terms of the *KGE* and *BIAS* criteria. Considering a large number of model assessment procedure (e.g., 255 covariate combinations for each sub-period in Weihe Basin,$C_8^1 + C_8^2 + C_8^3 + C_8^4 + C_8^5 + C_8^6 + C_8^7 + C_8^8 = 255$), only a small part of the results is presented hereafter for the sake of brevity. For each covariate combination, totally 5000 parameter sets were obtained from the calibration procedure and only the *KGE*-best for each covariate combination is shown here. For example, Table 4 presented the streamflow simulation performance for 8 different cases (C0–C7) during the sub-period 1 (SP1) in Weihe Basin. The case C0 was designed to represent the original GR4J model, where the value of parameter x_1 is invariant over time. Under cases C1–C7, the external covariate P_1, PET_1 and $NDVI_0$ were incorporated into the time-varying parameter $x_{1,t}$. The corresponding equations for time-varying parameter $x_{1,t}$ are also presented in Table 4. It should be noted that three forms of link functions including 'linear', 'exponential' and 'logarithmic' were used to link the external covariates with the time-varying parameter. Three link functions lead to similar model performance, while the 'linear' was always the best and simplest one. So only the 'linear' is shown here.

It can be found that the streamflow simulation performance significantly improves when parameter x_1 was allowed to change over time. For example, the *KGE* value is 0.752 when only the covariate P_1 was considered when compared to 0.727 for the original GR4J model. The difference in *BIAS* criteria (from −16.1% to −6.3%) also indicates a better water balance simulation of the GR4J model with time-varying parameter. Furthermore, the highest *KGE* value (0.761) was achieved when P_1, PET_1 and $NDVI_0$ were incorporated into the time-varying parameter.

Table 4. Comparison of streamflow simulation performance of the GR4J model (C0) and the GR4J-T model (C1–C7) in Weihe Basin during the sub-period 1 (SP1, from 1981 to 1985).

Case	Covariate			Equation for Time−Varying Parameter $x_{1,t}$	KGE (-)	BIAS (%)
	P_1	PET_1	$NDVI_0$			
C0				$x_{1,t} = 320.1$	0.727	−16.1
C1	✓			$x_{1,t} = 290.3 + 0.28P_{1,t}$	0.752	−6.3
C2		✓		$x_{1,t} = 306.5 + 0.54PET_{1,t}$	0.749	−8.1
C3			✓	$x_{1,t} = 296.8 + 99.8NDVI_{0,t}$	0.751	−11.2
C4	✓	✓		$x_{1,t} = 271.8 + 0.19P_{1,t} + 0.27PET_{1,t}$	0.755	−6.2
C5	✓		✓	$x_{1,t} = 312.5 + 0.17PET_{1,t} + 60.6NDVI_{1,t}$	0.759	−6.4
C6		✓	✓	$x_{1,t} = 298.4 + 0.25P_{1,t} + 41.8NDVI_{0,t}$	0.751	−9.3
C7	✓	✓	✓	$x_{1,t} = 303.4 + 0.18P_{1,t} + 0.28PET_{1,t} + 32.6NDVI_{0,t}$	0.761	−6.1

4.3. Streamflow Simulation Performance of the GR4J and GR4J-T Model

Similarly, the model assessment procedure was repeated for the rest sub-periods (SP2-SP26). Table 5 presents the streamflow simulation performance of the GR4J model and the GR4J-T model for all 26 sub-periods in Weihe Basin. Note that the equations for time-varying parameter $x_{1,t}$ corresponds to the covariate combination case with the best streamflow simulation performance in terms of the KGE criteria during the corresponding sub-period, i.e., the KGE-best one. The GR4J-T model significantly outperforms the original GR4J model in terms of the KGE criteria and achieve a better water balance simulation in most cases. This is not unexcepted as previous studies have proven the advantage of applying time-varying parameter in streamflow simulation [27,28].

Table 5. Comparison of streamflow simulation performance of the GR4J model and the GR4J-T model in Weihe Basin for all sub-periods (SP1–SP26).

Sub-Period	GR4J Model			GR4J-T Model		
	Value of x_1	KGE (-)	BIAS (%)	Equation for Time−Varying Parameter $x_{1,t}$	KGE (-)	BIAS (%)
SP1	320.1	0.727	−16.1	$x_{1,t} = 303.4 + 0.18P_{1,t} + 0.28PET_{1,t} + 32.6NDVI_{0,t}$	0.761	−6.1
SP2	443.2	0.617	−17.3	$x_{1,t} = 278.6 + 0.16P_{1,t} + 0.17PET_{1,t} + 57.7NDVI_{0,t}$	0.631	−14.5
SP3	324.1	0.571	−19.1	$x_{1,t} = 309.2 + 0.25P_{1,t} - 0.14PET_{1,t} - 28.6NDVI_{0,t}$	0.572	−11.6
SP4	263.5	0.509	−26.2	$x_{1,t} = 251.5 + 0.35P_{1,t} - 0.19PET_{1,t} + 38.6NDVI_{0,t}$	0.591	−20.2
SP5	447.3	0.611	−16.2	$x_{1,t} = 399.5 + 0.27P_{1,t} - 0.15PET_{1,t} - 4.3NDVI_{0,t}$	0.635	−13.2
SP6	375.5	0.389	−22.2	$x_{1,t} = 259.1 + 0.38P_{1,t} - 0.21PET_{1,t} + 44.6NDVI_{0,t}$	0.481	−20.3
SP7	367.4	0.392	−22.9	$x_{1,t} = 292.7 + 0.29P_{1,t} - 0.19PET_{1,t} + 44.2NDVI_{0,t}$	0.514	−27.9
SP8	300.9	0.442	−25.7	$x_{1,t} = 251.5 + 0.35P_{1,t} + 0.15PET_{1,t} + 21.7NDVI_{0,t}$	0.575	−26.4
SP9	256.3	0.341	−23.3	$x_{1,t} = 205.6 + 0.45P_{1,t} - 0.37PET_{1,t} + 61.9NDVI_{0,t}$	0.504	−19.6
SP10	414.5	0.403	−18.1	$x_{1,t} = 283.5 + 0.48P_{1,t} - 0.12PET_{1,t} + 17.1NDVI_{0,t}$	0.442	−15.2
SP11	417.6	0.392	−26.4	$x_{1,t} = 233.5 + 0.17P_{1,t} - 0.09PET_{1,t} - 7.6NDVI_{0,t}$	0.438	−26.6
SP12	244.1	0.483	−25.2	$x_{1,t} = 241.5 + 0.37P_{1,t} - 0.49PET_{1,t} + 48.2NDVI_{0,t}$	0.536	−19.7
SP13	182.4	0.230	−42.9	$x_{1,t} = 194.1 + 0.67P_{1,t} + 0.34PET_{1,t} - 27.9NDVI_{0,t}$	0.432	−40.3
SP14	340.4	0.499	−23.6	$x_{1,t} = 305.8 + 0.47P_{1,t} - 0.15PET_{1,t} - 34.8NDVI_{0,t}$	0.535	−17.9
SP15	424.1	0.659	−13.7	$x_{1,t} = 341.5 + 0.53P_{1,t} - 0.09PET_{1,t} - 13.6NDVI_{0,t}$	0.713	−7.8
SP16	208.1	0.601	−28.2	$x_{1,t} = 221.6 + 0.19P_{1,t} + 0.09PET_{1,t} - 6.3NDVI_{0,t}$	0.622	−26.4
SP17	295.5	0.582	−18.4	$x_{1,t} = 329.4 + 0.54P_{1,t} - 0.03PET_{1,t} - 28.1NDVI_{0,t}$	0.639	−10.1
SP18	315.3	0.489	−24.5	$x_{1,t} = 300.7 + 0.52P_{1,t} - 0.35PET_{1,t} - 12.2NDVI_{0,t}$	0.543	−17.5

Table 5. *Cont.*

Sub-Period	GR4J Model			GR4J-T Model		
	Value of x_1	KGE (-)	BIAS (%)	Equation for Time–Varying Parameter $x_{1,t}$	KGE (-)	BIAS (%)
SP19	378.7	0.734	−23.3	$x_{1,t} = 308.3 - 0.13P_{1,t} + 0.52PET_{1,t} - 23.6NDVI_{0,t}$	0.763	−16
SP20	429.1	0.779	−21.5	$x_{1,t} = 229.8 + 0.57PET_{1,t} - 29.2NDVI_{0,t}$	0.801	−17.4
SP21	339.5	0.732	−4.6	$x_{1,t} = 325.1 - 0.51P_{1,t} + 0.27PET_{1,t} + 29.1NDVI_{0,t}$	0.771	−1.5
SP22	460.1	0.746	−4.7	$x_{1,t} = 334.5 - 0.13P_{1,t} + 0.25PET_{1,t} - 12.1NDVI_{0,t}$	0.765	−3.2
SP23	382.8	0.729	−15.3	$x_{1,t} = 281.9 - 0.17P_{1,t} - 0.09PET_{1,t} + 22.4NDVI_{0,t}$	0.772	−13.8
SP24	334.5	0.563	−24.6	$x_{1,t} = 336.7 + 0.34P_{1,t} - 0.13PET_{1,t} + 15.2NDVI_{0,t}$	0.654	−10.6
SP25	223.1	0.522	−37.5	$x_{1,t} = 284.1 + 17.5P_{1,t} + 0.09PET_{1,t} + 11.5NDVI_{0,t}$	0.583	−22.3
SP26	426.5	0.586	−18.7	$x_{1,t} = 316.2 + 0.51P_{1,t} - 0.28PET_{1,t} + 33.8NDVI_{0,t}$	0.691	−13.8

The case of Tuojiang is more complicated since two parameters (x_1 and x_3) are assumed to be time-varying. Therefore, four scenarios were considered for each sub-period, including: (1) both parameters are constant, (2) x_1 is constant and x_3 is time-varying, (3) x_1 is time-varying and x_3 is constant, (4) both parameters are time-varying. However, only the results of the first and fourth scenario are presented here in Table 6, because: (1) the calibration procedure is not the main purpose of this study, (2) the second and third scenario show less significant improvement in term of *KGE* when compared to the fourth scenario. For Tuojiang Basin, it is clear that the GR4J-T outperforms the original GR4J model, though the improvement in terms of *KGE* is not as significant as the case of Weihe Basin. This may partly be attributed to the fact that the original GR4J can achieve a satisfying performance in most sub-periods for Tuojiang Basin.

Table 6. Comparison of streamflow simulation performance of the GR4J model and the GR4J-Model in Tuojiang Basin for all sub-periods (SP1–SP26).

Sub-Period	GR4J Model			GR4J-T Model		
	Values of x_1 and x_3	KGE (-)	BIAS (%)	Equations for $x_{1,t}$ and $x_{3,t}$	KGE (-)	BIAS (%)
SP1	$x_1 = 153.5$ $x_3 = 129.3$	0.706	−4.7	$x_{1,t} = 179.9 + 0.23PET_{1,t} - 52.3NDVI_{0,t}$ $x_{3,t} = 125.9 + 0.28PET_{1,t} - 50.1NDVI_{0,t}$	0.729	−3.8
SP2	$x_1 = 210.1$ $x_3 = 225.3$	0.619	−5.2	$x_{1,t} = 221.5 + 0.14P_{2,t} + 0.18PET_{1,t} - 50.1NDVI_{0,t}$ $x_{3,t} = 185.2 + 0.34PET_{1,t} - 41.5NDVI_{0,t}$	0.654	−3.2
SP3	$x_1 = 119.1$ $x_3 = 180.2$	0.677	−4.4	$x_{1,t} = 287.7 - 0.23PET_{1,t} - 66.3NDVI_{0,t}$ $x_{3,t} = 156.9 + 0.18PET_{1,t} - 31.7NDVI_{0,t}$	0.701	−4.2
SP4	$x_1 = 191.5$ $x_3 = 255.9$	0.725	−4.5	$x_{1,t} = 233.8 + 0.37PET_{1,t} + 7.2NDVI_{0,t}$ $x_{3,t} = 289.1 + 0.12PET_{1,t} - 33.2NDVI_{0,t}$	0.738	−1.9
SP5	$x_1 = 183.2$ $x_3 = 293.9$	0.696	−4.9	$x_{1,t} = 253.2 - 0.14PET_{1,t} - 38.7NDVI_{0,t}$ $x_{3,t} = 217.9 - 0.41PET_{1,t} - 48.3NDVI_{0,t}$	0.702	−6.3
SP6	$x_1 = 230.7$ $x_3 = 110.1$	0.738	−5.2	$x_{1,t} = 286.8 - 0.56PET_{1,t} + 19.6NDVI_{0,t}$ $x_{3,t} = 131.2 - 39.1NDVI_{0,t}$	0.754	−4.1
SP7	$x_1 = 239.2$ $x_3 = 124.6$	0.724	−2.2	$x_{1,t} = 263.1 - 0.27PET_{1,t} - 35.1NDVI_{0,t}$ $x_{3,t} = 217.9 - 0.24PET_{1,t} - 18.5NDVI_{0,t}$	0.743	−0.7
SP8	$x_1 = 201.3$ $x_3 = 281.2$	0.725	−2.5	$x_{1,t} = 267.8 + 0.18PET_{1,t} - 36.5NDVI_{0,t}$ $x_{3,t} = 237.5 + 0.24P_{2,t} - 0.31PET_{1,t} - 18.9NDVI_{0,t}$	0.752	−4.4
SP9	$x_1 = 259.7$ $x_3 = 115.8$	0.771	−3.5	$x_{1,t} = 161.5 + 0.32PET_{1,t} + 56.6NDVI_{0,t}$ $x_{3,t} = 237.9 + 0.14P_{2,t} - 0.28PET_{1,t} - 45.7NDVI_{0,t}$	0.792	−0.1
SP10	$x_1 = 234.6$ $x_3 = 122.7$	0.756	−5.1	$x_{1,t} = 191.3 + 0.64PET_{1,t} - 18.9NDVI_{0,t}$ $x_{3,t} = 187.2 + 0.19P_{2,t} - 72.1NDVI_{0,t}$	0.776	−4.8
SP11	$x_1 = 217.4$ $x_3 = 220.8$	0.759	−2.3	$x_{1,t} = 285.1 - 0.22P_{2,t} - 24.3NDVI_{0,t}$ $x_{3,t} = 188.2 + 0.12P_{2,t} + 0.28PET_{1,t} - 35.1NDVI_{0,t}$	0.771	−7.7

Table 6. *Cont.*

Sub-Period	GR4J Model			GR4J-T Model		
	Values of x_1 and x_3	KGE (-)	BIAS (%)	Equations for $x_{1,t}$ and $x_{3,t}$	KGE (-)	BIAS (%)
SP12	$x_1 = 137.3$ $x_3 = 199.3$	0.764	−4.9	$x_{1,t} = 216.5 + 0.36PET_{1,t} - 62.7NDVI_{0,t}$ $x_{3,t} = 244.7 + 0.34PET_{1,t} - 46.8NDVI_{0,t}$	0.779	−8.6
SP13	$x_1 = 261.8$ $x_3 = 161.6$	0.726	−7.5	$x_{1,t} = 307.8 + 0.18P_{2,t} - 0.42PET_{1,t} - 42.9NDVI_{0,t}$ $x_{3,t} = 155.8 + 0.52PET_{1,t} - 29.1NDVI_{0,t}$	0.739	−10.1
SP14	$x_1 = 240.4$ $x_3 = 152.5$	0.728	−2.1	$x_{1,t} = 271.9 - 0.19PET_{1,t} - 56.3NDVI_{0,t}$ $x_{3,t} = 141.4 + 0.43PET_{1,t} - 42.7NDVI_{0,t}$	0.736	−2.5
SP15	$x_1 = 240.1$ $x_3 = 177.2$	0.725	−6.7	$x_{1,t} = 234.7 - 0.16PET_{1,t} - 36.1NDVI_{0,t}$ $x_{3,t} = 185.5 - 0.16PET_{1,t} - 24.2NDVI_{0,t}$	0.728	−8.2
SP16	$x_1 = 265.3$ $x_3 = 178.9$	0.691	0.8	$x_{1,t} = 292.1 - 0.42PET_{1,t} - 40.9NDVI_{0,t}$ $x_{3,t} = 248.4 - 61.6NDVI_{0,t}$	0.703	3.4
SP17	$x_1 = 239.6$ $x_3 = 123.4$	0.721	−3.7	$x_{1,t} = 209.8 - 0.43PET_{1,t} + 22.3NDVI_{0,t}$ $x_{3,t} = 144.3 + 0.21P_{2,t} - 0.18PET_{1,t} + 18.3NDVI_{0,t}$	0.727	−2.3
SP18	$x_1 = 236.7$ $x_3 = 152.2$	0.745	−3.9	$x_{1,t} = 267.4 - 0.36PET_{1,t}$ $x_{3,t} = 195.9 - 0.52PET_{1,t} + 18.3NDVI_{0,t}$	0.751	−3.0
SP19	$x_1 = 210.5$ $x_3 = 138.8$	0.728	−9.1	$x_{1,t} = 188.3 + 0.13PET_{1,t} - 51.1NDVI_{0,t}$ $x_{3,t} = 137.8 + 0.15P_{2,t} - 0.28PET_{1,t} + 15.4NDVI_{0,t}$	0.739	−9.9
SP20	$x_1 = 221.5$ $x_3 = 138.4$	0.717	−11.9	$x_{1,t} = 107.6 + 0.12P_{2,t} + 97.3NDVI_{0,t}$ $x_{3,t} = 153.2 - 0.16PET_{1,t} - 42.4NDVI_{0,t}$	0.747	−7.4
SP21	$x_1 = 234.4$ $x_3 = 162.3$	0.795	−7.1	$x_{1,t} = 219.4 - 0.28PET_{1,t} + 75.1NDVI_{0,t}$ $x_{3,t} = 143.1 - 0.31PET_{1,t} - 42.4NDVI_{0,t}$	0.812	−5.2
SP22	$x_1 = 165.4$ $x_3 = 211.6$	0.714	−8.8	$x_{1,t} = 138.7 + 0.56PET_{1,t} - 62.3NDVI_{0,t}$ $x_{3,t} = 200.6 - 0.34PET_{1,t} - 39.4NDVI_{0,t}$	0.745	−6.6
SP23	$x_1 = 303.5$ $x_3 = 185.2$	0.749	−6.6	$x_{1,t} = 339.1 - 0.38PET_{1,t} - 60.8NDVI_{0,t}$ $x_{3,t} = 139.9 + 0.35PET_{1,t} - 62.4NDVI_{0,t}$	0.766	−7.1
SP24	$x_1 = 285.9$ $x_3 = 112.5$	0.688	−9.4	$x_{1,t} = 138.7 + 0.34PET_{1,t} + 41.6NDVI_{0,t}$ $x_{3,t} = 225.9 - 0.45PET_{1,t} - 68.5NDVI_{0,t}$	0.716	−7.5
SP25	$x_1 = 224.5$ $x_3 = 168.7$	0.678	−9.4	$x_{1,t} = 136.8 + 0.18PET_{1,t} + 57.9NDVI_{0,t}$ $x_{3,t} = 257.1 - 0.42PET_{1,t} - 61.3NDVI_{0,t}$	0.681	−6.9
SP26	$x_1 = 185.9$ $x_3 = 129.7$	0.687	−9.7	$x_{1,t} = 200.1 - 0.58PET_{1,t} + 72.6NDVI_{0,t}$ $x_{3,t} = 244.9 + 0.18P_{2,t} - 0.48PET_{1,t} - 71.2NDVI_{0,t}$	0.708	−7.4

The observed streamflow and the uncertainty bounds associated with model parameter for both models during part of SP1 (1982–1983) are presented in Figure 6 (for Weihe Basin) and Figure 7 (for Tuojiang Basin) for an illustration purpose. The uncertainty bounds associated with model parameter were calculated as follows [59]. For both the GR4J and GR4J-T model, the streamflow simulation for a certain sub-period was repeated using the 5000 parameter sets, and 5000 simulated hydrographs were therefore obtained. Then the likelihood value of each parameter set was assigned to the respective simulated hydrograph. At each time step of the simulation, the posterior distributions of simulated streamflow can be calculated based on the 5000 simulated streamflow and the corresponding likelihood values. The upper and lower uncertainty bounds are here defined as the 2.5% and 97.5% quantiles of the posterior distribution, respectively. Then the 95% predictive uncertainty bounds associated with model parameter for both models were obtained. More detail of the calculation of the predictive uncertainty bounds can be found in the work of Blasone et al. [59]. The red and grey shaded areas indicate the prediction uncertainty associated with the GR4J model and GR4J-T model, respectively. It can be found that most parts of the observed streamflow series lie within the uncertainty bounds for both models, indicating a good simulation performance. In addition, the original GR4J model tends to underestimate the low flows in Weihe Basin, as presented in Figure 6b, where the simulated streamflow series is constantly lower than the observed in November of 1983. However, the average relative bandwidth of the uncertainty bounds associated with the GR4J-T model is wider, which indicates a higher parameter uncertainty. This agrees with the finding of Wallner and Haberlandt [25], who showed that time-varying model parameters can improve the model performance at the cost of possible growth in parameter uncertainty. The growth in parameter uncertainty may also relate to the choice of the external time-varying covariates when applying time-varying parameter, which may be an interesting topic that deserves further exploration.

Figure 6a presents the constant value of the parameter x_1 and its temporal dynamic when treated as time-varying for Weihe Basin during part of SP1 (1982–1983). The constant value was obtained from the *KGE*-best parameter set of the GR4J model, while the temporal dynamic of $x_{1,t}$ is presented in the form of boxplot using the 5000 resultant parameter sets of the GR4J-T model. It can be seen that the time-varying parameter $x_{1,t}$ increases in the growing season and decreases as the dormant season begins, with its values around the constant value of the parameter x_1. The case is the same for the parameter x_1 and x_3 in Tuojiang Basin, as shown in Figure 7a,b.

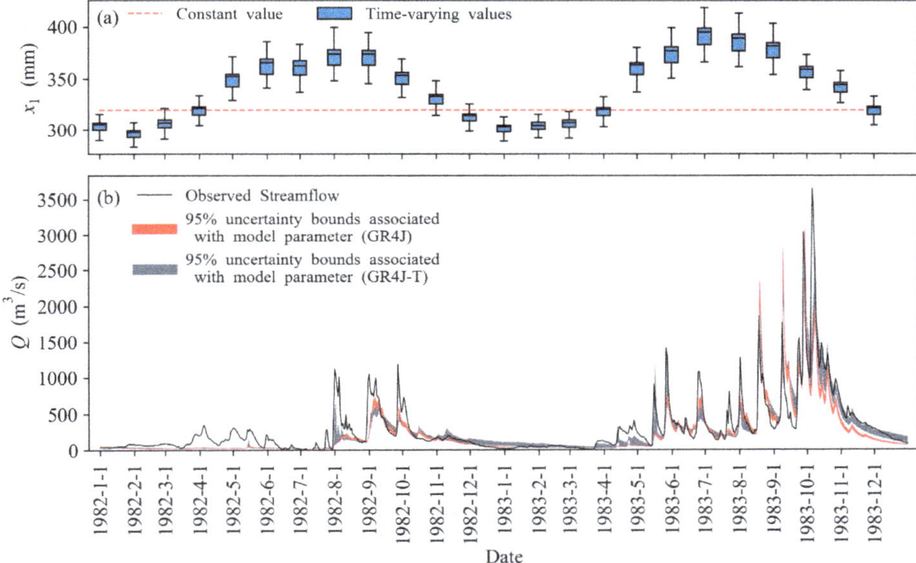

Figure 6. Temporal comparison of (**a**) the values of the parameter x_1 and (**b**) the simulated daily streamflow of the GR4J model and the GR4J-T model for Weihe Basin during SP1 (only the period from 1 January 1982 to 31 December 1983 are presented here). The uncertainty bounds associated with model parameter are derived from the streamflow simulation results using the corresponding 5000 parameter sets.

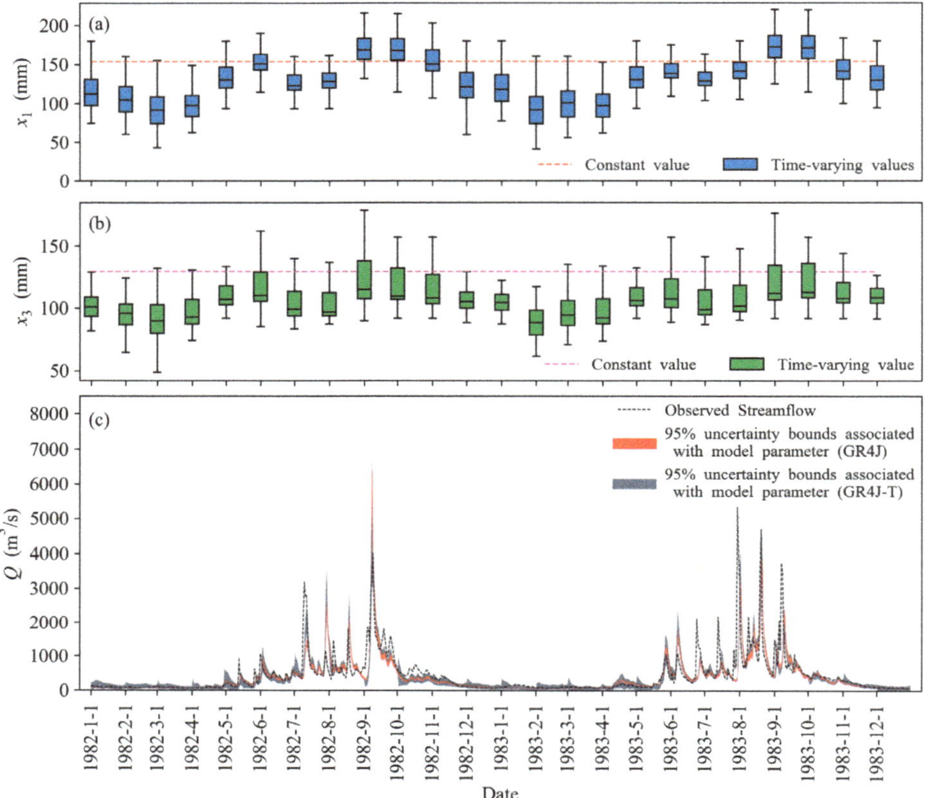

Figure 7. Temporal comparison of the values of the parameter (**a**) x_1 and (**b**) x_3, and (**c**) the simulated daily streamflow of the GR4J model and the GR4J-T model for Tuojiang Basin during SP1 (only the period from 1 January 1982 to 31 December 1983 are presented here). The uncertainty bounds associated with model parameter are derived from the streamflow simulation results using the corresponding 5000 parameter sets.

The increased value of the parameter x_1 during the growing season indicates larger catchment storage (e.g., interception) and, therefore, a weaker response of the watershed to precipitation inputs. As a result, the same hydrological inputs of precipitation may lead to less streamflow, therefore lower runoff ratio when compared to the dormant season. Likewise, the relatively low value of the parameter x_1 during the dormant season may lead to a stronger response to the precipitation inputs and may help to reduce the possibility to underestimate the low flow, as mentioned above. The parameter x_3, which controls the routing storage of GR4J model, shows a similar trend with x_1 in the case of Tuojiang. Similarly, the relatively low values of the parameter x_3 during dormant season, could contribute to a more rapid hydrological response and therefore a better representation of the low flow.

The improved streamflow simulation performance can be attributed to: (1) the dynamics in model parameter, which may help to compensate the possible deficiency of model structure, (2) the increased degree of freedom of model parameters. For the GR4J-T model applied in Weihe Basin, the number of parameters that required calibration is 7. The trade-off between a simple model structure and good model performance could be an interesting topic that deserves further researches, especially under changing environments.

4.4. Parameter Transferability of the GR4J and GR4J-T Model

To test the parameter transferability of the GR4J and GR4J-T Model, the abovementioned 5000 parameter sets obtained from each sub-period (donor period) were then validated at its independent sub-periods (receiver periods). For example, Figure 8 compares the streamflow simulation performance of the GR4J model and GR4J-T model under 10 different sub-periods using parameter sets obtained from SP1 in Weihe Basin. The corresponding *PTC* criteria are also presented. As a reference, the red solid line represents the highest value of *KGE* ($KGE^c_{R \to R}$) achieved by the original GR4J model using parameter from each receiver period, while the blue solid line for the GR4J-T model ($KGE^t_{R \to R}$). The blue boxes represent the *KGE* values ($KGE^c_{D \to R}$) achieved by the original GR4J model using the 5000 parameter sets obtained from SP1, and the green boxes represent those of the GR4J-T model ($KGE^t_{D \to R}$). It is clear that when parameter sets were transferred from the donor period to the receiver periods, the model performance significantly decreased for both the GR4J model and GR4J model. However, the maximum value of the green boxes ($KGE^t_{D \to R}$) is constantly higher than those of the blue boxes ($KGE^c_{D \to R}$), as shown in Figure 8. Besides, the *PTC* criteria, which is the difference between the average value of *KGE* achieved using 5000 parameter sets by both models over each receiver period (see Equation (6)), is above 0 in most cases, which indicates a better temporal transferability of the GR4J-T model. The case is the same for Tuojiang Basin (see Figure 9), though the corresponding *PTC* values are quite close to 0.

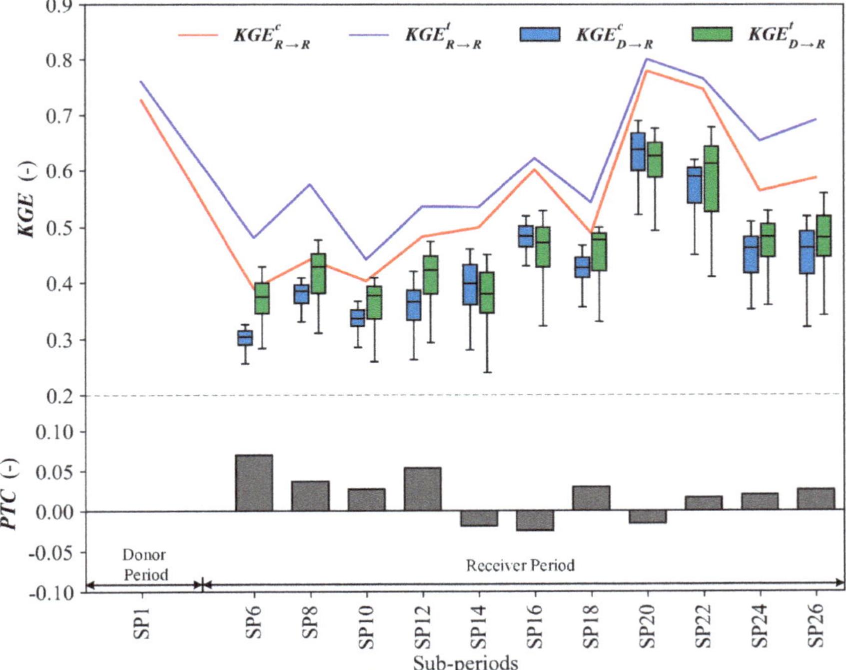

Figure 8. Comparison of streamflow simulation performance in terms of *KGE* when parameter sets calibrated during P1 are transferred to other sub-periods (SP6–SP26) for the GR4J model and the GR4J-T model in Weihe Basin. The boxplots represent the *KGE* values using the 5000 parameter sets obtained from SP1. The *PTC* values indicate the difference of the average simulation performance between the GR4J model and the GR4J-T model during the validation procedure.

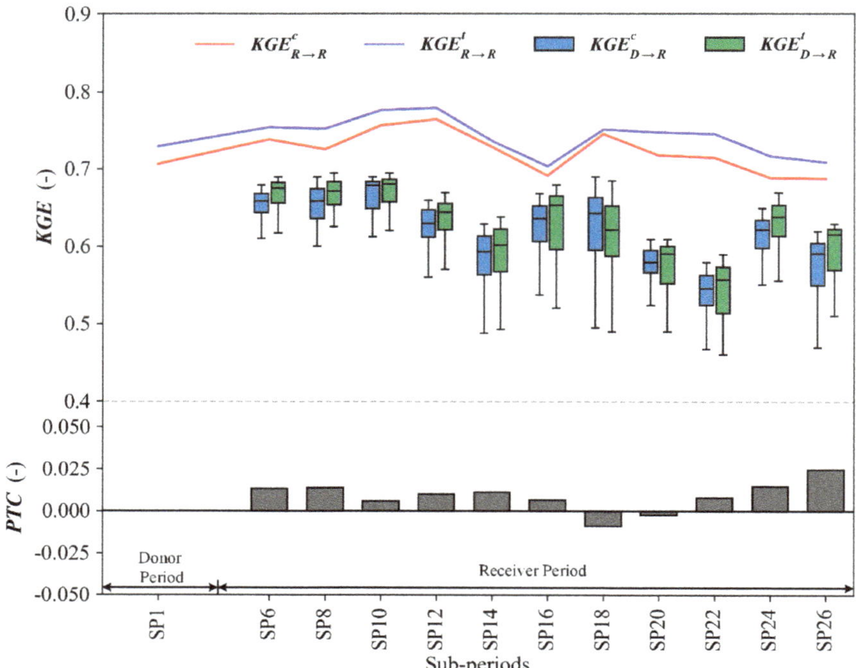

Figure 9. Comparison of streamflow simulation performance in terms of *KGE* when parameter sets calibrated during P1 are transferred to other sub-periods (SP6–SP26) for the GR4J model and the GR4J-T model in Tuojiang Basin.

To comprehensively compare the temporal transferability of the constant parameter and the time-varying parameter, the *PTC* criteria for every possible case is also presented in Figure 10. The blank area indicates the cases where the transferability test is inapplicable (e.g., between adjacent sub-periods or among sub-periods that overlap each other). The labels on the y-axis indicate the 26 donor periods while the labels on the x-axis indicate the receiver periods. Each cell in the figure stands for a transferability test and its color represents the corresponding *PTC* value. The red cells, indicating positive *PTC* values, stands for the cases where the GR4J-T model shows a better temporal transferability over the GR4J model, while the blue cells are in the opposite cases. It is clear that the *PTC* criteria have a positive value in most cases. This further demonstrates that the GR4J-T model has advantage in term of temporal transferability over the original GR4J model.

Note that the loss in model performance when the model parameter was transferred among sub-period with different climate and watershed conditions is inevitable even for the hydrological model with time-varying parameter. Here, it was proven that a better transferability can be achieved under changing environments when applying the time-varying parameter. Though in practice, applying the time-varying parameter is not an easy job, as a more complicated calibration procedure is necessary. However, it is still worth identifying the sensitive parameter and find out how it may change over time, which may help to improve the model structure and to provide new sights for better understanding the hydrological processes.

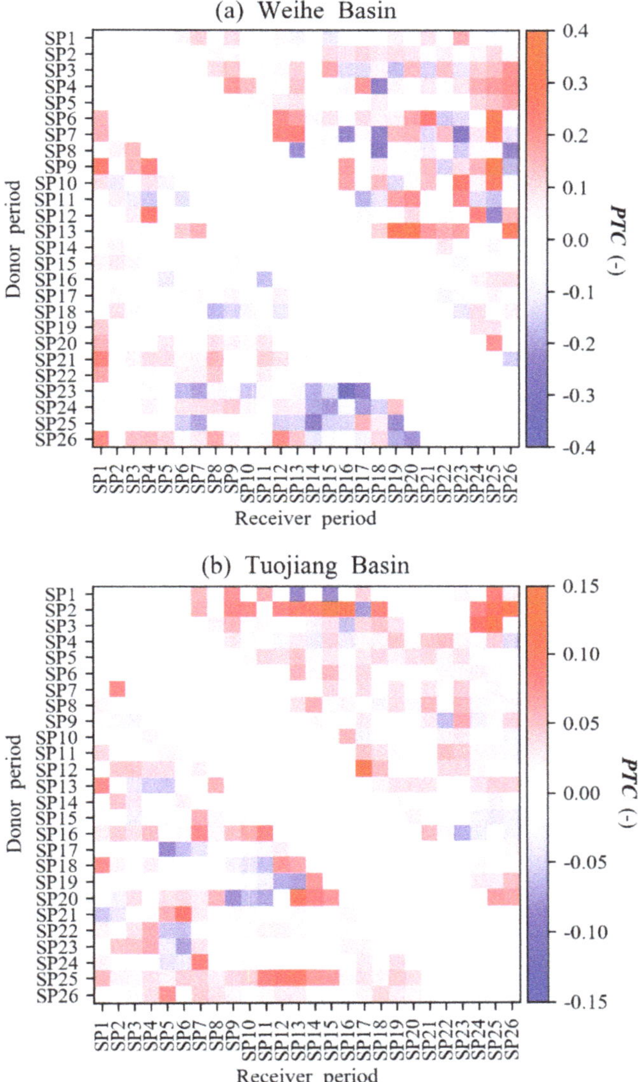

Figure 10. *PTC* values of the parameter transferability test for (**a**) Weihe Basin and (**b**) Tuojiang Basin.

5. Conclusions

This study investigated the parameter transferability under changing environments of the original GR4J model and its modified version, in which the parameter was allowed to vary over time. To set up the GR4J model with the time-varying parameter, a sensitivity analysis was applied to identify the most sensitive parameter, which was then treated as time-varying and assumed to be a function of several external covariates. Both models were calibrated and validated in a series of sub-periods with different climatic and watershed conditions. The investigation was carried out for Weihe Basin and Tuojiang Basin in western China. The main findings of this study are as follows:

(1) A better streamflow simulation performance in terms of the *KGE* criteria was achieved by the GR4J model with time-varying parameters (x_1 for Weihe Basin, x_1 and x_3 for Tuojiang Basin) during most calibration sub-periods for both basins, though at the cost of slight growth in parameter uncertainty,

(2) The GR4J model with time-varying parameter shows a better transferability among sub-periods with different climate and watershed conditions when compared to the original GR4J model for both basins.

Although the time-invariance of model parameters is one of the basic criteria of a high-quality hydrologic model, very few (if any) models can achieve this due to their inherent limitations. Numerous previous studies have proved that hydrological models with time-varying parameters can achieve a better streamflow simulation performance during the calibration period. This study further demonstrates the advantage of the GR4J model with time-varying parameter in terms of temporal transferability among different sub-periods. This may provide new insights for improving the structure of existing hydrological models and for building new models.

Considering the availability of the dataset for both basins, only six climate-related covariates (P_1, P_2, P_3, PET_1, PET_2, and PET_3) and 2 watershed-related covariates ($NDVI_0$ and $NDVI_1$) were accounted for to describe the dynamic variation of the selected parameter. More covariates should be investigated in further researches. Besides, the key factor that leads to the alternation in flow regimes also differs for different catchments. Thus, more emphases should be placed on the careful identification of proper external covariates when applying time-varying parameter.

Author Contributions: Conceptualization, L.Z. and L.X.; data curation, L.Z., J.C. and J.-S.K.; formal analysis, L.Z., L.X., and D.L.; funding acquisition, L.X.; investigation, L.Z.; methodology, L.Z. and L.X.; project administration, L.X.; resources, L.X. and J.C.; software, L.Z., D.L., and J.C.; supervision, L.X.; validation, L.Z. and J.-S.K.; visualization, L.Z.; writing – original draft, L.Z.; writing – review and editing, L.X., D.L., J.C., and J.-S.K.

Funding: This research is supported by the National Natural Science Foundation of China (Grant Nos. 41890822 and 51525902), the Research Council of Norway (FRINATEK Project 274310), and the Ministry of Education "111 Project" Fund of China (B18037), all of which are greatly appreciated.

Acknowledgments: Sincere thanks are due to the editor and two anonymous reviewers for all the remarks and suggestions that are very helpful and constructive for improving our manuscript.

Conflicts of Interest: The authors declare no conflict of interest.

References

1. Duan, Q.; Sorooshian, S.; Gupta, V.K. Effective and efficient global optimization for conceptual rain-runoff models. *Water Resour. Res.* **1992**, *28*, 1015–1031. [CrossRef]
2. Vrugt, J.A.; Gupta, H.V.; Bouten, W.; Sorooshian, S. A Shuffled Complex Evolution Metropolis algorithm for optimization and uncertainty assessment of hydrologic model parameters. *Water Resour. Res.* **2003**, *39*, 1201. [CrossRef]
3. Milly, P.; Betancourt, J.; Falkenmark, M.; Hirsch, R.M.; Kundzewicz, Z.W.; Lettenmaier, D.P.; Stouffer, R.J. Stationarity is Dead: Whither Water Management. *Science* **2008**, *319*, 573–574. [CrossRef]
4. Vansteenkiste, T.; Tavakoli, M.; Ntegeka, V.; De Smedt, F.; Batelaan, O.; Pereira, F.; Willems, P. Intercomparison of hydrological model structures and calibration approaches in climate scenario impact projections. *J. Hydrol.* **2014**, *519*, 743–755. [CrossRef]
5. Merz, R.; Parajka, J.; Blöschl, G. Time stability of catchment model parameters: Implications for climate impact analyses. *Water Resour. Res.* **2011**, *47*, w02531. [CrossRef]
6. Saft, M.; Peel, M.C.; Western, A.W.; Perraud, J.; Zhang, L. Bias in streamflow projections due to climate-induced shifts in catchment response. *Geophys. Res. Lett.* **2016**, *43*, 1574–1581. [CrossRef]
7. Xu, C.; Chen, H.; Guo, S. Hydrological Modeling in a Changing Environment: Issues and Challenges. *J. Water Resour. Res.* **2013**, *2*, 85–95. [CrossRef]
8. Silberstein, R.P.; Aryal, S.K.; Braccia, M.; Durrant, J. Rainfall-runoff model performance suggests a change in flow regime and possible lack of catchment resilience. In Proceedings of the 20th International Congress on Modelling and Simulation (MODSIM2013), Adelaide, Australia, 1–6 December 2013; pp. 2409–2415.

9. Omer, A.; Wang, W.G.; Basheer, A.K.; Yong, B. Integrated assessment of the impacts of climate variability and anthropogenic activities on river runoff: A case study in the Hutuo River Basin, China. *Hydrol. Res.* **2017**, *48*, 416–430. [CrossRef]

10. Le Lay, M.; Galle, S.; Saulnier, G.M.; Braud, I. Exploring the relationship between hydroclimatic stationarity and rainfall-runoff model parameter stability: A case study in West Africa. *Water Resour. Res.* **2007**, *43*, 7. [CrossRef]

11. Choi, H.T.; Beven, K. Multi-period and multi-criteria model conditioning to reduce prediction uncertainty in an application of TOPMODEL within the GLUE framework. *J. Hydrol.* **2007**, *332*, 316–336. [CrossRef]

12. Broderick, C.; Matthews, T.; Wilby, R.L.; Bastola, S.; Murphy, C. Transferability of hydrological models and ensemble averaging methods between contrasting climatic periods. *Water Resour. Res.* **2016**, *52*, 8343–8373. [CrossRef]

13. Singh, R.; van Werkhoven, K.; Wagener, T. Hydrological impacts of climate change in gauged and ungauged watersheds of the Olifants basin: A trading-space-for-time approach. *Hydrol. Sci. J.* **2014**, *59*, 29–55. [CrossRef]

14. Lacombe, G.; Ribolzi, O.; de Rouw, A.; Pierret, A.; Latsachak, K.; Silvera, N.; Pham Dinh, R.; Orange, D.; Janeau, J.; Soulileuth, B.; et al. Contradictory hydrological impacts of afforestation in the humid tropics evidenced by long-term field monitoring and simulation modelling. *Hydrol. Earth Syst. Sci.* **2016**, *20*, 2691–2704. [CrossRef]

15. De Vos, N.J.; Rientjes, T.H.M.; Gupta, H.V. Diagnostic evaluation of conceptual rainfall-runoff models using temporal clustering. *Hydrol. Process.* **2010**, *24*, 2840–2850. [CrossRef]

16. Gharari, S.; Hrachowitz, M.; Fenicia, F.; Savenije, H.H.G. An approach to identify time consistent model parameters: Sub-period calibration. *Hydrol. Earth Syst. Sci.* **2013**, *17*, 149–161. [CrossRef]

17. Guo, D.; Johnson, F.; Marshall, L. Assessing the Potential Robustness of Conceptual Rainfall-Runoff Models Under a Changing Climate. *Water Resour. Res.* **2018**, *54*, 5030–5049. [CrossRef]

18. Ekström, M.; Gutmann, E.D.; Wilby, R.L.; Tye, M.R.; Kirono, D.G.C. Robustness of hydroclimate metrics for climate change impact research. *Wiley Interdiscip. Rev. Water* **2018**, *5*, e1288. [CrossRef]

19. Motavita, D.F.; Chow, R.; Guthke, A.; Nowak, W. The comprehensive differential split-sample test: A stress-test for hydrological model robustness under climate variability. *J. Hydrol.* **2019**, *573*, 501–515. [CrossRef]

20. Fowler, K.; Coxon, G.; Freer, J.; Peel, M.; Wagener, T.; Western, A.; Woods, R.; Zhang, L. Simulating Runoff Under Changing Climatic Conditions: A Framework for Model Improvement. *Water Resour. Res.* **2018**, *54*, 9812–9832. [CrossRef]

21. Vormoor, K.; Heistermann, M.; Bronstert, A.; Lawrence, D. Hydrological model parameter (in) stability-"crash testing" the HBV model under contrasting flood seasonality conditions. *Hydrol. Sci. J.* **2018**, *63*, 991–1007. [CrossRef]

22. Kim, S.S.H.; Hughes, J.D.; Chen, J.; Dutta, D.; Vaze, J. Determining probability distributions of parameter performances for time-series model calibration: A river system trial. *J. Hydrol.* **2015**, *530*, 361–371. [CrossRef]

23. Paik, K.; Kim, J.H.; Kim, H.S.; Lee, D.R. A conceptual rainfall-runoff model considering seasonal variation. *Hydrol. Process.* **2005**, *19*, 3837–3850. [CrossRef]

24. Du, J.; Rui, H.; Zuo, T.; Li, Q.; Zheng, D.; Chen, A.; Xu, Y.; Xu, C.Y. Hydrological Simulation by SWAT Model with Fixed and Varied Parameterization Approaches Under Land Use Change. *Water Resour. Manag.* **2013**, *27*, 2823–2838. [CrossRef]

25. Wallner, M.; Haberlandt, U. Non-stationary hydrological model parameters: A framework based on SOM-B. *Hydrol. Process.* **2015**, *29*, 3145–3161. [CrossRef]

26. Pathiraja, S.; Marshall, L.; Sharma, A.; Moradkhani, H. Hydrologic modeling in dynamic catchments: A data assimilation approach. *Water Resour. Res.* **2016**, *52*, 3350–3372. [CrossRef]

27. Westra, S.; Thyer, M.; Leonard, M.; Kavetski, D.; Lambert, M. A strategy for diagnosing and interpreting hydrological model nonstationarity. *Water Resour. Res.* **2014**, *50*, 5090–5113. [CrossRef]

28. Deng, C.; Liu, P.; Wang, W.; Shao, Q.; Wang, D. Modelling time-variant parameters of a two-parameter monthly water balance model. *J. Hydrol.* **2019**, *573*, 918–936. [CrossRef]

29. Sadegh, M.; AghaKouchak, A.; Flores, A.; Mallakpour, I.; Nikoo, M.R. A Multi-Model Nonstationary Rainfall-Runoff Modeling Framework: Analysis and Toolbox. *Water Resour. Manag.* **2019**, *33*, 3011–3024. [CrossRef]

30. Pathiraja, S.; Anghileri, D.; Burlando, P.; Sharma, A.; Marshall, L.; Moradkhani, H. Time-varying parameter models for catchments with land use change: The importance of model structure. *Hydrol. Earth Syst. Sci.* **2018**, *22*, 2903–2919. [CrossRef]

31. Xiong, M.; Liu, P.; Cheng, L.; Deng, C.; Gui, Z.; Zhang, X.; Liu, Y. Identifying time-varying hydrological model parameters to improve simulation efficiency by the ensemble Kalman filter: A joint assimilation of streamflow and actual evapotranspiration. *J. Hydrol.* **2019**, *568*, 758–768. [CrossRef]

32. Rabassa, P.; Beck, C. Superstatistical analysis of sea-level fluctuations. *Phys. A Stat. Mech. Its Appl.* **2015**, *417*, 18–28. [CrossRef]

33. Yalcin, G.C.; Rabassa, P.; Beck, C. Extreme event statistics of daily rainfall: Dynamical systems approach. *J. Phys. A Math. Theor.* **2016**, *49*, 154001. [CrossRef]

34. Perrin, C.; Michel, C.; Andréassian, V. Improvement of a parsimonious model for streamflow simulation. *J. Hydrol.* **2003**, *279*, 275–289. [CrossRef]

35. Sobol, I.M. Global sensitivity indices for nonlinear mathematical models and their Monte Carlo estimates. *Math. Comput. Simul.* **2001**, *55*, 271–280. [CrossRef]

36. Herman, J.D.; Reed, P.M.; Wagener, T. Time-varying sensitivity analysis clarifies the effects of watershed model formulation on model behavior. *Water Resour. Res.* **2013**, *49*, 1400–1414. [CrossRef]

37. Patil, S.D.; Stieglitz, M. Comparing spatial and temporal transferability of hydrological model parameters. *J. Hydrol.* **2015**, *525*, 409–417. [CrossRef]

38. Arsenault, R.; Brissette, F.; Martel, J. The hazards of split-sample validation in hydrological model calibration. *J. Hydrol.* **2018**, *566*, 346–362. [CrossRef]

39. Coron, L.; Andréassian, V.; Perrin, C.; Lerat, J.; Vaze, J.; Bourqui, M.; Hendrickx, F. Crash testing hydrological models in contrasted climate conditions: An experiment on 216 Australian catchments. *Water Resour. Res.* **2012**, *48*, w05552. [CrossRef]

40. Coron, L.; Andréassian, V.; Perrin, C.; Bourqui, M.; Hendrickx, F. On the lack of robustness of hydrologic models regarding water balance simulation: A diagnostic approach applied to three models of increasing complexity on 20 mountainous catchments. *Hydrol. Earth Syst. Sci.* **2014**, *18*, 727–746. [CrossRef]

41. Gupta, H.V.; Kling, H.; Yilmaz, K.K.; Martinez, G.F. Decomposition of the mean squared error and NSE performance criteria: Implications for improving hydrological modelling. *J. Hydrol.* **2009**, *377*, 80–91. [CrossRef]

42. Feyen, L.; Vrugt, J.A.; Nualláin, B.Ó.; van der Knijff, J.; De Roo, A. Parameter optimisation and uncertainty assessment for large-scale streamflow simulation with the LISFLOOD model. *J. Hydrol.* **2007**, *332*, 276–289. [CrossRef]

43. Metropolis, N.; Rosenbluth, A.W.; Rosenbluth, M.N.; Teller, A.H.; Teller, E. Equation of state calculations by fast computing machines. *J. Chem. Phys.* **1953**, *21*, 1087–1092. [CrossRef]

44. Hastings, W.K. *Monte Carlo Sampling Methods Using Markov Chains and Their Applications*; Oxford University Press: Oxford, UK, 1970.

45. Ajami, N.K.; Duan, Q.; Sorooshian, S. An integrated hydrologic Bayesian multimodel combination framework: Confronting input, parameter, and model structural uncertainty in hydrologic prediction. *Water Resour. Res.* **2007**, *43*, w01403. [CrossRef]

46. Gelman, A.; Rubin, D.B. Inference from iterative simulation using multiple sequences. *Stat. Sci.* **1992**, *7*, 457–472. [CrossRef]

47. Li, S.; Xiong, L.; Li, H.; Leung, L.R.; Demissie, Y. Attributing runoff changes to climate variability and human activities: Uncertainty analysis using four monthly water balance models. *Stoch. Environ. Res. Risk Assess.* **2015**, *30*, 251–269. [CrossRef]

48. Xiong, L.; Du, T.; Xu, C.Y.; Guo, S.; Jiang, C.; Gippel, C.J. Non-Stationary Annual Maximum Flood Frequency Analysis Using the Norming Constants Method to Consider Non-Stationarity in the Annual Daily Flow Series. *Water Resour. Manag.* **2015**, *29*, 3615–3633. [CrossRef]

49. Jiang, C.; Xiong, L.; Wang, D.; Liu, P.; Guo, S.; Xu, C.Y. Separating the impacts of climate change and human activities on runoff using the Budyko-type equations with time-varying parameters. *J. Hydrol.* **2015**, *522*, 326–338. [CrossRef]

50. Zou, L.; Xia, J.; She, D. Analysis of Impacts of Climate Change and Human Activities on Hydrological Drought: A Case Study in the Wei River Basin, China. *Water Resour. Manag.* **2018**, *32*, 1421–1438. [CrossRef]

51. Su, X.L.; Kang, S.Z.; Wei, X.M.; Xing, D.W.; Cao, H. Impact of climate change and human activity on the runoff of Wei River basin to the Yellow River. *J. Northwest A F Univ. (Nat. Sci. Ed.)* **2007**, *35*, 153–159.

52. Luan, X.; Wu, P.; Sun, S.; Li, X.; Wang, Y.; Gao, X. Impact of Land Use Change on Hydrologic Processes in a Large Plain Irrigation District. *Water Resour. Manag.* **2018**, *32*, 3203–3217. [CrossRef]

53. National Meteorological Information Center of China. Available online: http://data.cma.cn/ (accessed on 30 July 2019).

54. Blaney, H.F.; Criddle, W.D. *Determining Consumptive Use and Irrigation Water Requirements, USDA Technical Bulletin 1275*; US Department of Agriculture: Washington, DC, USA, 1962.

55. Fensholt, R.; Rasmussen, K. Analysis of trends in the Sahelian 'rain-use efficiency' using GIMMS NDVI, RFE and GPCP rainfall data. *Remote Sens. Environ.* **2011**, *115*, 438–451. [CrossRef]

56. Fensholt, R.; Proud, S.R. Evaluation of Earth Observation based global long-term vegetation trends—Comparing GIMMS and MODIS global NDVI time series. *Remote Sens. Environ.* **2012**, *119*, 131–147. [CrossRef]

57. Mann, H.B. Non-parametric tests against trend. *Econometrica* **1945**, *13*, 245–259. [CrossRef]

58. Kendall, M.G. *Rank Correlation Measures*; Charles Griffin: London, UK, 1975.

59. Blasone, R.; Madsen, H.; Rosbjerg, D. Uncertainty assessment of integrated distributed hydrological models using GLUE with Markov chain Monte Carlo sampling. *J. Hydrol.* **2008**, *353*, 18–32. [CrossRef]

Article

Study on the Single-Multi-Objective Optimal Dispatch in the Middle and Lower Reaches of Yellow River for River Ecological Health

Tao Bai [1,*], Xia Liu [1], Yan-ping HA [2], Jian-xia Chang [1], Lian-zhou Wu [1], Jian Wei [1] and Jin Liu [3]

[1] State Key Laboratory of Eco-hydraulics in Northwest Arid Region of China, Xi'an University of Technology, Xi'an 710048, China
[2] Xi'an Thermal Power Research Institute Co., Ltd., Xi'an 710048, China
[3] The Pearl River Hydraulic Research Institute, Guangzhou 510611, China
* Correspondence: baitao@xaut.edu.cn; Tel.: +86-029-82312036

Received: 23 February 2020; Accepted: 17 March 2020; Published: 24 March 2020

Abstract: Given the increasingly worsening ecology issues in the lower Yellow River, the Xiaolangdi reservoir is chosen as the regulation and control target, and the single and multi-objective operation by ecology and power generation in the lower Yellow River is studied in this paper. This paper first proposes the following three indicators: the ecological elasticity coefficient (f1), the power generation elasticity coefficient (f2), and the ecological power generation profit and loss ratio (k). This paper then conducts a multi-target single dispatching study on ecology and power generation in the lower Yellow River. A genetic algorithm (GA) and an improved non-dominated genetic algorithm (NSGA-II) combining constraint processing and feasible space search techniques were used to solve the single-objective model with the largest power generation and the multi-objective optimal scheduling model considering both ecology and power generation. The calculation results show that: (1) the effectiveness of the NSGA-IIcombined with constraint processing and feasible spatial search technology in reservoir dispatching is verified by an example; (2) compared with the operation model of maximizing power generation, the power generation of the target model was reduced by 0.87%, the ecological guarantee rate was increased by 18.75%, and the degree of the impact of ecological targets on the operating results was quantified; (3) in each typical year, the solution spatial distribution and dimensions of the single-target and multi-target models of change are represented by the Pareto-front curve, and a multi-objective operation plan is generated for decision makers to choose; (4) the f1, f2, and k indicators are selected to analyze the sensitivity of the five multi-objective plans and to quantify the interaction between ecological targets and power generation targets. Ultimately, this paper discusses the conversion relationship and finally recommends the best equilibrium solution in the multi-objective global equilibrium solution set. The results provide a decision-making basis for the multi-objective dispatching of the Xiaolangdi reservoir and have important practical significance for further improving the ecological health of the lower Yellow River.

Keywords: multi-objective optimal operation model; feasible search space; Pareto-front optimal solution set; loss–benefit ratio of ecology and power generation; elasticity coefficient

---

1. Introduction

Rivers are the most important life support system for mankind—they are the key to maintaining the material and energy cycle of a region or river basin. On the other hand, the hydrological situation of rivers affects all aspects of natural ecosystems. Good hydrological conditions are important for maintaining the health of rivers and maintaining aquatic life. Ecosystem integrity plays a very important role. As the mother river of China, the Yellow River is the largest water supply source in Northwest China and North China. It undertakes water supply tasks for 15% of the country's cultivated land diversion irrigation, 12% of the population, and more than 50 large and medium-sized cities. As we all know, the Yellow River is the river with the highest sand content in the world. Its largest sediment is located in the middle reaches and flows through the Loess Plateau. The large amount of sediment carried in the Yellow River caused great damage to the national economy and ecological environment. Thus, there is an urgent need to mitigate this problem by means of reservoir management. The construction of cascade reservoirs in the Yellow River basin has made great economic benefits, resulting in serious ecological problems at the same time. The ecological problems in the middle and lower reaches of the Yellow River are mainly reflected in the following aspects: ① The reservoir operation has a great effect on the morphological changes of the river. The water and sediment regulation of the Xiaolangdi dam is beneficial to river erosion, but the erosion amplitudes are becoming smaller [1–3]. It is predicted that the river downstream of Xiaolangdi will quickly be desilted after 2020, and the amount of desilted river will reach the pre-reservoir level by around 2028. By then, the Zhongshui River Channel will be difficult to maintain, and the ecological flow in some sections will be destroyed [4,5]. ② The seasonal basic ecological flow in the river channel cannot be guaranteed, and the species diversity in the river channel is severely damaged [6,7]. ③ The water quality of the downstream river water deteriorates [8], the oxygen content of the water body decreases, and BOD, COD, and other indicators exceed the normal threshold, resulting in the self-healing and dirt-holding capacity being significantly reduced.

The multi-objective optimal dispatching of reservoirs has become a hot spot in the research of reservoir operation in China. In the past, the study on the optimal operation of reservoirs has been carried out by power generation [9], flood control [10,11], irrigation [12], and other single target to maximize the reservoir economic benefits. With the rapid development of social economy, the reservoir urgently needs to improve the utilization of water and play its comprehensive benefit. The operation models and the optimization algorithms from single target [13,14] to multi-objective [15–26] have become the future overall trend of reservoir optimization. The operation model was based on a single-objective model of power generation originally. Robin W. [27] established a single-objective model of power generation and constructed a reservoir [28] that established an optimal power generation operation model, which is simple and easy to implement and provides an effective method for the optimal operation of cascade hydropower stations. In today's world, the reservoir has to undertake more and more tasks, so the reservoir began to use the multi-objective model to guide the actual operation. Authors such as Huang Cao [29] established a power generation, flood control, and ecological multi-objective joint operation model—a model that can better reflect the interrelationships and changes between these objectives. Yang Na [30] established a multi-objective reservoir operation model that takes into account the profiting requirements of a reservoir and the ecological requirements of a river. The optimization results show that the scheduling method of preference for eco-environmental targets is more favorable for the implementation of the water diversion project of the South-to-North Water Transfer Project. Jin Xin [31] constructed a multi-objective ecological operation model for water supply reservoirs, which can provide theoretical support for the ecological operation decision of the northern reservoir for water supply. In 1962, domestic scholars began to apply a genetic algorithm as a classical single-objective solution algorithm [32] to reservoir operation. At the beginning of this century, Chang Jianxia [33] coded an algorithm in decimal mode that is now widely used in reservoir operation. The genetic algorithm can also transform the multi-objective into a single-objective solution by constraint processing [34]. However, with the emergence of dimensionality, the genetic algorithm (GA) also derives the evolutionary algorithm for solving multi-objective models, which is the non-dominated sorting genetic algorithm (NSGA-II) [35–39]. NSGA-II is an improvement on the basis of GA. At the beginning of the twenty-first century, a multi-objective evolutionary algorithm is applied to the field of reservoir operation [40], and Yuan Ruan [41] use NSGA-II to solve multi-objective optimization scheduling model. This example analysis shows that the algorithm has a good accuracy in solving the multi-objective problem. Zhu Jie [42] uses improved NSGA-II to obtain a better dispatching scheme for the Zhanghe reservoir. NSGA-II has shown its superiority in the application of multi-objective reservoir scheduling.

However, the existing NSGA-II stochastic search algorithm still has some shortcomings, such as the instability of results and a long computation time. At the same time, in the current research on the ecological regulation of reservoirs, the ecological demand was mostly a fixed process [43–49], or even a fixed value, and there was no distinction between different incoming water years, which is not a comprehensive consideration of the ecological needs of the basin. In view of this, this paper set different ecological needs in each typical year and conducted both a single- and multi-objective optimization scheduling of the Xiaolangdi reservoir, and studied the impact of different ecological objectives on the operation mode of the reservoir. The used genetic algorithm and improved NSGA-II algorithm based on constraint processing and the spatial optimization technology-multi-objective model to obtain the single-objective optimal solution and its distribution characteristics in the multi-objective Pareto-front optimal solution set, revealing the multi-objective scheduling mutual conversion rules between targets. This paper chooses index factors to quantify the ecological and mutual feedback conversion relationship between power generation targets and finally recommends the use of a global optimal balance solution in a multi-target solution set to minimize the impact of the reservoir on the surrounding environment adverse effects [50]. The research results provide a decision-making basis for the dispatchers of the Xiaolangdi reservoir and the Xixiayuan reservoir. This paper has important practical significance and application value for improving the ecological environment of the lower Yellow River and improving the comprehensive utilization of reservoirs.

2. Research Area and Data

The Yellow River originated from Maqu in the Yoguzonglie basin at the northern foot of the Bayankala Mountains on the Qinghai-Tibet Plateau, and finally flowed into the Bohai Sea in Shandong Province. The basin area is 795,000 km^2, and the total length of the main stream is 5464 km, which is the second longest river in China. The Xiaolangdi Reservoir has great regulation and storage capacity, and plays an important role in ecological protection of the Yellow River and sediment control in the river. This paper selects Xiaolangdi as the research object, whose location is on the main stream of the Yellow River, north of Luoyang. The location overview shown is in Figure 1. The main parameters of the Xiaolangdi reservoir are shown in Table 1.

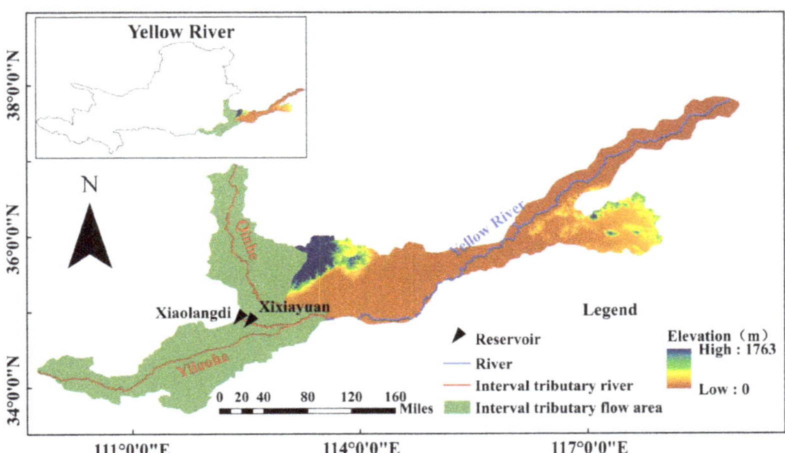

Figure 1. Location overview of Xiaolangdi.

Table 1. Main parameters of Xiaolangdi.

Installed Capacity (MW)	Guaranteed Output (MW)	Total Storage Capacity (Billion m^3)	Normal Water Level (m)	Dead Water Level (m)	Ecological Guarantee Rate (%)	Maximum Overflow (m^3/s)	Adjustment Performance
1800	354	126.5	275	230	90	1776	Year

Long series runoff collected during the historical period from 1961 to 2009 (hydrological year) of the Xiaolangdi Reservoir was used as input data, and the data was input according to the reservoir optimization model. The monthly average natural runoff of the Xiaolangdi Long-series is shown in Figure 2.

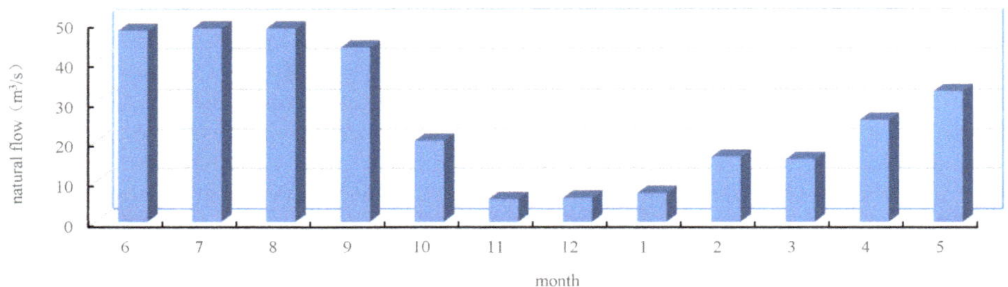

Figure 2. Monthly average runoff process of the Xiaolangdi reservoir.

This article integrates the collected industrial, agricultural, domestic, and ecological water demand processes downstream of the Xiaolangdi reservoir to obtain the comprehensive water demand process downstream of the Xiaolangdi reservoir, and uses it as a constraint for optimal scheduling research. The comprehensive water demand process is shown in Figure 3.

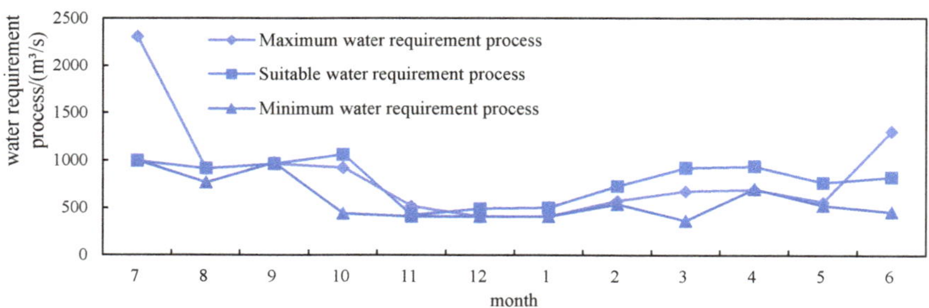

Figure 3. Integrated water requirement process in the lower reaches of Xiaolangdi.

In Figure 3, the minimum ecological water requirement is the flow process needed to maintain the downstream ecosystem without degradation during the minimum water requirement. The suitable ecological water requirement during the process of water demand is the flow process to maintain the suitable habitat of the lower reaches and ensure the normal survival and reproduction of the downstream species; the maximum ecological water requirement is the flow process to maintain the balance of river scouring and deposition and to restore the capacity of river pollution.

3. Modeling

Model 1: Maximize power generation

$$MaxE = \sum_{t=1}^{T} kQ_o(t)H(t) \times \Delta t \tag{1}$$

where E is the total power generation of hydropower stations during dispatching period, and t and T are the time serial numbers and the total number of periods, respectively. In addition, k is the comprehensive output coefficient for power station; $Q_o(t)$ and $H(t)$ are the power discharge and water head in the t-th period of the power station, respectively.

Model 2: Maximize power generation while meeting the ecological flow requirement

The objective function and constraints are the same as Model 1. In addition, the ecological flow requirement needs to be considered:

$$Q_d(t) \leq Q_o(t) \tag{2}$$

where $Q_d(t)$ is the ecological flow requirement of the downstream channel during the t-th period.

Model 3: Multi-objective optimal operation considering both ecology and power generation

$$Obj_1: \qquad MaxE = \sum_{t=1}^{T} kQ_o(t)H(t) \times \Delta t \tag{3}$$

$$Obj_2: \qquad MinW = \sum_{t=1}^{T} Q_s(t) \times \Delta t \tag{4}$$

where W and $Q_s(t)$ are the total water shortage in dispatching periods and water shortage flow in the t-th period. The constraints are the same as Model 1.

The constraints for models one to three are as follows:

(1) water balance constraints

$$V(i,t) = V(i,t-1) + [QI(i,t-1) - QO(i,t-1)] \times \Delta t \tag{5}$$

where $V(i,t)$ and $V(i,t-1)$ are the initial storages of the ith reservoir at times t and $t-1$, respectively. $QI(i,t-1)$ and $QO(i,t-1)$ are the inflow and outflow of the ith reservoir at time $t-1$, respectively. t is the time interval.

(2) Outflow constraints

$$QO_{min}(i,t) \leq QO(i,t) \leq QO_{min}(i,t) \tag{6}$$

$$QO(i,t) = QI(i,t) - [V(i,t+1) - V(i,t)]/\Delta t \tag{7}$$

where $QO_{min}(i,t)$ and $QO_{min}(i,t)$ are the minimum and maximum allowable outflows of the ith reservoir at time t, respectively. $QO(i,t)$ and $QI(i,t)$ are the outflow and inflow of the ith reservoir at time t, respectively. $V(i,t+1)$ and $V(i,t)$ are the initial and final storages of the ith reservoir at times $t+1$ and t, respectively.

(3) water level constraints

$$Z_{min}(i,t) \leq Z(i,t) \leq Z_{max}(i,t) \tag{8}$$

where Z_{min} and Z_{max} are the minimum and the maximum water levels of the ith reservoir at time t, respectively.

(4) Hydropower outputs constraints

$$N_{min}(i,t) \leq N(i,t) \leq N_{max}(i,t) \tag{9}$$

where $N_{min}(i, t)$ and $N_{max}(i, t)$ are the minimum and maximum hydropower outputs of the ith reservoir at time t, respectively. In general, N_{min} is the guaranteed output and N_{max} is the installed capacity.

4. Methodology

4.1. Single-Objective Solution

Genetic algorithm (GA) is a common method for solving the single-objective optimal dispatching model [51]. This article intends to use the basic genetic algorithm, such as references, to solve the single-objective Model 1 and Model 2.

4.2. Multi-Objective Solution

There are many methods to solve the multi-objective optimization model of reservoir, such as the constraint method, the weight coefficient method, the multi-objective evolutionary algorithm, and so on. The constraint method is to transform the multi objective into a single objective, and the weight coefficient method is to set the weight of each target and transform it into a single objective. Indeed, the two objective problems are transformed into a single-objective solution, while it is impossible to reveal the transformation rules among the targets from the global equilibrium solution to clarify the sensitive relationship among the targets. In this multi-objective optimal dispatching model, which considers both ecology and power generation, the key of each dispatching objective is the process of outflow. Specifically, the expected discharge process of power generation target has strong pulse in non-flood season, reducing the outflow volume, to obtain high water level and maximum power generation; the expected outflow process in Model 3 is similar to ecological water demand, and the pulse is weakened. In the non-flood season, the steady outflow of the reservoir can be maintained to meet the gentle ecological base flow of the lower reaches, so that the difference between them is reduced, and the minimum amount of water shortage is obtained. Hence, one can see that the ecology and power generation objectives are contradictory and counter-productive contradictions. Thus, this paper intends to use the non-dominated ranking genetic algorithm (NSGA-II), which is based on constraint processing to optimize the feasible search space, and treats each dispatching objective equally to make it survive the competition in the process of optimization so as to obtain the global equilibrium solution.

In order to improve the accuracy and efficiency of the NSGA-II, this paper optimizes the NSGA-II by optimizing the feasible search space. The reservoir multi-objective optimization scheduling model has a highly nonlinear characteristic, and different treatment methods should be adopted for different constraints. In this paper, the constraints are divided into the convertible constraints and the non-convertible constraints, in which the convertible constraints are used to optimize the initial search space. The convertible constraints are water level constraints, discharge constraints, ecological water demand constraints. The non-convertible constraints include flow balance constraints, water balance constraints, and power output constraints.

The specific feasible search space optimization steps are as follows:

*Step*1: The minimum and maximum water level of the reservoir form the initial search space, as shown in Figure 1;

*Step*2: In the convertible constraints, the constraint of reservoir capacity and the flow constraint of reservoir outflow are converted to the upper and lower limits of the water level, and it is intersected with the water level constraints of the reservoir in order to eliminate the search space that does not satisfy the convertible constraint conditions.

*Step*3: The initial population of each target is generated in the feasible search space, and the optimization is finished by the constraint processing and the search space optimization.

The optimized feasible search space is shown in Figure 4. Among them, Z represents the water level, the unit is meter, T represents time in seconds. Z(T) represents the water level of the reservoir during the T period, Z^0 represents the initial water level of the reservoir, Z_{max} represents the highest water level, and Z_{min} represents the lowest water level.

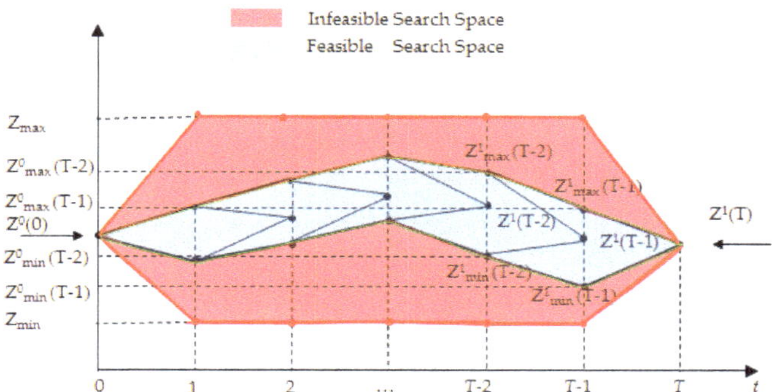

Figure 4. Optimization of the feasible search space.

The optimization of feasible search space: the infeasible solution space is removed from the initial search space, which, in a sense, not only does not reduce the diversity of the population, but also improves the quality of initial population, which makes easier to converge.

When the initial population is optimally calculated, the stage solutions of all the generations of each group can meet the transformable constraints, only to determine the non-transitive constraints. When the non-transformable constraint is not satisfied, the penalty function will be adopted to reduce the fitness function value, thereby reducing the chance that the individual will inherit to the next generation and ensure the optimal path for the optimization. The optimized NSGA-II, on the one hand, optimizes the initial search space to the feasible search space, eliminates the infeasible search space, reduces the search range of the global equilibrium solution, and improves the optimization speed; on the other hand, all the solutions of all generations in the optimal search satisfy the convertible constraints, and only need to determine whether the non-convertible constraints are satisfied or not, thus avoiding the cumbersome and redundant judgment of all the constraints. After the NSGA-II is optimized, it not only avoids falling into the local optimal solution, but also accelerates the convergence rate and improves the search efficiency and the precision of the algorithm. The operation flow chart of the NSGA-II in which the feasible search space was optimized is shown as Figure 5.

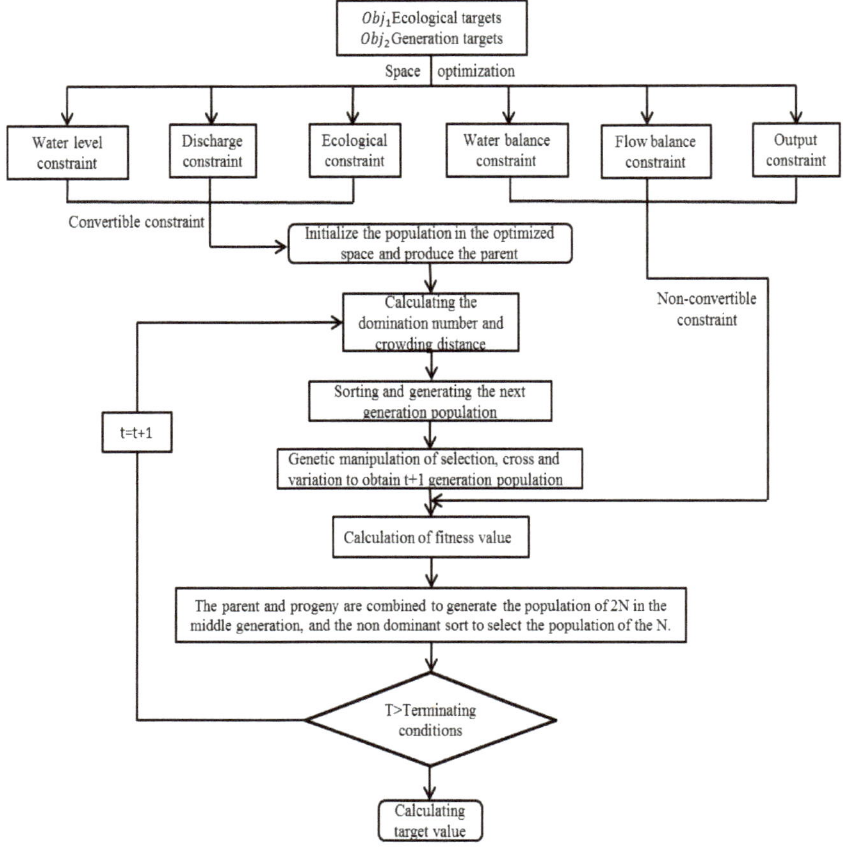

Figure 5. The operation flow chart of the non-dominated genetic algorithm (NSGA-II).

5. Results and Discussion

5.1. Illustrate the Effectiveness of the NSGA-II

The parameters of the genetic algorithm, multi-objective particle swarm optimization (MOPSO), NSGA-II, and optimized NSGA-II are set in Table 2. Verifying the validity and stability of NSGA-II based on constrained processing is a crucial part of optimizing the feasible search space. Since NSGA-II belongs to a posteriori algorithm, the effectiveness of the multi-objective reservoir scheduling is verified by selecting the 1981 data, and the comparison results are shown in Table 3.

Table 2. Parameter setting of the genetic algorithm (GA) and the improved NSGA-II.

Parameter	GA	MOPSO	NSGA-II	Optimized NSGA-II [12]
Population size	300	300	300	300
Selection probability	0.5	-	0.5	0.5
Cross probability	0.8	-	0.8	0.8
Mutation probability	0.05	-	0.05	0.05
Maximum number of iterations	200	200	200	200

Table 3. Comparison of statistical parameters of each target with different calculation times.

Algorithm	Target	1		2		3		4		5		Mean Value		Standard Deviation		Time
		Lower Limit	Upper Limit	Lower Limit	Upper Limit	Lower Limit	Upper Limit	Lower Limit	Upper Limit	Lower Limit	Upper Limit	Lower Limit	Upper Limit	Lower Limit	Upper Limit	
GA	Power generation	-	103.15	-	103.17	-	103.15	-	103.15	-	103.13	-	103.15	-	103.15	1.63
	Water shortage	-	5.00	-	5.07	-	5.05	-	5.06	-	5.05	-	5.04	-	5.04	
MOPSO	Power generation	102.90	103.15	102.82	103.14	102.83	103.15	102.85	103.15	102.82	103.13	102.85	103.14	102.85	103.14	2.63
	Water shortage	0.01	5.06	0.00	5.04	0.00	5.09	0.00	5.16	0.00	5.01	0.00	5.07	0.00	5.07	
NSGA-II	Power generation	102.80	103.17	102.82	103.06	102.78	103.16	102.78	103.17	102.80	103.13	102.80	103.14	102.80	103.14	2.65
	Water shortage	0.00	5.04	0.00	3.35	0.00	5.06	0.00	5.06	0.00	5.06	0.00	4.71	0.00	4.71	
Improved NSGA-II	Power generation	102.83	103.14	102.81	103.17	102.77	103.16	102.83	103.15	102.80	103.13	102.80	103.15	102.80	103.15	2.40
	Water shortage	0.00	5.06	0.00	5.06	0.00	5.05	0.00	5.06	0.00	5.06	0.00	5.06	0.00	5.06	

Notes: Power generation/10^8kW·h; Water shortage/10^8m^3; Time/s.

In order to determine the accuracy and stability of the GA, MOPSO, NSGA-II, and the improved NSGA-II algorithm, take the result of running five times as an example, in which the different indexes of each algorithm are obtained. The upper and lower limits of the optimal value of power generation and water shortage, the mean value, the standard deviation, and the average running time are shown in Table 3. It can be seen that:

(1) The upper limit of the power generation of optimized NSGA-II is 10.315 billion kW·h. Compared with NSGA-II and MOPSO, however, both of which improved by 1 million kW·h, there is almost no change relative to the GA. The lower limit of the power generation of optimized NSGA-II is the same as that of NSGA-II and MOPSO.

(2) From the view of ecological water shortage, the lower limit of ecological deficiency is consistent. The upper limit of ecological water shortage for optimized NSGA-II is 506 million m^3, which is higher than NSGA-II but lower than MOPSO.

(3) From the stability of results, the maximum and minimum standard deviation of ICS optimization results is 0.03, which is lower than the other three algorithms. In addition, the first results of MOPSO significantly premature convergence, the lower limit of the amount of water is non-zero. Meanwhile, the second result of NSGA-II has local convergence, and the power generation does not reach the maximum. Therefore, the results of optimized NSGA-II are relatively stable.

(4) From the optimization time, NSGA-II is longer than GA, and different optimization strategies will make the algorithm optimization time longer. While optimized NSGA-II is shorter than NSGA-II, which reflects that the improved strategy proposed in this paper accelerates the population convergence under the premise of ensuring a certain accuracy. Overall, the optimized NSGA-II is superior to GA, NSGA-II, and MOPSO, indicating that the optimization strategy is effective.

5.2. Analysis of the Impact of Ecological Goals on Dispatching Results

For the single-objective model with the maximum power generation as the target, a strong ecological constraint is added to form Model 2. The maximum degree of influence of ecological goals on the scheduling results can be found by comparing Model 1 with Model 2. Among the long series of results, the total power generation was 8.759 billion kW · h, which was 0.87% lower than the 8.836 billion kW · h of Model 1. However, the ecological guarantee rate of the Huayuankou section increased by 18.75%, which met the planned 90% guarantee rate. It shows that using ecological goals as a constraint can significantly increase the guarantee rate of ecological water demand, but the power generation benefits are slightly lost.

Model 2 is a single-objective model with maximum power generation after adding strong ecological constraints, and the impact of ecological objectives on the dispatching results can be compared between Model 1 and Model 2. Select 1981–1982, 1960–1961, and 1972–1973 as wet, normal, dry. Among these dates, July–October is the flood season, November–March is the dry season, and April–June is the water-supply period. The process of reservoir water level, reservoir discharge, and the power generation of Model 1 and Model 2 are compared and analyzed, as shown in Figure 6, Figure 7, and Figure 8.

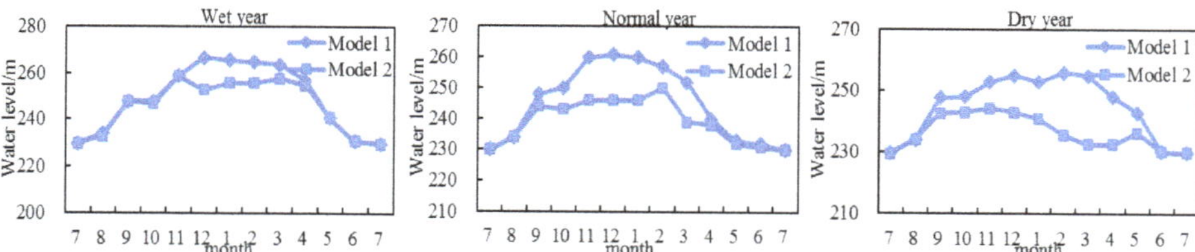

Figure 6. Water level change of the Xiaolangdi reservoir in each typical year.

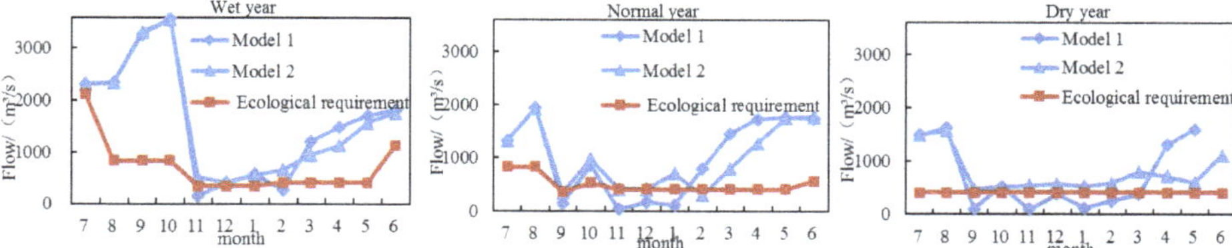

Figure 7. Variation of discharge from the Xiaolangdi reservoir in each typical year.

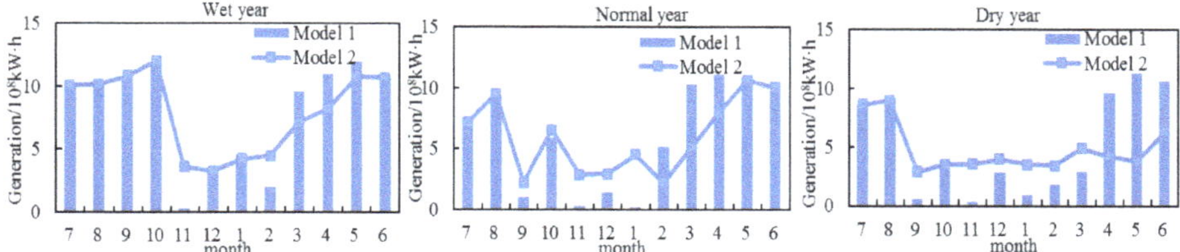

Figure 8. Variation of the monthly generating capacity of Xiaolangdi in each typical year.

(1) The water level of each month in Model 2 is basically lower than Model 1, especially in the non-flood period of the wet year, the normal, and the dry year, and the water level of the reservoir is obviously maintained at the low water level. This is because, in order to meet the downstream ecological water storage, the reservoir needs to increase the discharge, which causes the water level to drop. Therefore, considering the ecological constraints, the water level during the water supply period of the Xiaolangdi dam during the dispatching period has been changed.

(2) Restricted by ecological constraints, the monthly outflow of the Xiaolangdi dam in Model 2 basically meets the ecological requirements. In Model 1, the discharge flow of individual months is lower than the ecological water demand and cannot meet the ecological requirements. This is because, in order to maximize the power generation, Model 1 needs to discharge as little and as much water as possible during the storage period and store it at the normal storage level as soon as possible. As a result, the amount of water discharged during the storage period cannot meet the ecological water demand. The destruction of the month occurred in the dry season. The number of damaging months is 2, 4, and 4, respectively, and the degree of damage for the wet year, normal year, and dry year are as follows.

(3) The power generation of Xiaolangdi Hydropower Station shows the same variation in all typical years: the flood season is large, the dry season is small, and the water supply period is large. In Model 2, the power generation in the dry season is obviously larger than Model 1. The main reasons are as follows: In Model 1, the destruction of ecological water demand are all in the dry season. In order to ensure the ecological flow, Model 2 runs at the low water level during the dry year, which increases the discharge and power generation.

Table 4 shows the annual electricity generation and the number of eco-friendly months of different models in each typical year. As can be seen from Table 4, the monthly number of ecological guarantees in each typical year of Model 2 was increased by 2, 4, and 4 months, while the power generation decreased by 0.33, 0.86 and 1.27 billion kW·h. The ecological target has a great impact on power generation during the year and has the greatest impact on power generation in the dry year.

Therefore, adding Model 2 with strong ecological constraints, there is a clear change in the operation process and the magnitude of the reservoir during the dispatch cycle, and Model 2 sacrifices more power generation to complete the ecological target.

Table 4. Analysis of power generation and ecological guarantee in each typical year.

	Wet Year		Normal Year		Dry Year	
	Model 1	**Model 2**	**Model 1**	**Model 2**	**Model 1**	**Model 2**
Power generation/10^8 kW·h	95.97	95.64	67.95	67.09	60.02	58.75
The number of guarantee ecology	10	12	8	12	8	12

5.3. Result Analysis of the Multi-Objective Model

In this paper, we use the optimized NSGA-II to solve the multi-objective Model 3. The results of the multi-objective optimization scheduling model are plotted as a Pareto-front curve, as shown in Figure 9.

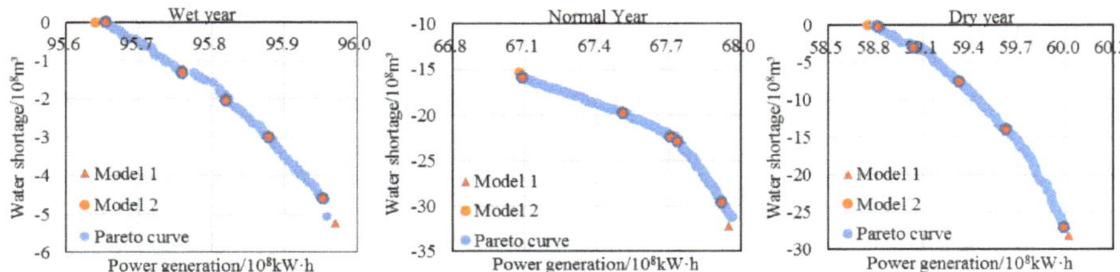

Figure 9. Multi-objective Pareto-front curves in each typical year.

In the Pareto-front curve of multi-objective optimal scheduling, the annual variation in the annual water power generation and the comprehensive water shortage in the wet year are 9.564–9.597 billion kW·h and 0–0.530 billion m^3, respectively; the annual variation in the annual water power generation and the comprehensive water shortage in normal year are 6.707–6.795 billion kW·h and 1.547–3.246 billion m^3, respectively; and the annual variation in the annual water power generation and the comprehensive water shortage in dry year are 5.875–6.002 billion kW·h and 0–2.835 billion m^3, respectively.

It can be seen that with the decrease in typical annual runoff that the decrease in hydropower generation and the increase in ecological comprehensive water shortage make the contradiction between ecological target and power generation target aggravating. The total water shortage in normal year is larger than that of the dry year, which is mainly caused by the annual runoff process and the integrated water demand process.

Due to the different water requirement processes in typical years, Model 1 and Model 2 are different in Pareto-front curves:

(1) The annual generation capacity of the model 1 is 9.597 billion kW·h, and the comprehensive water shortage is 0.530 billion m^3, which is located at the right extreme value of the Pareto-front curve; the annual generation capacity of Model 2 is 9.564 billion kW·h, and the comprehensive water shortage is 0 billion m^3, which is located at the left extreme value of the Pareto-front curve.

(2) The annual generation capacity of Model 1 and Model 2 in the normal water year is 6.795 billion kW·h and 6.707 billion kW·h, respectively, and the comprehensive water shortage is 3.246 billion m^3 and 1. 547 million m^3, respectively.

(3) The annual generation capacity of Model 1 and Model 2 in the dry water year is 6.002 billion kW·h and 5.875 billion kW·h, respectively, and the comprehensive water shortage is 2.835 billion m^3 and 0 million m^3, respectively.

(4) With the increase in the total amount of incoming water, the power generation in the dry year, the normal year, and the wet year increase sequentially, and the water shortage also increases in turn. There is a lot of water in the normal year, but it is restricted by the constraints of poor distribution

in the water year, and the ecological water demand is greater than in the dry year, which causes the maximum water shortage to increase.

It can be seen that the single-objective calculation results fall on the Pareto-front curve of the global equilibrium solution. With the decrease in incoming water, the power generation decreased drastically, the water deficit increased significantly, and the contradiction between the target of ecological comprehensive water demand and the power generation goal was aggravated.

On the Pareto-front curve of the multi-objective global equilibrium solution, we select the power generation index, and generate Schemes 1 to 5 shown from small to large (the position is shown in Figure 9). The outflow of Xiaolangdi in each typical year is shown in Figures 10–12. The scheduling results of each scheme are shown in Table 5.

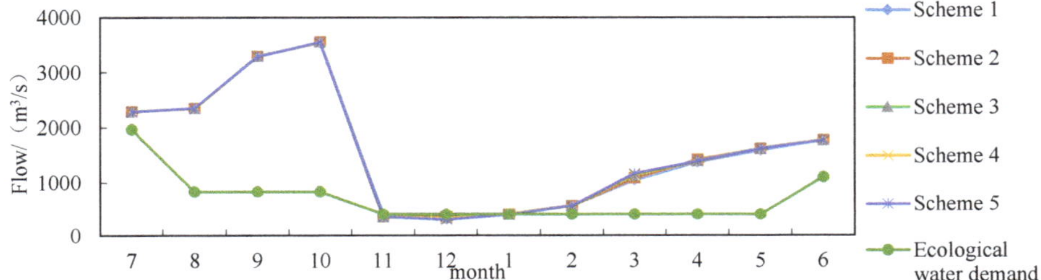

Figure 10. Comparative analysis of the outflow of Xiaolangdi in the wet year.

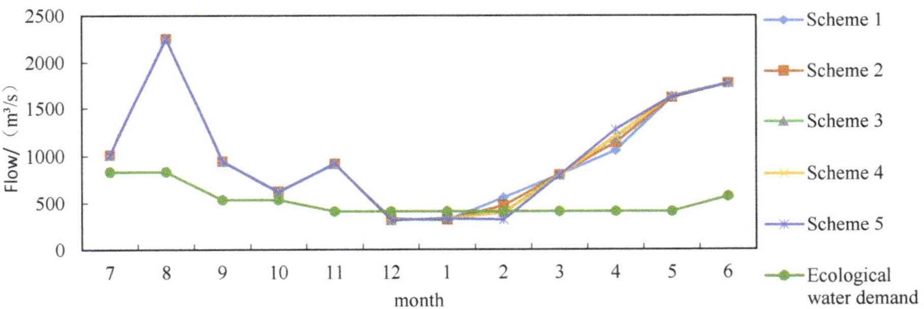

Figure 11. Comparative analysis of the outflow of Xiaolangdi in the normal year.

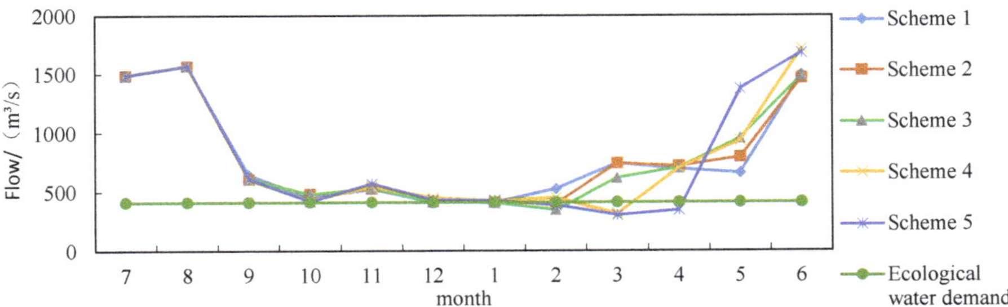

Figure 12. Comparative analysis of the outflow of Xiaolangdi in the dry year.

Table 5. Comparative analysis of various schemes in typical years.

Scheme	Wet Year		Normal Year		Dry Year	
	Power Generation/10^8kW·h	Eco-Water Shortage/10^8m^3	Power Generation/10^8kW·h	Eco-Water Shortage/10^8m^3	Power Generation/10^8kW·h	Eco-Water Shortage/10^8m^3
Scheme 1	95.65	0	67.09	16.01	58.82	0.28
Scheme 2	95.76	1.34	67.51	19.88	59.04	3.38
Scheme 3	95.82	2.08	67.71	22.6	59.33	7.9
Scheme 4	95.88	3.05	67.73	22.98	59.63	14.32
Scheme 5	95.95	4.62	67.92	29.57	60	27.36

Table 5 shows that in wet year, the multi-objective scheme can meet the ecological requirements under the condition of minimum power generation. From Schemes 1–5, the power generation increased from 9.565 billion kW·h to 9.595 billion kW·h, and the outflow decreased during the dry year. From the ecological destruction duration, the monthly number of guarantees is reduced from 12 months to 9 months; from the depth of ecological destruction, the ecological water shortage increased from 0 to 0.462 billion m^3, and Schemes 2–5 failed to meet the ecological requirements.

Compared with the wet year, the power generation Schemes 1–5 consist of 6.709–5 kW·h increased to 6.792 billion kW·h, from the diachronic perspective of ecological destruction, which dropped from 11 months to 8 months. From the depth of ecological damage, the ecological water shortage of 1.601 billion m^3 increased to 2.957 billion m^3.

Like the normal year, the power generation of each scheme is reduced to a minimum value of 5.875 billion kW·h, which can basically guarantee the minimum ecological runoff process in the dry year. The ecological flow of Scheme 2 to Scheme 5 has been destroyed in varying degrees. Under the condition of a continuous decrease in incoming water, the contradiction between power generation target and ecological target is aggravated day by day.

From the above analysis, the following conclusions are drawn:

(1) Due to the difference in incoming water and ecological water demand, the Pareto-front optimal solution set shows the solution spatial distribution and the dimension change of each typical single-multi-objective model. The results of the single-objective Model 1 or Model 2 are basically on the curve, and the results of Model 1 are close to the maximum target of power generation, and Model 2's results are close to the minimum target of water shortage.

(2) The location of the single-objective result in the curve shows that it has some limitations. The multi-objective model considers the power generation and ecological targets and gives a multi-objective global equilibrium solution set. The multi-objective model has superiority that can fully meet the ecological flow requirements. It provides a multi-objective dispatching solution for decision-makers in reservoir operation and watershed management.

(3) In the multi-objective model, except for Scheme 1, other schemes cannot meet the downstream ecological objective. The safety of ecological flow in the lower reaches of the Yellow River is serious, and as the inflow decreases, the amount of power generation will be correspondingly reduced, the number of ecological assurances and the amount of ecological water deficit will increase, and the ecological targets will be harder to meet. The contradiction between ecological and power generation targets is increasingly aggravated—it is bound to sacrifice the efficiency of power generation in exchange for ecological benefits.

5.4. Sensitivity Analysis

In order to reveal the law of mutual transformation between power generation and ecology, this paper analyzes the sensitivity of various dispatching schemes in each typical year. Using the dimensionless method, three new indicators are creatively proposed: the ratio of power generation to the installed capacity is defined as the coefficient of elasticity (f_1); the ratio of the difference in the total water requirement and water deficit to the total water demand is defined as the ecological elastic coefficient (f_2), as shown in Table 5.

The increase in power generation benefit when reducing the 1% ecological benefit is the ratio of ecological- and power-loss benefits (k), just like the Formula (4). The greater the k, the smaller the impact of ecological benefits on power generation benefits, and the greater the overall benefits of the scheme.

$$k(i) = \{f_1(i) - f_1(i-1)\}/\{f_2(i) - f_2(i-1)\} \tag{10}$$

Based on the Scheme 1, the k of each typical year can be obtained as shown in Figures 12 and 13.

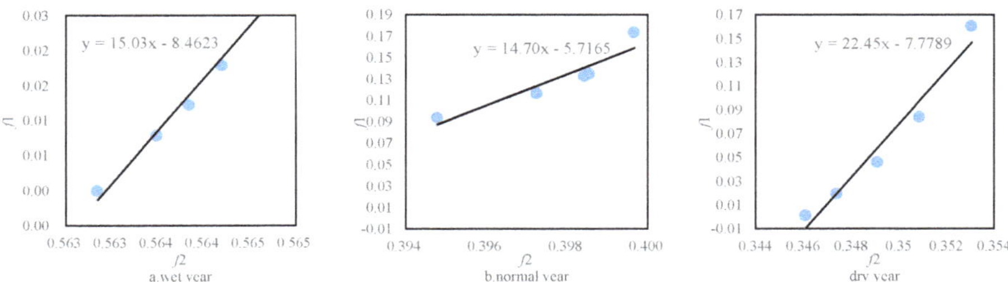

Figure 13. Correlation between power generation and ecological elasticity in each typical year.

It can be seen from Figure 14 that the linear fitting relationship between f_1 and f_2 in each typical year is better, and the linear slopes of the typical years are 15.03, 14.70, and 22.45, respectively, showing a consistent and significant increasing trend. In response to the sensitivity of power generation and ecology, the normal year is the smallest, followed by the wet year and the dry year. That is, with the decrease in inflow, the amount of power generation, eco-elastic coefficient, and the number of ecological guarantee months are all reduced, the ecological water deficit is greatly increased, the restrictive relationship between power generation and ecological targets is strengthened, and the sensitivity between the two is enhanced. The decrease in the ecological elasticity coefficient will make the restoration of ecological environment more severe; the reservoir should take the ecological as the main goal in the dry year and take the phased measures to alleviate the ecological deterioration situation.

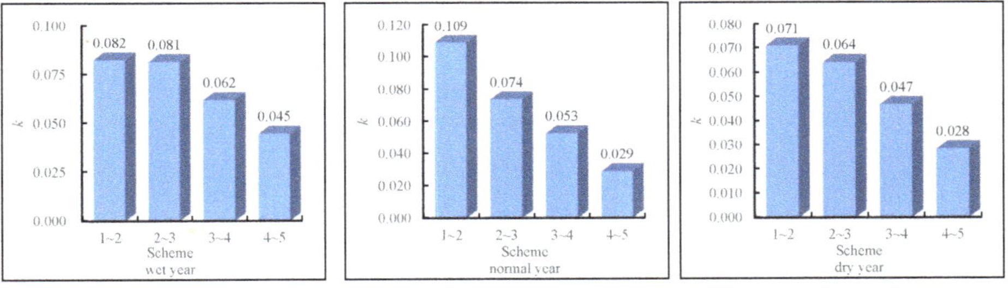

Figure 14. k value of different scheme intervals in each typical year.

It can be seen from Figure 13 that from Scheme 1 to Scheme 5, the k value in each typical year shows a continuous decreasing trend, and the maximum value of k in each typical year is 0.082, 0.109, and 0.071, respectively. The corresponding optimal mediation scheme all are Schemes 1–2 intervals. The recommended best coordination solution can maximize the comprehensive benefit.

6. Conclusions

In this paper, the multi-objective optimization algorithm is applied to solve the different models by GA and optimized NSGA-II. The results of the single-objective and multi-objective calculation are compared and analyzed, and the following conclusions are obtained.

(1) Compared with the GA, MOPSO, and NSGA-II algorithms, the feasibility and effectiveness of the optimized NSGA-II are verified.

(2) With the single-objective model of maximum power generation, the total power generation is 8.84 billion kW·h, and the guarantee rate of ecological water demand is 74.43%; the single-objective model considering ecological constraints has a total power generation of 8.759 billion kW·h, and increases the ecological guarantee rate to 93.18%, which basically meets the ecological requirements. The comparison of the results of the two models reveals the degree of impact of ecological goals on dispatching.

(3) The single-objective typical annual optimal value basically falls on the multi-objective Pareto-front curve, which is in good agreement with the Pareto-front curve, further demonstrating the accuracy and reliability of each model and algorithm. With the decrease in incoming water, the ecological security situation in the lower reaches of the Yellow River is poor. The multi-objective model gives a global equilibrium solution set of power generation and ecology, which provides the best coordination scheme for the decision makers of reservoir actual operation and river basin management.

(4) In the multi-objective power generation and ecological sensitivity, the weakest year is in the dry year, and the strongest in the wet year, which is the contradiction between the ecological and the power generation in the dry year. To improve the same ecological benefits, it is bound to sacrifice greater power generation benefits. As the situation of ecological security in the dry years is more severe, stage-by-stage measures should be taken to ease the deterioration of the ecological environment.

In this paper, the different ecological water demand processes of each typical year is regarded as the ecological target. The influence of water quality factors such as water and sediment from alluvial water and algae on the ecological flow in Xiaolangdi is neglected. The next step is to establish an ecological and discharge flow response model to determine the process of integrated ecological water demand for the lower reaches of the Yellow River, and then improve the multi-objective optimization scheduling model.

Author Contributions: T.B, J.-x.C. and J.L. provided the writing ideas and supervised the study; Y.-p.H. conceived and designed the methods; T.B. and X.L. wrote the paper, and all the authors were responsible for data processing and data analysis. All authors have read and agreed to the published version of the manuscript.

Funding: This research received no external funding.

Acknowledgments: This study is supported by the National Key R&D Program of China (2017YFC0405900), the China Postdoctoral Science Foundation (2017M623332XB), the Postdoctoral Research Funding Project of Shaanxi Province (2017BSHYDZZ53), the Basic Research Plan of Natural Science in Shaanxi Province (2018JQ5145), the Natural Science Foundation of China (51879213), Guangdong Province Water Conservancy Science and Technology Innovation Project (2017-09).The authors would like to thank the Editors and anonymous Reviewers for their valuable and constructive comments on this manuscript.

Conflicts of Interest: The authors declare no conflict of interest.

References

1. Fu, C.L.; Li, Q.Y.; Wang, Q.B.; Li, Q. Influence of water and sediment regulation of the Xiaolangdi reservoir on the channel scouring and silting of lower reaches of the Yellow River. *J. Water Resour. Water Eng.* **2012**, *23*, 173–175.
2. Jin, W.; Chang, J.; Wang, Y.; Bai, T. Long-term water-sediment multi-objectives regulation of cascade reservoirs: A case study in the Upper Yellow River, China. *J. Hydrol.* **2019**, *577*, 123978. [CrossRef]
3. Bai, T.; Wu, L.; Chang, J.; Huang, Q. Multi-objective optimal operation model of cascade reservoirs and its application on water and sediment regulation. *Water Resour. Manag.* **2015**, *29*, 2751–2770. [CrossRef]
4. Wan, Z.; An, C.; Yan, Z. Scouring Effect and Trend Predicting of Xiaolangdi Reservoir on Downstream River. *Yellow River* **2015**, *34*, 6–8.
5. Pan, B.; Guan, Q.; Liu, Z.; Gao, H. Analysis of channel evolution characteristics in the hobq desert reach of the yellow river (1962–2000). *Glob. Planet. Chang.* **2015**, *135*, S0921818115300989. [CrossRef]
6. Han, T. Problems and countermeasures for the conservation of aquatic biological resources in Xiaolangdi Reservoir. *Henan Fish.* **2014**, *98*, 11–12, 24.

7.   Cui, B.; Chang, X.; Shi, W. Abrupt Changes of Runoff and Sediment Load in the Lower Reaches of the Yellow River. *Water Resour.* **2014**, *41*, 252–260. [CrossRef]

8.   Shi, J.; Zhou, Y.; Mu, Y.; Zhang, N. Impact of Xiaolangdi Reservoir operation on the distribution of downstream water pollution. *Yellow River* **2006**, *10*, 41–42, 54.

9.   Liao, X.; Zhu, X.; Tian, W. Analysis on Increasing Power Generation Benefit of the Xiaolangdi Power Station. *Water Conserv. Sci. Technol. Econ.* **2014**, *20*, 100–102.

10.   Turgeon, A. A decomposition method for the long-term scheduling of reservoirs in series. *Water Resour. Res.* **1981**, *17*, 1565–1570. [CrossRef]

11.   Niewiadomska-Szynkiewicz, E.; Malinowski, K.; Karbowski, A. Predictive Methods for Real-Time Control of Flood Operation of a Multi-Reservoir System: Methodology and Comparative Study. *Water Resour. Res.* **1996**, *32*, 2885–2896. [CrossRef]

12.   Tian, Y.; Zhang, Z.; Chen, J. Real-time joint irrigation scheduling of reservoirs in central Liaoning. *Water Resour. Manag Technol.* **1994**, *1*, 47–49.

13.   Zhao, T.; Zhao, J.; Zhao, T. Optimizing Operation of Water Supply Reservoir: The Role of Constraints. *Math. Probl. Eng.* **2014**, *2014*, 1–15. [CrossRef]

14.   Ikhar, P.R.; Regulwar, D.G.; Kamodkar, R.U. Optimal reservoir operation using soil and water assessment tool and genetic algorithm. *ISH J. Hydraul. Eng.* **2017**, *24*, 249–257. [CrossRef]

15.   Wang, X.; Zhou, J.; Ou Yang, S.; Zhang, Y.-C. TGC eco-friendly generation multi-objective optimal dispatch model and its solution algorithm. *J. Hydraul. Eng.* **2013**, *44*, 154–163.

16.   Tabari, M.M.R.; Soltani, J. Multi-Objective Optimal Model for Conjunctive Use Management Using SGAs and NSGA-II Models. *Water Resour.* **2013**, *27*, 37–53. [CrossRef]

17.   Abido, M.A. Multiobjective Evolutionary Algorithms for Electric Power Dispatch Problem. *IEEE Trans. Evol. Comput.* **2006**, *10*, 315–329. [CrossRef]

18.   Ahmadi, M.; Bozorg Haddad, O.; Marino, M.A. Extraction of flexible multi-objective real-time reservoir operation rules. *Water Resour. Manag.* **2014**, *28*, 131–147. [CrossRef]

19.   Shokri, A.; Haddad, O.B.; Mari, O.M.A. Algorithm for Increasing the Speed of Evolutionary Optimization and its Accuracy in Multi-objective Problems. *Water Resour. Manag.* **2013**, *27*, 2231–2249. [CrossRef]

20.   Afshar, A.; Sharifi, F.; Jalali, M.R. Applying the non-dominated archiving multi-colony ant algorithm for multi-objective optimization: Application to multi-purpose reservoir operation. *Engng. Optimiz.* **2009**, *41*, 313–325. [CrossRef]

21.   Alizadeh, M.R.; Nikoo, M.R.; Rakhshandehroo, G.R. Hydro-Environmental Management of Groundwater Resources: A Fuzzy-Based Multi-Objective Compromise Approach. *J. Hydrol.* **2017**, *551*, 540–554. [CrossRef]

22.   Afshar, A.; Shojaei, N.; Sagharjooghifarahani, M. Multiobjective Calibration of Reservoir Water Quality Modeling Using Multiobjective Particle Swarm Optimization (MOPSO). *Water Resour. Manag.* **2013**, *27*, 1931–1947. [CrossRef]

23.   Reddy, M.J.; Kumar, D.N. Optimal Reservoir Operation Using Multi-Objective Evolutionary Algorithm. *Water Resour. Manag.* **2006**, *20*, 861–878. [CrossRef]

24.   Bai, T.; Chang, J.X.; Chang, F.J.; Huang, Q.; Wang, Y.M.; Chen, G.S. Synergistic Gains from the Multi-Objective Optimal Operation of Cascade Reservoirs in the Upper Yellow River Basin. *J. Hydrol.* **2015**, *523*, 758–767. [CrossRef]

25.   Reddy, M.J.; Nagesh Kumar, D. Multi-Objective Particle Swarm Optimization for Generating Optimal Trade-Offs in Reservoir Operation. *Hydrol. Process.* **2010**, *21*, 2897–2909. [CrossRef]

26.   Wang, X.; Chang, J.; Meng, X.; Wang, Y. Research on Multi-Objective Operation Based on Improved Nsga-ii for the Lower Yellow River. *J. Hydraul. Eng.* **2017**, *48*, 135–145.

27.   Robin, W.; Mohd, S. Evaluation of genetic algorithm for optimal reservoir system operation. *J. Water Resour. Plan. Manag.* **1999**, *125*, 25–33.

28.   Wang, X.; Li, C. Research and application of genetic algorithm in short-term optimal power generation dispatching. *J. Yangtze River Sci. Acad. (Eng. Ed.)* **2003**, *20*, 28–31.

29.   Huang, C.; Wang, Z.; Li, S.; Chen, S.-L. A multi-reservoir operation optimization model and application in the upper Yangtze River Basin, I. *Principle and solution of the model. J. Hydraul. Eng.* **2014**, *45*, 1009–1018.

30.   Yang, N.; Mei, Y.; Yu, L. Multi-objective reservoir optimal operation model considering natural flow regime and its application. *J. Hehai Univ. (Nat. Sci.)* **2013**, *41*, 86–89.

31. Jin, X.; Hao, C.; Wang, G.; Wang, L.H. Multi-objective ecological operation of water supply reservoir. *South-North Transf. Water Sci. Technol.* **2015**, *13*, 463–466.

32. Shi, Z.; Luo, Y.; Qiu, J. Optimal dispatch of cascaded hydropower stations using Matlab genetic algorithm toolbox. *Electr. Power Autom. Equip.* **2005**, *25*, 30–33.

33. Chang, J.; Huang, Q.; Wang, Y. Optimal Operation of Hydropower Station Reservoir by Using an Improved Genetic Algorithm. *J. Hydroelectr. Eng.* **2001**, *3*, 1–20.

34. You, J.; Ji, C.; Fu, X. New method for solving multi-objective problem based on genetic algorithm. *J. Hydraul. Eng.* **2003**, *34*, 64–69.

35. Deb, K.; Pratap, A.; Agarwal, S.; Meyarivan, T. A fast and elitist multiobjective genetic algorithm: NSGA-II. *IEEE Trans. Evol. Comput.* **2002**, *6*, 182–197. [CrossRef]

36. Kim, T.; Heo, J.H.; Jeong, C.S. Multi-reservoir system optimization in the Han River basin using multi-objective genetic algorithms. *Hydrol. Process* **2006**, *20*, 2057–2075. [CrossRef]

37. Li, H.; Zhang, Q. Multiobjective Optimization Problems With Complicated Pareto Sets, MOEA/D and NSGA-II. *IEEE Trans. Evol. Comput.* **2009**, *13*, 284–302. [CrossRef]

38. Chang, L.C.; Chang, F.J. Multi-objective evolutionary algorithm for operating parallel reservoir system. *J. Hydrol.* **2009**, *377*, 12–20. [CrossRef]

39. Tsai, W.P.; Chang, F.J.; Chang, L.C.; Herricks, E.E. AI techniques for optimizing multi-objective reservoir operation upon human and riverine ecosystem demands. *J. Hydrol.* **2015**, *530*, 634–644. [CrossRef]

40. Chen, L.; Mc, P.J.; Yeh, W.W.G. A diversified multi-objective GA for optimizing reservoir Rule curves. *Adv. Water Resour.* **2007**, *30*, 1082–1093. [CrossRef]

41. Yun, R.; Dong, Z.; Wang, H. Multiobjective optimization of a reservoir based on NSGA2. *J. Shandong Univ. (Eng. Sci.)* **2010**, *40*, 125–128.

42. Zhu, J.; Chen, S.; Wan, B.; Huang, S.T.; Tu, J.J.; Fu, X. The Application of NSGA-II for Medium or Long-term Multi-Objective Optimal Scheduling in Zhanghe Reservoir. *China Rural Water Hydropower* **2013**, *9*, 60–62, 66.

43. Vieira, J.; Cunha, M.C.; Luis, R. Integrated Assessment of Water Reservoir Systems Performance with the Implementation of Ecological Flows under Varying Climatic Conditions. *Water Resour. Manag.* **2018**, *32*, 5183–5205. [CrossRef]

44. Ziyu, D.; Guohua, F.; Xin, W.; Tan, Q. A novel operation chart for cascade hydropower system to alleviate ecological degradation in hydrological extremes. *Ecol. Model.* **2018**, *384*, 10–22.

45. Niu, W.J.; Feng, Z.K.; Cheng, C.T. Optimization of variable-head hydropower system operation considering power shortage aspect with quadratic programming and successive approximation. *Energy* **2018**, *143*, 1020–1028. [CrossRef]

46. Williams, P.J.; Kendall, W.L. A guide to multi-objective optimization for ecological problems with an application to cackling goose management. *Ecol. Model.* **2017**, *343*, 54–67. [CrossRef]

47. Xia, Y.; Feng, Z.; Niu, W.; Qin, H. Simplex quantum-behaved particle swarm optimization algorithm with application to ecological operation of cascade hydropower reservoirs. *Appl. Soft Comput.* **2019**, *84*, 105715. [CrossRef]

48. Yang, Z.; Yang, K.; Su, L.; Hu, H. The multi-objective operation for cascade reservoirs using MMOSFLA with emphasis on power generation and ecological benefit. *J. Hydroinf.* **2019**, *21*, 257–278. [CrossRef]

49. Bai, T.; Wei, J.; Chang, F.J.; Yang, W.; Huang, Q. Optimize multi-objective transformation rules of water-sediment regulation for cascade reservoirs in the Upper Yellow River of China. *J. Hydrol.* **2019**, *577*, 123987. [CrossRef]

50. Qi, Z. Analysis on Environmental Impact of Xiaolangdi Reservoir Operation. *Yellow River* **2011**, *33*, 97–98.

51. Lei, Y.; Zhang, S. *Matlab Genetic Algorithm Toolbox and Its Application*; Xidian University Press: Xi'an, China, 2014.

Article

Multi-Dimensional Interval Number Decision Model Based on Mahalanobis-Taguchi System with Grey Entropy Method and Its Application in Reservoir Operation Scheme Selection

Changming Ji [1], Xiaoqing Liang [1,2], Yang Peng [1], Yanke Zhang [1,*], Xiaoran Yan [1] and Jiajie Wu [1]

[1] School of Renewable Energy, North China Electric Power University, Beijing 102206, China;
cmji@ncepu.edu.cn (C.J.); l_x_q1688@163.com (X.L.); pengyang@ncepu.edu.cn (Y.P.);
yanxiaoran1014@163.com (X.Y.); 1172111017@ncepu.edu.cn (J.W.)

[2] Department of Physics and Hydropower Engineering, Gansu Normal University for Nationalities,
Hezuo 747000, Gansu, China

* Correspondence: ykzhang2008@163.com

Received: 15 January 2020; Accepted: 28 February 2020; Published: 3 March 2020

Abstract: In decision-making with interval numbers, there are problems such as how to reduce the loss of decision information to improve decision accuracy and the difficulty of using interval numbers for sorting. On the basis of fully considering the subjective and objective weights of indexes, the grey entropy method (GEM) is improved by taking advantage of the Mahalanobis-Taguchi System (MTS) in which the orthogonal design has few tests but much obtained information, and the Mahalanobis distance can reflect the correlation between indexes. Then, the signal-to-noise ratio is integrated with the improved degree of balance and approach, and a multi-dimensional interval number decision model based on MTS and GEM is put forth. This model is applied to selecting the optimal scheme of controlling the Pankou reservoir's water level in flood season. Compared with the decision results of other methods, the optimal scheme selected by the proposed model can achieve greater benefits within an acceptable risk range and thus better coordinate the balance between risk and benefit, which verifies the feasibility and validity of the model.

Keywords: Mahalanobis-Taguchi System; grey entropy method; signal-to-noise ratio; degree of balance and approach; interval number

---

1. Introduction

Multi-attribute decision-making refers to the process of sorting and selecting among a group of alternatives by using the obtained information [1]. Multi-attribute decision-making methods include Technique for Order Preference by Similarity to an Ideal Solution (TOPSIS) [2], set pair analysis (SPA) [3], grey relation analysis (GRA) [4], grey target method (GTM) [5], etc. Under the influence of many factors, the objective things in nature are complex and changeable (e.g., runoff and flood in the hydrological and water resources system), which show great uncertainty like randomness, fuzziness and greyness [6], in addition to the fuzziness of human thinking. In recent years, the multi-attribute decision-making methods with uncertain decision information, especially uncertain attribute values and attribute weights, have become a hot research issue. In general, the priority and also difficulty in solving this kind of problem is to transform uncertain decision-making into definite decision-making. During this process, on the one hand, we should try to avoid or reduce the loss of decision information. The more information loss, the larger deviation caused in the selection of the optimal scheme, which may fail to achieve the expected objectives in scheme implementation and incur risk events. On the other hand, the cumbersome calculation of interval numbers should be reduced, so that the technique is more operable in a practical application. For instance, when using the methods in references [7,8], the comparison of interval numbers only considers their upper and lower boundaries, which loses too much information even though the decision-making efficiency is enhanced. Moreover, the tedious calculation in rough set method [9] and Vlsekriterijumska Optimizacija I Kompromisno Resenje (VIKOR) method [10] makes interval number ranking more difficult.

At present, grey analysis, set pair analysis, stochastic analysis and fuzzy analysis are the four main kinds of hydrologic uncertainty analysis methods. In information theory, a system with some known information and also some unknown information is a grey system, such as the hydrologic and water resources system. Grey relation analysis is a multi-factor analysis method in the grey system theories [11]. Its basic idea is to determine the closeness of relation between a comparison sequence and a reference sequence according to their geometric similarity. Grey relation analysis can overcome the shortcomings of regression analysis, variance analysis and other factor analysis methods in traditional mathematical statistics [12], and has the advantages of a small sample size and simple calculation procedure. Gao and Zhang [13] first used grey relation analysis for scheme selection. However, in the relation coefficient series obtained from the alternatives and the reference scheme, the larger relation coefficient often plays a decisive role in determining the relation degree, while the information implied by other relation coefficients tends to be neglected. As an improvement, Zhang [14] put forward the grey entropy method (GEM) based on the idea that an alternative scheme is better when it is evenly closer to the reference scheme. This method combines the relation degree in grey relation analysis with the balance degree defined in grey entropy by multiplication, and thus, develops the degree of balance and approach for scheme decision-making, which in some way remedies grey relation analysis. The GEM has been widely used in logistics [15], transportation [16], tourism [17] and other industries, with satisfactory results. Grey entropy is a concept that combines grey system theory and information entropy theory. As an effective statistical measurement for information uncertainty [18], it is consistent with the physical meaning of Shannon entropy. Wang et al. [19] defined the grey distance entropy of the real grey number and the interval grey number, respectively, took it as the measurement for the proximity degree of the two grey numbers, and discussed the multi-attribute decision-making methods based on grey distance entropy and TOPSIS. Liu et al. [20] deemed that in the grey entropy method the attribute value is a real number and the attribute weight is not considered, thus they extended the application scope to interval numbers. At present, grey entropy is mainly used to determine the objective weight of indexes in uncertain decision-making, and there are not many in-depth theoretical studies. The comparison of the above decision-making methods is shown in Table 1.

Table 1. Comparison of the five multi-attribute decision-making methods.

Method	Advantage	Shortage
TOPSIS	Alternative schemes are evaluated by both how close they are to the positive ideal scheme and how far away from the negative ideal scheme.	The Euclidean distance does not consider the correlation between indexes, producing indistinct decision result, "reverse order" and other problems.
SPA	Alternative schemes are evaluated by their identical-discrepant-contrary degree to the optimal scheme, preventing the "reverse order" problem of TOPSIS.	The uncertainty of the discrepancy coefficient may lead to decision risk.
GRA	Alternative schemes are evaluated by their geometric proximity to the optimal scheme.	The Euclidean distance is used to calculate the correlation coefficient, which neglects the correlation between indexes; larger correlation coefficients determine the correlation degree, which causes information loss.
GTM	Alternative schemes are evaluated by their distance from the optimal scheme (off-target distance).	The single bull's-eye may cause the decision result to be indistinct; the off-target distance is calculated using the Euclidean distance, which neglects correlation between indexes, and index weight is not considered.
GEM	Alternative schemes are evaluated by how evenly close they are to the optimal scheme, which remedies GRA in some way.	The Euclidean distance is used to calculate the correlation coefficient, which neglects the correlation between indexes; the calculation of correlation degree does not consider index weight.

The combination of the Mahalanobis-Taguchi System (MTS) theory and grey entropy can provide a new idea for uncertain multi-attribute decision-making. The Mahalanobis-Taguchi System [21,22], proposed by Japanese engineer Taguchi G. in the 1990s, is a pattern recognition method for unbalanced data. Scholars are becoming gradually more familiar with MTS and constantly try to explore its application or its combination with other theories in different fields. Buenviaje et al. [23] obtained medical patterns from historical data sets through MTS. Huang et al. [24] utilized the data mining function of MTS, combined it with the artificial neural network (ANN) and came up with the MTS-ANN algorithm. Zeng et al. [25] studied how to make the risk decision on power transformer maintenance by using MTS and the grey cumulative prospect theory. Chang et al. [26] employed the three key tools of MTS, namely, orthogonal table, Mahalanobis distance and signal-to-noise ratio, to tackle the problem of multi-attribute decision-making with interval numbers on the basis of TOPSIS. The orthogonal table [27] is a direct test method for multi-factor system optimization. It can be expressed in the form of $L_a(b^N)$, where a is the number of tests, b the number of levels of each factor and N the number of factors that can be arranged at most in the orthogonal table. The orthogonal table is a prepared set of standard tables, from which the suitable one is chosen in actual application according to the number of factors and the number of levels of each factor. The orthogonal table designs a small number of tests and obtains comprehensive information, which can effectively reduce the loss of information. The Mahalanobis distance [28], proposed by Indian statistician Mahalanobis, is a covariance distance that, compared with Euclidean distance, can better reflect the correlation between attributes. The concept of signal-to-noise ratio (SNR) [29] originates from signal transmission and is defined as the ratio of signal power to noise power. Taguchi G. redefined the SNR, regarding the square (μ^2) and the variance (σ^2) of the expected value of an index (non-negative and continuous) as the signal power and the noise power, respectively. The SNR can be divided into three types: nominal-the-better, smaller-the-better and larger-the-better. The first one means that the closer to the expected value when it is positive, the better; the second one means that the smaller the expected value when it is 0, the better; the third one means that the larger the expected value when it is $+\infty$, the better. The signal-to-noise ratio can be used to measure the volatility of indexes and thereby ensure the accuracy of decision results.

Like other pattern recognition methods, MTS also uses distance to measure the similarity between samples; however, instead of using Euclidean distance, it uses Mahalanobis distance, which is more suitable to distinguish sample similarity. As the theory of MTS has only been developed for the past 20 years and there are few studies on its integration with grey analysis, more work needs to be done to deepen this field of theoretical research and expand its application.

To address the uncertain multi-attribute decision-making problem in which both attribute weight and attribute value are interval numbers, we improved the grey entropy method based on the treatment of interval numbers in reference [26], and put forward a multi-dimensional interval number decision model based on the Mahalanobis-Taguchi System with the grey entropy method (MTS-GEM). The model is applied to the selection of the optimal scheme of controlling the Pankou reservoir's water level in flood season, which can facilitate reservoir operation research.

2. Multi-Dimensional Interval Number and Mahalanobis-Taguchi System

2.1. Orthogonal Test of Multi-Dimensional Interval Number

Interval number generally refers to the normal interval number in the form of $\widetilde{a} = \left[a^L, a^U\right]$ (a^L, $a^U \in R$ and $a^U \geq a^L$), where a^L and a^U are, respectively, the lower and upper bounds of the interval number. Real numbers can be regarded as interval numbers whose lower bounds are equal to their upper bounds. For the basic operation rules of any two interval numbers $\widetilde{a} = \left[a^L, a^U\right]$ and $\widetilde{b} = \left[b^L, b^U\right]$, see reference [30].

Suppose that there are n factors $\widetilde{x}_j$ (j=1,2, ... ,n, and $n \leq N$) in a test, which are uniformly distributed within their variation range, and the number of levels of each factor is b. Because n factors can constitute a hypercube in an *n*-dimensional space, the result of one orthogonal test corresponds to a point on the hypercube.

Take the orthogonal test of three factors as an example. Let the lower bound value of interval number be level 1 and the upper bound value of interval number be level 2, then the orthogonal table of $L_4(2^3)$ shown in Table 2 can be used for the orthogonal test, and the results are shown in Figure 1.

Table 2. Orthogonal table of $L_4(2^3)$.

Test	Factor		
	$\tilde{x}_1$	$\tilde{x}_2$	$\tilde{x}_3$
One	1	1	1
Two	1	2	2
Three	2	1	2
Four	2	2	1

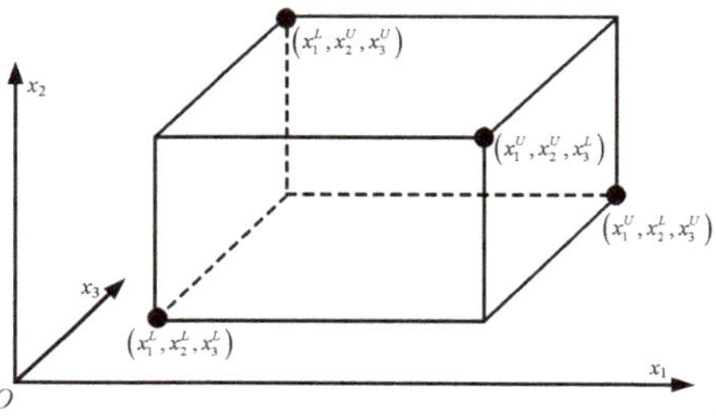

Figure 1. Orthogonal test results of three-dimensional interval numbers.

2.2. Signal-to-Noise Ratio and Mahalanobis Distance

Let D be the population of a non-negative continuous index, whose signal-to-noise ratio η^* is calculated as follows [31]:

$$\eta^* = \mu^2/\sigma^2, \tag{1}$$

where μ^2 and σ^2 are the mean square and the variance of population D, respectively.

The three types of signal-to-noise ratio for measuring index volatility are calculated as follows.

1   Nominal-the-better (NB) signal-to-noise ratio

According to the knowledge of mathematical statistics, the unbiased estimators of variance σ^2 and mean square μ^2 of population D are as follows:

$$\hat{\sigma}^2 = \frac{1}{a-1}\sum_{g=1}^{a}\left(D_g - \overline{D}^2\right),\ \hat{\mu}^2 = \overline{D}^2 - \frac{\hat{\sigma}^2}{a}, \tag{2}$$

where $\overline{D} = \frac{1}{a}\sum_{g=1}^{a}D_g$ and a is the number of samples.

Put $\hat{\sigma}^2$ and $\hat{\mu}^2$ into Equation (1) and conduct the common logarithmic transformation on η^*, then the NB signal-to-noise ratio η^{NB} can be obtained:

$$\eta^{NB} = 10\log_{10}\frac{\hat{\mu}^2}{\hat{\sigma}^2}, \tag{3}$$

2   Smaller-the-better (SB) signal-to-noise ratio

For an SB index, the smaller μ^2 and σ^2 are, the better, meaning that the smaller $\mu^2 + \sigma^2 = E(D^2)$ is, the better. The unbiased estimator of D^2 is $\hat{D}^2 = \frac{1}{a}\sum_{g=1}^{a} D_g^2$. Take $(\hat{D}^2)^{-1}$ as the signal-to-noise ratio, and the SB SNR η^{SB} is calculated as follows:

$$\eta^{SB} = -10\log_{10}\left(\frac{1}{a}\sum_{g=1}^{a} D_g{}^2\right), \tag{4}$$

3   Larger-the-better (LB) signal-to-noise ratio

If D stands for an LB index, then D^{-1} stands for an SB index. From the above derivation process of the η^{SB} formula, the LB SNR η^{LB} can be obtained:

$$\eta^{LB} = -10\log_{10}\left(\frac{1}{a}\sum_{g=1}^{a} \frac{1}{D_g{}^2}\right), \tag{5}$$

Let Z be a population of an n-dimensional real number space, $\mu = (\mu_1\mu_2 \cdots \mu_n)^T$ and Σ^{-1} be the mean vector and the inverse of covariance matrix of Z, respectively. If $x = (x_1x_2 \cdots x_n)^T$ is one sample of the n-dimensional real number space, then the Mahalanobis distance d from sample x to population Z is:

$$d^2 = (x-\mu)^T \sum^{-1} (x-\mu), \tag{6}$$

In scheme decision-making, an alternative scheme is better if its Mahalanobis distance to the reference scheme is smaller, hence, we adopted the SB SNR. If the reference scheme's sample x and population Z are taken as the input of MTS, then the Mahalanobis distance d between the alternative scheme and the reference scheme is the responding output.

3. Improved Grey Entropy Method by Mahalanobis-Taguchi System

3.1. Grey Entropy

Grey entropy is the entropy of a grey number [32,33]. Grey number refers to a number set with incomplete and uncertain information, in which the numbers are possible values. For a discrete grey number $q = \{q_i|i = 1, 2, \cdots, n\}, i \in J$ (where J is a finite set and $q_i \geq 0$, $\sum_{i=1}^{n} q_i = 1$), $H(q) = -\sum_{i=1}^{n} q_i \ln q_i$ is called the corresponding grey entropy. When $q_i = 0$, $q_i \ln q_i = 0$. The more equal $q_1, q_2, \cdots, q_n$ are, the greater the grey entropy $H(q)$ is, thereby, $H(q)_{max} = \ln n$, and $B(q) = H(q)/H(q)_{max}$ is defined as the balance degree of the grey number.

3.2. The Comparison of Grey Entropy and Information Entropy

Information entropy refers to the entropy of a signal source [34]. A discrete signal source can be expressed as $X : \begin{bmatrix} x_1 & x_2 & \cdots & x_i & \cdots & x_n \\ p_1 & p_2 & \cdots & p_i & \cdots & p_n \end{bmatrix}, i \in J$ (where J is an infinite set, the probability of random variable x_i is p_i and $P(X = x_i \cap X = x_j) = 0, i \neq j$, $\sum_{i=1}^{n} p_i = 1$). Then, $H(X) = -\sum_{i=1}^{n} p_i \ln p_i$ is called the information entropy of the signal source.

Accordingly, grey entropy and information entropy have the following similarities and differences.

- Similarities

    1. With the same form of calculation formula, grey entropy and information entropy share some characteristics, such as symmetry, non-negativity, additivity, convexity and extremum property.
    2. The physical meanings of grey entropy and information entropy are essentially the same. The former is to measure the fluctuation degree of a grey number, while the latter is to describe the uncertainty of a signal source.

- Differences

1. Grey entropy is defined in a finite information space, whereas information entropy is defined in an infinite information space.
2. Grey entropy is a kind of non-probability entropy with greyness, that is, q_i is a possible value. On the contrary, information entropy is a type of probability entropy with certainty, that is, p_i is a certain value.

3.3. Fundamentals of Grey Entropy Method

GEM was originally used for scheme decision-making with multi-dimensional real numbers. Suppose an initial decision matrix $X = \left[x_{ij}\right]_{m \times n}$ ($I = 1,2, \ldots ,m; j=1,2, \ldots ,n$) composed of m schemes and n indexes. Through weighted standardization, X is converted into $C = \left[c_{ij}\right]_{m \times n}$, and the alternative scheme is $c_i = (c_{i1}, c_{i2}, \cdots , c_{in})$. Given the reference scheme $p_k = (p_{k1}, p_{k2}, \cdots , p_{kn})$ ($k = 1,2, \ldots ,z$), the correlation degree G_{ki} between the alternative scheme and the reference scheme is calculated as follows:

$$G_{ki} = \sum\nolimits_{j=1}^{n} r_{ki}^{(j)} / n,$$
(7)

$$r_{ki}^{(j)} = \frac{\min\limits_{i}\min\limits_{j}\left|p_k^{(j)} - c_i^{(j)}\right| + \zeta \max\limits_{i}\max\limits_{j}\left|p_k^{(j)} - c_i^{(j)}\right|}{\left|p_k^{(j)} - c_i^{(j)}\right| + \zeta \max\limits_{i}\max\limits_{j}\left|p_k^{(j)} - c_i^{(j)}\right|},$$
(8)

where $r_{ki}^{(j)}$ is the correlation coefficient of index j between c_i and p_k and ζ is the distinguishing coefficient and generally set at 0.5.

The correlation coefficient sequence of c_i and p_k is $r_{ki} = \left\{r_{ki}^{(j)} \middle| j = 1, 2, \cdots , n\right\}$, whose balance degree J_{ki} is calculated as follows:

$$J_{ki} = -\frac{1}{\ln n} \sum\nolimits_{j=1}^{n} q_{ki}^{(j)} \ln q_{ki}^{(j)},$$
(9)

where $q_{ki}^{(j)} = r_{ki}^{(j)} / \sum_{j=1}^{n} r_{ki}^{(j)}$.

The degree of balance and approach of c_i and p_k is calculated as follows:

$$\gamma_{ki} = G_{ki} \times J_{ki},$$
(10)

The greater the degree of balance and approach is, the similar the alternative scheme is to the reference scheme.

It can be seen from Equation (8) that the correlation coefficient reflects the distance between two points. When using the Euclidean distance for calculation, the correlation between indexes, which usually exits in reality, is ignored, making the calculation results of the correlation degree and the balance degree unreasonable. Moreover, there is interaction between indexes, either controllable or uncontrollable. Additionally, the uncontrollable interaction affects the stability of the responding output. Consequently, although the balance degree is an indicator of the volatility of correlation coefficient series, it cannot embody the variability caused by the uncontrollable interaction between indexes. The merit of MTS is that it can not only use the orthogonal table to deal with multi-dimensional real numbers or multi-dimensional interval numbers, taking the interaction among indexes into account, but also use the Mahalanobis distance considering the correlation between indexes to calculate the correlation coefficient. Therefore, using MTS to improve GEM can enhance the decision-making performance of GEM.

4. Multi-Dimensional Interval Number Decision Model Based on MTS-GEM

4.1. Development of the Weighted Standardized Decision Matrix

Suppose that there are m alternative schemes and n indexes, and they constitute the initial interval number decision matrix $\widetilde{X} = \left[\widetilde{x}_{ij}\right]_{m \times n}$ (i=1,2,$\ldots$,m; j=1,2,$\ldots$,n) where $\widetilde{x}_{ij} = \left[x_{ij}^L, x_{ij}^U\right]$. As the dimensions of the indexes are often different, we used Equations (11) and (12) to nondimensionalize different types of indexes, and get the standardized interval number decision matrix $\widetilde{B} = \left[\widetilde{b}_{ij}\right]_{m \times n}$, where $\widetilde{b}_{ij} = \left[b_{ij}^L, b_{ij}^U\right]$.

For a benefit index:

$$\widetilde{b}_{ij} = \left[\frac{x_{ij}^L - \min\limits_i x_{ij}}{\max\limits_i x_{ij} - \min\limits_i x_{ij}}, \frac{x_{ij}^U - \min\limits_i x_{ij}}{\max\limits_i x_{ij} - \min\limits_i x_{ij}}\right], \tag{11}$$

For a cost index:

$$\widetilde{b}_{ij} = \left[\frac{\max\limits_i x_{ij} - x_{ij}^U}{\max\limits_i x_{ij} - \min\limits_i x_{ij}}, \frac{\max\limits_i x_{ij} - x_{ij}^L}{\max\limits_i x_{ij} - \min\limits_i x_{ij}}\right], \tag{12}$$

where $\min\limits_i x_{ij} = \min(\min\limits_i x_{ij}^L, \min\limits_i x_{ij}^U)$ and $\max\limits_i x_{ij} = \max(\max\limits_i x_{ij}^L, \max\limits_i x_{ij}^U)$.

To make the weight distribution of each index reasonable, the integration of expert scoring (subjective weight) and entropy weight method [35] (objective weight) is used to determine the combined weight of each index.

1. Subjective weight $\widetilde{S} = \left[\widetilde{s}_j\right]_{1 \times n}$ where $\widetilde{s}_j = \left[s_j^L, s_j^U\right]$

Assuming that there are v experts who participate in the weight scoring of n indexes, the scoring matrix is $\widetilde{Y} = \left[\widetilde{y}_{lj}\right]_{v \times n}$ (l=1,2,$\ldots$,v; j=1,2,$\ldots$,n) where $\widetilde{y}_{lj} = \left[y_{1j}^L, y_{1j}^U\right]$, then $\widetilde{s}_j = \left[\min\limits_l y_{1j}^L, \max\limits_l y_{1j}^U\right]$.

2. Objective weight $\widetilde{T} = \left[\widetilde{t}_j\right]_{1 \times n}$, where $\widetilde{t}_j = \left[t_j^L, t_j^U\right]$

First, the entropy of the interval number index is calculated based on the standardized interval number decision matrix $\widetilde{B}$:

$$E_j^L = -\frac{1}{\ln m}\sum_{i=1}^m \left[\left(\frac{b_{ij}^L}{\sum_{i=1}^m b_{ij}^L}\right)\ln\left(\frac{b_{ij}^L}{\sum_{i=1}^m b_{ij}^L}\right)\right]$$
$$E_j^U = -\frac{1}{\ln m}\sum_{i=1}^m \left[\left(\frac{b_{ij}^U}{\sum_{i=1}^m b_{ij}^U}\right)\ln\left(\frac{b_{ij}^U}{\sum_{i=1}^m b_{ij}^U}\right)\right] \tag{13}$$

where E_j^L and E_j^U are the information entropy of the interval number index's lower bound and upper bound, respectively; when $b_{ij}^L = 0$ or $b_{ij}^U = 0$, $\left(\frac{b_{ij}^L}{\sum_{i=1}^m b_{ij}^L}\right)\ln\left(\frac{b_{ij}^L}{\sum_{i=1}^m b_{ij}^L}\right) = 0$ or $\left(\frac{b_{ij}^U}{\sum_{i=1}^m b_{ij}^U}\right)\ln\left(\frac{b_{ij}^U}{\sum_{i=1}^m b_{ij}^U}\right) = 0$.

Then, calculate the objective weight of the interval number index $\widetilde{t}_j$:

$$\widetilde{t}_j = \left[t_j^L, t_j^U\right] = \left[\min\left(\frac{1 - E_j^L}{\sum_{j=1}^n (1 - E_j^L)}, \frac{1 - E_j^U}{\sum_{j=1}^n (1 - E_j^U)}\right), \max\left(\frac{1 - E_j^L}{\sum_{j=1}^n (1 - E_j^L)}, \frac{1 - E_j^U}{\sum_{j=1}^n (1 - E_j^U)}\right)\right], \tag{14}$$

3. Combined weight $\widetilde{W} = \left[\widetilde{w}_j\right]_{1 \times n}$, where $\widetilde{w}_j = \left[w_j^L, w_j^U\right]$

Considering both the subjective weight $\widetilde{S}$ and the objective weight $\widetilde{T}$, we obtained the combined weight of the interval number index $\widetilde{W} = \left[\widetilde{w}_j\right]_{1 \times n}$ where $\widetilde{w}_j = \left[\beta s_j^L + (1 - \beta)t_j^L, \beta s_j^U + (1 - \beta)t_j^U\right]$ and is an empirical factor that reflects the preference of decision makers between subjective experience and objective data [36].

With the standardized decision matrix $\widetilde{B}$ and the indexes' combined weight $\widetilde{W}$, we used the multiplication algorithm of interval number to obtain the weighted standardized decision matrix $\widetilde{C} = \left[\widetilde{c}_{ij}\right]_{m \times n}$, where $\widetilde{c}_{ij} = \widetilde{w}_j \times \widetilde{b}_{ij}$. Additionally, based on TOPSIS, the positive ideal scheme $\widetilde{p}_+$ and the negative ideal scheme $\widetilde{p}_-$ were determined:

$$
\begin{aligned}
\widetilde{p}_+ &= \left\{ \left[\max_i c^L_{i1}, \max_i c^U_{i1}\right], \left[\max_i c^L_{i2}, \max_i c^U_{i2}\right], \cdots, \left[\max_i c^L_{in}, \max_i c^U_{in}\right] \right\} \\
\widetilde{p}_- &= \left\{ \left[\min_i c^L_{i1}, \min_i c^U_{i1}\right], \left[\min_i c^L_{i2}, \min_i c^U_{i2}\right], \cdots, \left[\min_i c^L_{in}, \min_i c^U_{in}\right] \right\}
\end{aligned}
\tag{15}
$$

4.2. Orthogonal Test of Schemes and Calculation of Derivative Indicators

According to the number of indexes, a two-level orthogonal table with $N \geq n$ was selected, where n indexes can be arranged in any n columns. For the interval number $\widetilde{a} = \left[a^L, a^U\right]$, take a^L as level 1 and a^U as level 2. The layout matrix C_i of the alternative scheme $\widetilde{c}_i$ ($i=1,2,\ldots,m$) is as follows:

$$
C_i = \begin{bmatrix} c_i^{(1)} \\ c_i^{(2)} \\ \vdots \\ c_i^{(a)} \end{bmatrix} = \begin{bmatrix} c_{i1}^{(1)} & c_{i2}^{(1)} & \cdots & c_{in}^{(1)} \\ c_{i1}^{(2)} & c_{i2}^{(2)} & \cdots & c_{in}^{(2)} \\ \vdots & \vdots & \ddots & \vdots \\ c_{i1}^{(a)} & c_{i2}^{(a)} & \cdots & c_{in}^{(a)} \end{bmatrix}_{a \times n} ,
\tag{16}
$$

where $c_i^{(g)}$ ($g=1,2,\ldots,a$) is the distribution point of scheme $\widetilde{c}_i$ and $c_{i1}^{(g)}, c_{i2}^{(g)}, \ldots, c_{in}^{(g)}$ are the components of $c_i^{(g)}$.

Similarly, the layout matrix of the positive ideal scheme P_+ and the layout matrix of the negative ideal scheme P_- can be obtained.

Here, we define derivative indicators as the signal-to-noise ratio and the degree of balance and approach based on square Mahalanobis distance obtained by the initial interval number indexes. According to Equation (6), calculate the square Mahalanobis distance between the alternative scheme's distribution point $c_i^{(g)}$ and the positive/negative ideal scheme, and we obtain $d^2\left[c_i^{(g)}, P_+\right]$ and $d^2\left[c_i^{(g)}, P_-\right]$.

1 Signal-to-noise ratio

The SB signal-to-noise of the scheme i to the positive (negative) ideal scheme is η_{+i} (η_{-i}):

$$
\begin{aligned}
\eta_{+i} &= -10\log_{10}\left\{ \sum_{g=1}^{a} d'^2\left[c_i^{(g)}, P_+\right] / a \right\} \\
\eta_{-i} &= -10\log_{10}\left\{ \sum_{g=1}^{a} d'^2\left[c_i^{(g)}, P_-\right] / a \right\}
\end{aligned}
\tag{17}
$$

where a is the number of orthogonal tests and $d'^2\left[c_i^{(g)}, P_+\right]$, $d'^2\left[c_i^{(g)}, P_-\right]$ is the square Mahalanobis distance standardized by Equation (18):

$$
\begin{aligned}
d'^2\left[c_i^{(g)}, P_+\right] &= \frac{\max\limits_g d^2\left[c_i^{(g)}, P_+\right] - d^2\left[c_i^{(g)}, P_+\right]}{\max\limits_g d^2\left[c_i^{(g)}, P_+\right] - \min\limits_g d^2\left[c_i^{(g)}, P_+\right]} \\
d'^2\left[c_i^{(g)}, P_-\right] &= \frac{d^2\left[c_i^{(g)}, P_-\right] - \min\limits_g d^2\left[c_i^{(g)}, P_-\right]}{\max\limits_g d^2\left[c_i^{(g)}, P_-\right] - \min\limits_g d^2\left[c_i^{(g)}, P_-\right]}
\end{aligned}
\tag{18}
$$

2 Improved degree of balance and approach

The correlation degree of the alternative scheme $\widetilde{c}_i$ and the positive (negative) ideal scheme is G_{+i} (G_{-i}):

$$G_{+i} = \sum_{g=1}^{a} r_{+i}^{(g)}/a, \, G_{-i} = \sum_{g=1}^{a} r_{-i}^{(g)}/a, \tag{19}$$

where $r_{+i}^{(g)}$ $(r_{-i}^{(g)})$ is the correlation coefficient of the alternative scheme $\widetilde{c}_i$ and the positive (negative) ideal scheme at distribution point g.

Based on the square Mahalanobis distance, the formula of correlation coefficient is revised as follows:

$$r_{+i}^{(g)} = \frac{\min\limits_{i} \min\limits_{g} d^2\left[c_i^{(g)}, P_+\right] + \zeta \max\limits_{i} \max\limits_{g} d^2\left[c_i^{(g)}, P_+\right]}{d^2\left[c_i^{(g)}, P_+\right] + \zeta \max\limits_{i} \max\limits_{g} d^2\left[c_i^{(g)}, P_+\right]}$$

$$r_{-i}^{(g)} = \frac{\min\limits_{i} \min\limits_{g} d^2\left[c_i^{(g)}, P_-\right] + \zeta \max\limits_{i} \max\limits_{g} d^2\left[c_i^{(g)}, P_-\right]}{d^2\left[c_i^{(g)}, P_-\right] + \zeta \max\limits_{i} \max\limits_{g} d^2\left[c_i^{(g)}, P_-\right]} \tag{20}$$

where the distinguishing coefficient ζ is set at 0.5.

The balance degree of the alternative scheme $\widetilde{c}_i$ and the positive (negative) ideal scheme is J_{+i} (J_{-i}):

$$J_{+i} = -\frac{1}{\ln a}\sum_{g=1}^{a} q_{+i}^{(g)} \ln q_{+i}^{(g)}$$
$$J_{-i} = -\frac{1}{\ln a}\sum_{g=1}^{a} q_{-i}^{(g)} \ln q_{-i}^{(g)} \tag{21}$$

where $q_{+i}^{(g)} = r_{+i}^{(g)}/\sum_{g=1}^{a} r_{+i}^{(g)}$ and $q_{-i}^{(g)} = r_{-i}^{(g)}/\sum_{g=1}^{a} r_{-i}^{(g)}$.

The degree of balance and approach of the alternative scheme $\widetilde{c}_i$ and the positive (negative) ideal scheme is γ_{+i} (γ_{-i}):

$$\gamma_{+i} = G_{+i} \times J_{+i}, \, \gamma_{-i} = G_{-i} \times J_{-i}, \tag{22}$$

4.3. Scheme Decision-Making

Based on the signal-to-noise ratio and the improved degree of balance and approach, the decision matrix Y is constructed as shown in Equation (23), in which the benefit indicator is η_{-i} and γ_{+i}, while the cost indicator is η_{+i} and γ_{-i}:

$$Y = \begin{bmatrix} \eta_{+1} & \eta_{-1} & \gamma_{+1} & \gamma_{-1} \\ \eta_{+2} & \eta_{-2} & \gamma_{+2} & \gamma_{-2} \\ \vdots & \vdots & \vdots & \vdots \\ \eta_{+m} & \eta_{-m} & \gamma_{+m} & \gamma_{-m} \end{bmatrix}_{m \times 4}, \tag{23}$$

Now, the decision-making problem with multi-dimensional interval numbers is changed into a decision-making problem with multi-dimensional real numbers [37]. The optimal scheme can be selected with the following multi-dimensional real vector space decision-making method (MRVSDM).

1. The n-dimensional real numbers are regarded as the points $A_1, \ldots, A_i, \ldots, A_m$ in an n-dimensional space with O as the origin, and then we can get the vectors $a_1 = \overrightarrow{OA_1}, \ldots, a_i = \overrightarrow{OA_i}, \ldots, a_m = \overrightarrow{OA_m}$.

2. Assuming that the reference scheme vector is p, $|a_i|$ and $|p|$ are the modules of vector a_i and vector p, respectively. Between vectors a_i and p, calculate their angle $\theta_i = (\hat{a_i, p}) = \arccos\left(\frac{a_i \cdot p}{|a_i||p|}\right)$ (where $a_i \cdot p$ is their product), as well as their mapping distance $MD_i = |a_i|\sin\theta_i$.

3. The set of mapping distance $MD = \{MD_i | i = 1, 2, \cdots, m\}$ can be obtained. According to the principle that the smaller MD_i is, the closer vector a_i is to vector p, the scheme that satisfies the objective $\min\limits_{i} MD_i$ is selected as the optimal scheme.

The diagram of decision-making in a three-dimensional real number vector space is shown in Figure 2. The input of the MTS-GEM model is the interval number indexes derived from different schemes. After orthogonal tests and calculation of the Mahalanobis distance, the signal-to-noise ratio, the improved degree of balance and approach, the output is the mapping distance from each scheme to the ideal scheme. The flow chart of this study is shown in Figure 3.

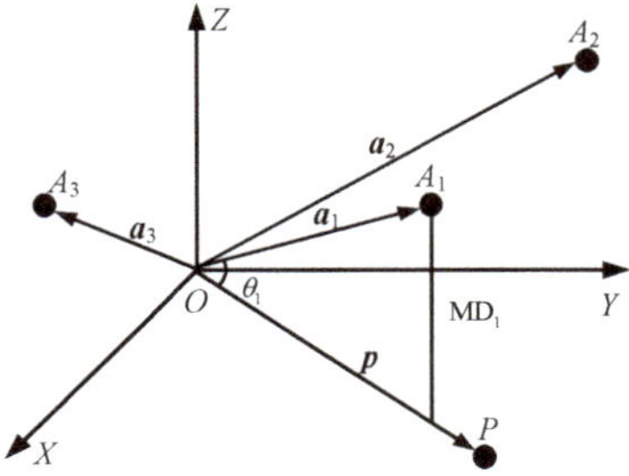

Figure 2. Diagram of decision-making in a three-dimensional real number vector space.

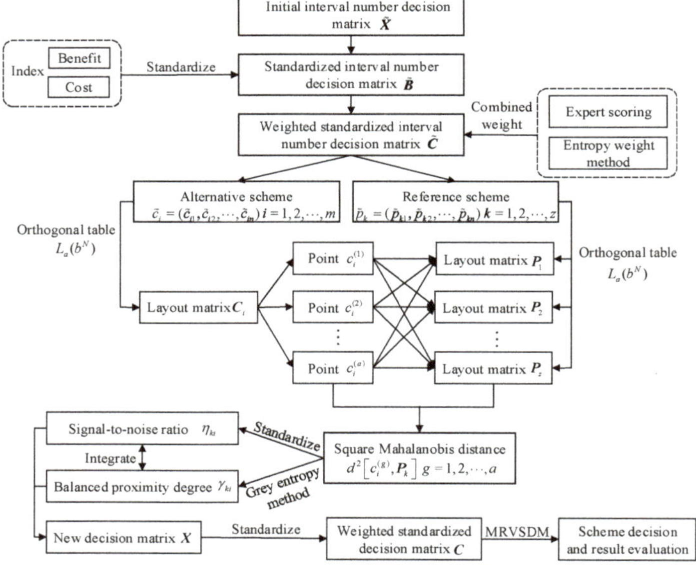

Figure 3. Flow chart of multi-dimensional interval number decision-making based on Mahalanobis-Taguchi System with grey entropy method.

5. Case Study

5.1. Initial Interval Number Decision Matrix and Its Weighted Standardization

Multi-objective reservoir operation is a multi-dimensional and complicated system engineering issue. Affected by runoff forecast, operation model, solution method and other factors, in the obtained non-inferior solution set, the attribute values are not always a precise real number but an interval number with uncertainty. The Pankou reservoir, located at the upstream of the Du River in China, is an annual-regulating reservoir with comprehensive utilization tasks of power generation, flood control, water supply and so forth. Its basic information is listed in Table 3, and its location map is shown in Figure 4.

Table 3. Basic information of the Pankou reservoir.

Item	Unit	Pankou
Dead water level	m	330
Flood control limit water level	m	347.6
Normal water level	m	355
Flood control high water level	m	358.4
Design flood water level	m	357.14
Spillway flood water level	m	360.82
Minimum storage capacity	10^8 m^3	8.5
Regulating storage capacity	10^8 m^3	11.2
Total storage capacity	10^8 m^3	23.38
Regulating performance	–	Annual regulating
Installed power capacity	MW	500
Average annual generated power	10^8 kW·h	10.474

Figure 4. The location of the Pankou reservoir.

According to the current regulations of the China Yangtze River Flood Control and Drought Relief Headquarters, for the Pankou reservoir, the upper limit of operation water level in flood season (June 20–August 20) is the flood control limit water level (347.6 m), while in other periods it is the normal water level (355 m); the lower limit of operation water level in all periods is the dead water level (330 m), and falling to the lower limit should be avoided during operation.

However, in the actual operation of the Pankou reservoir, it can barely store water to the normal water level in most years, undermining such benefits as power generation and water supply. In order to reasonably adjust the upper limit of water level in flood season and improve the utilization rate of flood resources, in reference [38] the three indexes of flood control risk rate, annual generated power and water storage at the end of flood season were used as the evaluation indexes for selecting the optimal scheme of water level upper limit in flood season. Flood control risk rate was obtained by means of flood stochastic simulation [39]. Firstly, the flood stochastic simulation model was used to simulate n (a large number) pieces of annual maximum flood inflow hydrographs, which can fully reflect the statistical characteristics of the reservoir's measured flood inflow. Then, a flood operation calculation was conducted to obtain the highest annual water level sequence, and the ratio of times that the water level limit was exceeded to n was taken as the flood control risk rate, the value of which needs to meet people's acceptable level of risk. Annual generated power refers to the total electricity generated by the hydropower station within a one-year operation cycle, which represents the power generation benefit of the hydropower station. The more annual generated power, the greater the annual power generation benefit. The water storage at the end of the flood season refers to the water storage between the dead water level and the particular water level at the end of flood season, which represents the water supply benefit of the reservoir. The more water storage at the end of flood season, the greater the water supply benefit.

The scheme setting is shown in Figure 5. Currently, Scheme 1 is being adopted as the operation strategy of the Pankou reservoir's upper water level limit in flood season.

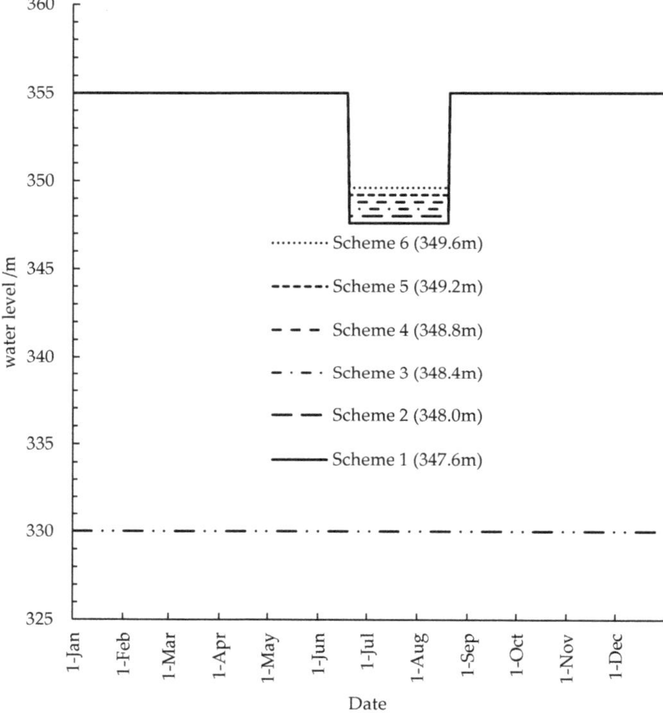

Figure 5. Scheme setting of the Pankou reservoir's upper water level limit in flood season.

For the flood control risk rate, the Monte Carlo method was used to randomly simulate 100 groups of 1000 floods corresponding to the 1000-year return period (0.1%). After flood operation calculation of each flood, the number of times that the design flood water level (357.14m) is exceeded in each group of 1000 floods were counted as $\lambda_\xi(\xi = 1,2,\cdots,100)$. The flood control risk rate of each group was $\lambda_\xi/1000$, and the flood control risk rate in the form of an interval number was $\left[\min_{1\leqslant\xi\leqslant100}(\lambda_\xi/1000), \max_{1\leqslant\xi\leqslant100}(\lambda_\xi/1000)\right]$. Similarly, the annual generated power and water storage at the end of flood season were obtained through reservoir operation calculation by using the monthly inflow data of 41 years from 1971 to 2011. Then, the minimum and maximum values of the 41 results formed the interval numbers. The initial interval number decision matrix $\widetilde{X}$ is shown in Table 4.

Table 4. Initial interval number decision matrix $\widetilde{X}$.

Scheme i	Flood Control Risk Rate $\tilde{x}_{i1}$/ %	Annual Generated Power $\tilde{x}_{i2}$/10^8 kW·h	Water Storage at the End of Flood Season $\tilde{x}_{i3}$/10^8 m^3
1	[1.812,4.371]	[6.531,14.392]	[4.887,11.200]
2	[1.882,4.408]	[7.086,14.983]	[5.411,11.200]
3	[1.993,4.421]	[7.687,15.639]	[6.033,11.200]
4	[2.211,4.689]	[8.275,16.136]	[6.751,11.200]
5	[2.534,5.028]	[8.699,16.641]	[7.523,11.200]
6	[2.977,5.594]	[9.320,17.343]	[8.431,11.200]

According to Equation (11)–(12), the benefit indexes (annual generated power, water storage at the end of flood season) and the cost index (flood control risk rate) in Table 3 are standardized, resulting in the standardized interval number decision matrix $\widetilde{B}$ shown in Table 5.

Table 5. Standardized interval number decision matrix $\widetilde{B}$.

Scheme i	Flood Control Risk Rate $\tilde{b}_{i1}$	Annual Generated Power $\tilde{b}_{i2}$	Water Storage at the End of Flood Season $\tilde{b}_{i3}$
1	[0.323,1.000]	[0.000,0.727]	[0.000,1.000]
2	[0.314,0.981]	[0.051,0.782]	[0.083,1.000]
3	[0.310,0.952]	[0.107,0.842]	[0.182,1.000]
4	[0.239,0.895]	[0.161,0.888]	[0.295,1.000]
5	[0.150,0.809]	[0.201,0.935]	[0.418,1.000]
6	[0.000,0.692]	[0.258,1.000]	[0.561,1.000]

According to Lynne [40] who derives the variance formula of interval number sample matrix based on uniform distribution, for $\widetilde{B} = \left[\tilde{b}_{ij}\right]_{m\times n}$, its variance $D(\tilde{b}_j)$ can be expressed as follows:

$$D(\tilde{b}_j) = \frac{1}{3m}\sum_{i=1}^{m}\left[(b_{ij}^L)^2 + b_{ij}^L x_{ij}^U + (b_{ij}^U)^2\right] - \frac{1}{4m^2}\left[\sum_{i=1}^{m}(b_{ij}^L + b_{ij}^U)\right]^2, \tag{24}$$

Based on the definition of the correlation coefficient of two random real number variables in the probability theory and mathematical statistics, the correlation coefficient of two interval variables $\rho(\tilde{b}_h, \tilde{b}_j)$ $(h, j=1,2,\ldots,n)$ is:

$$\rho(\tilde{b}_h, \tilde{b}_j) = \frac{\mathrm{Cov}(\tilde{b}_h, \tilde{b}_j)}{\sqrt{D(\tilde{b}_h,)}\sqrt{D(\tilde{b}_j)}} = \frac{D(\tilde{b}_h + \tilde{b}_j) - D(\tilde{b}_h) - D(\tilde{b}_j)}{2\sqrt{D(\tilde{b}_h)}\sqrt{D(\tilde{b}_j)}}, \tag{25}$$

From Equation (24)–(25), the correlation coefficient matrix of $\widetilde{B} = \left[\tilde{b}_{ij}\right]_{m\times n}$ can be obtained, as shown in Table 6.

Table 6. Correlation coefficient matrix of the standardized interval number decision matrix.

Index	Flood Control Risk Rate $\tilde{b}_1$	Annual Generated Power $\tilde{b}_2$	Water Storage at the End of Flood Season $\tilde{b}_3$
Flood control risk rate $\tilde{b}_1$	1.000	0.613	0.573
Annual generated power $\tilde{b}_2$	0.613	1.000	0.971
Water storage at the end of flood season $\tilde{b}_3$	0.573	0.971	1.000

It is shown in Table 5 that there is positive correlation among the three indexes. The correlation between the flood control risk rate and annual generated power or water storage at the end of flood season is slightly weak. The correlation between annual generated power and water storage at the end of flood season is significant in that their correlation coefficient is 0.971. Thus, it is necessary to consider the correlation between the indexes in the decision-making process of the reservoir water level scheme in flood season.

There are five experts to grade the importance of the indexes, and the subjective weight of the interval number indexes obtained from the scoring matrix is $\widetilde{S}$ =

$\{[0.320, 0.397]\ [0.292, 0.360]\ [0.235, 0.300]\}$. According to Equation (13), the information entropy weight is $t_j^L = (0.252, 0.347, 0.401)$ and $t_j^U = (0.573, 0.427, 0)$. Thus, the objective weight of the interval number indexes is $\widetilde{T} = \{[0.252, 0.573]\ [0.347, 0.427]\ [0, 0.401]\}$. Let the empirical factor β be 0.5, then the combined weight is $\widetilde{W} = \{[0.286, 0.485]\ [0.320, 0.394]\ [0.118, 0.351]\}$. From $\widetilde{B}$ and $\widetilde{W}$, the weighted standardized decision matrix $\widetilde{C}$ is obtained, as shown in Table 7.

Table 7. Weighted standardized interval number decision matrix $\widetilde{C}$.

Scheme i	Flood Control Risk Rate $\tilde{c}_{i1}$	Annual Generated Power $\tilde{c}_{i2}$	Water Storage at the End of Flood Season $\tilde{c}_{i3}$
1	[0.092, 0.485]	[0.000, 0.286]	[0.000, 0.351]
2	[0.090, 0.476]	[0.016, 0.308]	[0.010, 0.351]
3	[0.089, 0.462]	[0.034, 0.332]	[0.021, 0.351]
4	[0.068, 0.434]	[0.052, 0.350]	[0.035, 0.351]
5	[0.043, 0.392]	[0.064, 0.368]	[0.049, 0.351]
6	[0.000, 0.336]	[0.083, 0.394]	[0.066, 0.351]

Through the matrix $\widetilde{C}$, based on TOPSIS and Equation (15), the positive (negative) ideal scheme $\widetilde{p}_+$ ($\widetilde{p}_-$) is determined as follows:

$$\widetilde{p}_+ = \{[0.092, 0.485]\ [0.083, 0.394]\ [0.066, 0.351]\}$$
$$\widetilde{p}_- = \{[0.000, 0.336]\ [0.000, 0.286]\ [0.000, 0.351]\}.$$

5.2. Orthogonal Test of the Schemes

Since there are three indexes, the orthogonal table is $L_4(2^3)$. The alternative scheme's layout matrix C_i (i=1,2, … ,6), the positive ideal scheme's layout matrix P_+ and the negative ideal scheme's layout matrix P_- are shown in Table 8.

Table 8. Layout matrixes of the schemes.

Matrix	Flood Control Risk Rate	Annual Generated Power	Water Storage at the End of Flood Season	Matrix	Flood Control Risk Rate	Annual Generated Power	Water Storage at the End of Flood Season
C_1	0.092	0.000	0.000	C_5	0.043	0.064	0.049
	0.092	0.286	0.351		0.043	0.368	0.351
	0.485	0.000	0.351		0.392	0.064	0.351
	0.485	0.286	0.000		0.392	0.368	0.049
C_2	0.090	0.016	0.010	C_6	0.000	0.083	0.066
	0.090	0.308	0.351		0.000	0.394	0.351
	0.476	0.016	0.351		0.336	0.083	0.351
	0.476	0.308	0.010		0.336	0.394	0.066
C_3	0.089	0.034	0.021	P_+	0.092	0.083	0.066
	0.089	0.332	0.351		0.092	0.394	0.351
	0.462	0.034	0.351		0.485	0.083	0.351
	0.462	0.332	0.021		0.485	0.394	0.066
C_4	0.068	0.052	0.035	P_-	0.000	0.000	0.000
	0.068	0.350	0.351		0.000	0.286	0.351
	0.434	0.052	0.351		0.336	0.000	0.351
	0.434	0.350	0.035		0.336	0.286	0.000

5.3. Scheme Decision-Making and Result Evaluation

The square Mahalanobis distance from each point in the layout matrix C_i to the positive ideal scheme's layout matrix P_+ and the negative ideal scheme's layout matrix P_- is worked out, as shown in Table 9.

Table 9. Square Mahalanobis distance.

Scheme i	$d^2[c_i^{(g)}, P_+]$				$d^2[c_i^{(g)}, P_-]$			
1	4.139	1.567	3.263	2.428	1.658	1.649	4.166	4.175
2	3.774	1.662	2.966	2.290	1.424	1.906	3.858	4.190
3	3.386	1.790	2.630	2.155	1.187	2.222	3.478	4.192
4	3.152	2.076	2.238	1.909	1.054	2.583	2.930	3.936
5	3.071	2.436	1.901	1.668	1.037	3.018	2.308	3.583
6	3.132	3.112	1.542	1.543	1.177	3.806	1.628	3.355

The square Mahalanobis distance is standardized by Equation (18) and the SNR indicator is obtained. The indicator of the degree of balance and approach is worked out by Equation (19)–(22). Accordingly, the decision matrix X is derived, as shown in Table 10.

Table 10. Derivative indicator decision matrix X.

Scheme i	η_{+i}	η_{-i}	γ_{+i}	γ_{-i}
1	2.998	3.010	0.753	0.652
2	2.829	2.894	0.775	0.660
3	2.508	2.784	0.801	0.669
4	1.869	2.633	0.824	0.686
5	2.429	2.446	0.841	0.705
6	2.984	3.010	0.838	0.710

In accordance with Equation (11)–(12), X is standardized and with the weight of each index being 0.25, the weighted standardized decision matrix C is obtained, as shown in Table 11.

Table 11. Weighted standardized derivative indicator decision matrix C.

Scheme i	η^*_{+i}	η^*_{-i}	γ^*_{+i}	γ^*_{-i}
1	0.000	0.250	0.000	0.250
2	0.037	0.199	0.063	0.216
3	0.109	0.150	0.136	0.177
4	0.250	0.083	0.202	0.103
5	0.126	0.000	0.250	0.022
6	0.003	0.250	0.241	0.000

$p = [0.25, 0.25, 0.25, 0.25]$ is taken as the reference scheme to calculate the mapping distance MD_i from the alternative scheme to the reference scheme.

To verify the feasibility and effectiveness of the method in this paper, the first comparison method of scheme ranking (method 1) is based on the closeness degree of the SNR defined by Equation (26), according to reference [40]:

$$\eta_i = \frac{\eta_{-i}}{\eta_{+i} + \eta_{-i}}, \tag{26}$$

On the basis of TOPSIS, the improved closeness degree of the degree of balance and approach is defined as follows:

$$\gamma_i = \frac{\gamma_{+i}}{\gamma_{+i} + \gamma_{-i}}, \tag{27}$$

When $\gamma_{-i} \to 0$, $\gamma_i \to 1$, the closer scheme i is to the positive ideal scheme, whereas when $\gamma_{+i} \to 0$, $\gamma_i \to 0$, the farther scheme i is to the positive ideal scheme. This ranking criterion is that the larger γ_i is, the better the corresponding scheme i is, which is the second method for comparison (method 2).

The schemes are sorted according to the different ranking criteria of method 1, method 2 and the method in this study. The results are shown in Table 12.

Table 12. Results of scheme ranking.

Scheme i	Method 1		Method 2		This Study'S Method	
	η_i	Ranking	γ_i	Ranking	MD_i	Ranking
1	0.5010	6	0.5359	6	0.2500	6
2	0.5057	3	0.5401	5	0.1585	3
3	0.5260	2	0.5449	2	0.0492	1
4	0.5848	1	0.5457	1	0.1378	2
5	0.5018	5	0.5440	3	0.1983	4
6	0.5021	4	0.5413	4	0.2443	5

It can be seen from Table 11 that Scheme 1 ranks the last in the decision results of all the three methods, meaning that the currently-adopted strategy is far from decent and other schemes need to be selected, which conforms to the status quo of the Pankou reservoir. This study's method chooses Scheme 3 as the optimal scheme, which raises the upper water level limit in flood season by 0.8 m compared with Scheme 1.

The reasons for the inconsistent scheme ranking results can be explained by the calculation process of the SNR and the improved degree of balance and approach in Section 4.2. The two indicators respectively reflect the output strength and the degree of balance and approach of geometric curves in the alternative and the reference schemes. If method 1 or method 2 is used alone, their decision results will be less likely to be adopted. The method proposed in this paper includes the advantages of both method 1 and method 2, and gets closer to the reference scheme by the mapping distance, producing more accurate and reasonable decision results. Therefore, scheme 3 is recommended as the optimal.

In Table 11, the closeness degree of the SNR and the closeness degree of the balance and approach degree are benefit indicators, whereas the mapping distance is a cost indicator, and therefore converted into a benefit indicator by the range method. After normalizing the three indicators, the decision results of the three methods are illustrated in Figure 6. It can be seen that compared with method 1 and method 2, this study's method produces more obviously distinct decision results, which proves that the proposed model can effectively mine the hidden rules in data, especially for the analysis of a system with deficient information.

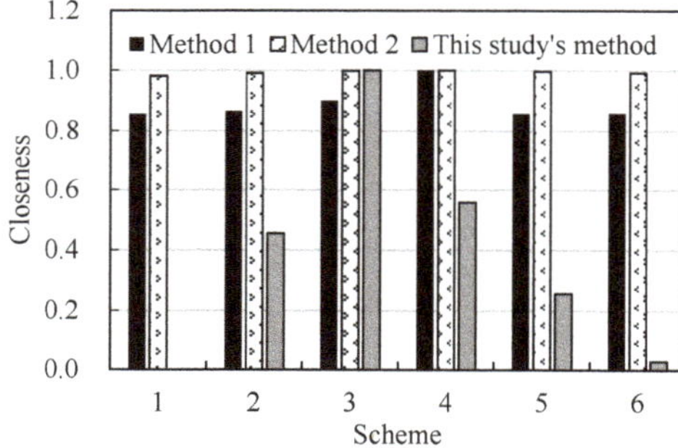

Figure 6. Comparison of closeness degree of each alternative scheme to the reference scheme.

6. Conclusions

In order to reduce the uncertainty of interval numbers as well as computation work and fully mine the implied information in multi-attribute decision-making, a multi-dimensional interval number decision model based on Mahalanobis-Taguchi System with grey entropy method (MTS-GEM) was proposed and verified by a case study. The principal conclusions are as follows.

1. MTS-GEM can effectively reduce the uncertainty created by interval numbers. In the model, the bounded uncertain n-dimensional interval number is expressed quantitatively as a hypercube in the n-dimensional space. Meanwhile, the alternative and reference schemes are all transformed into finite vertices of the hypercube, which realizes the transformation from an interval number decision vector to a real number decision vector.

2. MTS-GEM can produce markedly distinctive decision results, which demonstrates the sufficiency of decision information contained in the model. The model not only considers the output response strength and the degree of balance and approach between alternative and reference schemes, but also uses the idea of further approaching the reference scheme by mapping distance, which elevates the accuracy and reliability of the decision results.

3. The case study of selecting the optimal scheme of controlling the Pankou reservoir's water level in flood season shows that the proposed method can pick out the best scheme that better coordinates risk and benefit, which further proves the comprehensive and excellent decision-making performance of the model.

Author Contributions: Conceptualization, C.J. and X.L.; Data curation, X.Y.; Formal analysis, X.L., Y.Z. and X.Y.; Methodology, C.J., X.L. and Y.P.; Validation, X.L., Y.P. and Y.Z.; Writing—original draft, X.L. and J.W.; Writing—review and editing, J.W. All authors have read and agreed to the published version of the manuscript.

Funding: This research received no external funding.

Acknowledgments: This study was financially supported by the National Key Research and Development Program of China (2016YFC0402309), the National Natural Science Foundation of China (51709105) and the Fundamental Research Funds for the Central Universities (2019MS031). The authors are grateful to the reviewers for their valuable comments and constructive suggestions.

Conflicts of Interest: The authors declare no conflict of interest.

References

1. Xu, Z. *Uncertain Multiple Attribute Decision Making: Methods and Applications*; Tsinghua University Press: Beijing, China, 2004.
2. Yoon, K.; Hwang, C.L. *Multiple Attribute Decision Making: Methods and Applications*; Springer: Berlin, Germany, 1981.
3. Liu, Y.; Wang, W. Set Pair Analysis and its application to optimization of standard schemes of urban flood control. *J. North China Univ. Water Res. Electr. Power (Nat. Sci.)* **2018**, *39*, 77–80.
4. Tu, E.Q.; Liao, X.Y.; Chen, Z.J.; Wang, H.M. Effects of rainfall on runoff from a small watershed in Three Gorges Reservoir area. *Bull. Soil Water Conserv.* **2010**, *30*, 7–11.
5. Chen, S.; Li, Z.; Xu, Q. Grey target theory based equipment condition monitoring and wear mode recognition. *Wear* **2006**, *260*, 438–449. [CrossRef]
6. Wang, W.; Li, Y.; Jin, J.; Ding, J. *Set Pair Analysis for Hydrology and Water Resources Systems*; Science Press: Beijing, China, 2010.
7. Gao, F. Possibility degree and comprehensive priority of interval numbers. *Syst. Eng. Theor. Pract.* **2013**, *33*, 2033–2040.
8. Hou, J.; Li, Y.; Chi, H. An new method of interval numbers ordering in uncertain multiple attribute decision-making. *Math. Pract. Theor.* **2017**, *47*, 24–27.
9. Li, L. Research on Multi-Attribute Decision Making of Interval-Valued Information System based on Rough Set. Master's Thesis, Nanjing University of Aeronautics and Astronautics, Nanjing, China, 1 March 2014.
10. Mohammad, K.S.; Majeed, H.; Kamran, S. Extension of VIKOR method for decision making problem with interval numbers. *Appl. Math. Model.* **2009**, *33*, 2257–2262.
11. Deng, J. *Grey Theory Basis*; Huazhong University of Science & Technology Press: Wuhan, China, 2002.
12. Wu, Q. *System Engineering*; Beijing Institute of Technology Press: Beijing, China, 2015.
13. Gao, X.; Zhang, Q. Gray relational grade rule of uncertain decision making. *Oil-Gas Field Surf. Eng.* **1995**, *14*, 57–58.
14. Zhang, Q.; Li, X.; Deng, J. Grey entropy method of uncertain decision making. *J. Manag. Sci.* **1995**, *6*, 37–39.

15. Shang, L.; Tan, Q. Emergency logistics supplier evaluation based on grey entropy model. *Stats. Decis.* **2013**, *3*, 45–47.
16. Fang, Y.; Li, G.; Du, L. Intercity rail transit system decision-making method based on grey entropy. *Syst. Eng.* **2015**, *33*, 152–158.
17. Guo, Q.; Li, W. Algorithm of tourist routes based on grey entropy decision model. *Comput. Eng. Desig.* **2017**, *38*, 1988–1991.
18. Xia, J. *Grey System Hydrology*; Huazhong University of Science and Technology Press: Wuhan, China, 1998.
19. Wang, P.; Fang, Z.; Liu, J.; Chi, Y. Study on grey distance entropy model with application in multiple attribute decision making. In Proceedings of the 2009 IEEE International Conference on Grey Systems and Intelligent Services, Nanjing, China, 10–12 November 2009; pp. 1051–1054.
20. Liu, R.; Gao, X.; Zhang, G. Research on the uncertain multi-attribute decision making methods and application based on grey entropy model with interval-type attribute values and weights. *Control Decis.* **2018**, 1–11.
21. Taguchi, G.; Jugulum, R. New Trends in Multivariate Diagnosis. *Sankhyā Indian J. Stat. Ser. B* **2000**, *62*, 233–248.
22. Wang, Z.; Lu, C.; Wang, Z.; Liu, H.; Fan, H. Fault diagnosis and health assessment for bearings using the Mahalanobis–Taguchi system based on EMD-SVD. *Trans. Inst. Meas. Control* **2013**, *35*, 798–807. [CrossRef]
23. Buenviaje, B.; Bischoff, J.E.; Roncace, R.A.; Willy, C.J. Mahalanobis–Taguchi System to Identify Preindicators of Delirium in the ICU. *IEEE Trans. Inf. Technol. B.* **2016**, *20*, 1205–1212. [CrossRef] [PubMed]
24. Huang, C.; Hsu, T.; Liu, C. The Mahalanobis–Taguchi system—Neural network algorithm for data-mining in dynamic environments. *Expert Syst. Appl.* **2009**, *36*, 5475–5480. [CrossRef]
25. Zeng, W.; Hao, Y.; Fan, R.; Yuan, J.; Li, C. Risk decision-making with intervals based on Mahalanobis-Taguchi System and grey cumulative prospect theory for maintenance of power transformers. *J. North China Electr. Power Univ.* **2015**, *42*, 100–110.
26. Chang, Z.; Cheng, L.; Liu, J. Multiple attribute decision making method with intervals on Mahalanobis-Taguchi System and TOPSIS method. *Syst. Eng. Theor. Pract.* **2014**, *34*, 168–175.
27. Zhang, L. *Orthogonal Method and Applied Mathematics*; Science Press: Beijing, China, 2009.
28. Mahalanobis, P.C. On the generalized distance in statistics. *P. Indian Natl. Sci. A.* **1936**, *2*, 49–55.
29. Chang, Z. Research progress of Mahalanobis-Taguchi system. *Control Decis.* **2019**, *34*, 2505–2516.
30. Zhou, Z.; Xing, Y.; Sun, H.; Cai, Y.; Yu, X. Interval game analysis of government subsidies on haze governance in Beijing-Tianjin-Hebei. *Syst. Eng. Theor. Pract.* **2017**, *37*, 2640–2648.
31. Wang, W. *The Design of the Test and Analysis*; Higher Education Press: Beijing, China, 2004.
32. Zhang, Q.; Qin, H.; Deng, J. New definition of grey number's grey grade. *J. Northeast Petrol. Univ.* **1996**, *20*, 89–92.
33. Zhang, Q.; Guo, X.; Deng, J. Grey relation entropy method of grey relation analysis. *Syst. Eng. Theor. Pract.* **1996**, *8*, 7–11.
34. Wu, N.; Yuan, S. *Maximum Entropy Method*; Hunan Science & Technology Press: Changsha, China, 1991.
35. Li, C.; Zhang, F.; Zhu, T.; Feng, T.; An, P. Evaluation and correlation analysis of land use performance based on entropy-weight TOPSIS method. *T. Chin. Soc. Agric. Eng.* **2013**, *29*, 217–227.
36. Peng, Y.; Ji, C.; Liu, F. Multi-objective optimization model for coordinative dispatch of water and sediment in cascade reservoirs. *J. Hydraul. Eng.* **2013**, *44*, 1272–1277.
37. Zhang, P.; Ji, C. In *Multi-Dimensional Vector Space Decision Model and Its Application for Multi-Objective Operation Risk Analysis of Reservoirs, Proceedings of the World Environmental and Water Resources Congress 2017: International Perspectives, History and Heritage, Emerging Technologies, and Student Papers, Sacramento, CA, USA, 21–25 May 2017*; Dunn, C.N., Van Weele, B., Eds.; Amer Soc Civil Engineers: New York, NY, USA, 2017; pp. 184–191.
38. Wang, L.; Yan, X.; Wang, B.; Yu, H.; Ji, C. Interval number grey target decision-making model based on multi-dimensional association sampling and its application. *Syst. Eng. Theor. Pract.* **2019**, *39*, 1610–1622.
39. Xiao, Y.; Guo, S.; Luo, Z.; Xiong, L.H.; Li, W. Estimation of hydrological risk rate for dams. *Eng. J. Wuhan Univ.* **2005**, *38*, 10–13.
40. Lynne, B.; Edwin, D. *Symbolic Data Analysis: Conceptual Statistics and Data Mining*; John Wiley & Sons Inc.: Chichester, UK, 2006.

Article

Evaluation of GloFAS-Seasonal Forecasts for Cascade Reservoir Impoundment Operation in the Upper Yangtze River

Kebing Chen [1], Shenglian Guo [1,*], Jun Wang [1], Pengcheng Qin [2], Shaokun He [1], Sirui Sun [3] and Matin Rahnamay Naeini [4,*]

1   State Key Laboratory of Water Resources and Hydropower Engineering Science, Wuhan University, Wuhan 430072, China; chenkb@whu.edu.cn (K.C.); wangjwd@whu.edu.cn (J.W.); he_shaokun@whu.edu.cn (S.H.)
2   Wuhan Regional Climate Centre, Hubei Meteorology Bureau, Wuhan 430074, China; qinpengcheng027@163.com
3   Middle Changjiang River Bureau of Hydrology and Water Resources Survey, Wuhan 430012, China; threesun123@icloud.com
4   Center for Hydrometeorology and Remote Sensing (CHRS) & Department of Civil and Environmental Engineering, University of California, Irvine, CA 92697, USA
*   Correspondence: slguo@whu.edu.cn (S.G.); rahnamam@uci.edu (M.R.N.)

Received: 1 November 2019; Accepted: 28 November 2019; Published: 1 December 2019

Abstract: Standard impoundment operation rules (SIOR) are pre-defined guidelines for refilling reservoirs before the end of the wet season. The advancement and availability of the seasonal flow forecasts provide the opportunity for reservoir operators to use flexible and early impoundment operation rules (EIOR). These flexible impoundment rules can significantly improve water conservation, particularly during dry years. In this study, we investigate the potential application of seasonal streamflow forecasts for employing EIOR in the upper Yangtze River basin. We first define thresholds to determine the streamflow condition in September, which is an important period for decision-making in the basin, and then select the most suitable impoundment operation rules accordingly. The thresholds are used in a simulation–optimization model to evaluate different scenarios for EIOR and SIOR by multiple objectives. We measure the skill of the GloFAS-Seasonal forecast, an operational global seasonal river flow forecasting system, to predict streamflow condition according to the selected thresholds. The results show that: (1) the 20th and 30th percentiles of the historical September flow are suitable thresholds for evaluating the possibility of employing EIOR; (2) compared to climatological forecasts, GloFAS-Seasonal forecasts are skillful for predicting the streamflow condition according to the selected 20th and 30th percentile thresholds; and (3) during dry years, EIOR could improve the fullness storage rate by 5.63% and the annual average hydropower generation by 4.02%, without increasing the risk of flooding. GloFAS-Seasonal forecasts and early reservoir impoundment have the potential to enhance hydropower generation and water utilization.

Keywords: Yangtze River; cascade reservoirs; impoundment operation; GloFAS-Seasonal; forecast evaluation

1. Introduction

The rapid population and economic growth in recent decades, along with climate change and variability, impose more stress on water resources and cascade reservoir systems. Reservoirs, as one of the most important components of the hydrologic system, play a significant role as water supply by altering natural streamflow across space and time [1,2], while mitigating the effect of extreme events [3,4]. Conventionally, during wet season, reservoir operators release water preferentially for flood control [5,6] while storing water before the end of wet season to meet the demand for hydropower generation, navigation, and water supply. For full replenishment of storage, reservoir operators and academic researchers have highlighted the importance of reservoir impoundment operation and impoundment rules in several studies [7–10].

Reservoir operation rules are often used to provide guidelines for reservoir operators to determine the amount of controlled discharge. Among these rules, the reservoir impoundment rules are designed to refill the reservoir and raise the storage water level. The New York City rule [11] is among the early guidelines for reservoir impoundment and provides the probability of spills rather than the amount of spill to minimize the water shortage [12]. Since the development of the New York City rule, different types of reservoir impoundment operation rules have been developed and employed for various reservoir systems [7,13–17]. In the Yangtze River, which is one of the largest rivers in the world by discharge volume with huge reservoir storage capacity, the impoundment operation is complex and challenging. Water managers and stakeholders employ predefined impoundment rules for reservoir systems. These fixed rules, so-called standard impoundment operation rules (SIOR), are derived based on the historical flow records [7,9,18]. The SIOR is designed to reduce the flood control risk during the impoundment period. However, these fixed rules are unable to fully replenish the storage of the reservoir during dry years, which could lead to water shortage. On the contrary, flexible early impoundment operation rules (EIOR) allow the reservoir operators to start the impoundment process earlier and avoid unnecessary spills. However, employing EIOR requires information about the streamflow forecast to alleviate the risk of flooding.

The recent advancements of the meteorological and hydrological forecast systems provide an unprecedented opportunity for employing flexible operation rules rather than fixed ones for reservoir systems [19–21]. Combining seasonal meteorological forecasts with hydrological models at continental-scale has provided several continental-scale seasonal hydro-meteorological forecasting systems [22–24], such as the European Flood Awareness System [25], the Australian Government Bureau of Meteorology Seasonal Streamflow Forecasts [26], and the National Hydrologic Ensemble Forecast Service, USA [27]. Several studies have demonstrated that a skillful streamflow forecast can enhance the efficiency of water allocation systems to manage the trade-off between hydropower, irrigation, municipal, and environmental services [28–31]. The potential for employing seasonal forecast in the Yangtze River basin has been investigated in several research studies, mostly through statistical techniques [32–34]. However, the potential application of the available seasonal forecasts for reservoir impoundment operation is not well understood in the Yangtze River basin. In this study, we evaluate the global seasonal river flow forecasting system (GloFAS-Seasonal) developed by the European Centre for Medium-Range Weather Forecasts (ECMWF) [35] for reservoir impoundment in the upper Yangtze River basin.

We follow two steps for our evaluation. First, we investigate different streamflow thresholds to evaluate the possibility of employing EIOR. These thresholds can be considered as an indicator for the dry condition which has adverse effects on reservoir impoundment operation. To find these thresholds, we analyze multiple impoundment operation scenarios for EIOR and SIOR using a simulation–optimization model. We test different percentiles of historical streamflow as thresholds to find the impoundment rule curves for these scenarios. These rule curves are derived using the non-dominated sorting genetic algorithm-II (NSGA-II) [36] through a multi-objective optimization process. We then analyze the objective function values to select the most suitable thresholds for employing EIOR or SIOR. Second, we apply these thresholds to evaluate the skill of the GloFAS-Seasonal streamflow forecast for selecting EIOR or SIOR. When the streamflow forecast is below the selected threshold, it means a dry condition. Hence, the EIOR provides a longer impoundment period during this condition and a more suitable impoundment approach for water conservation. We employ different scores to show the skill of the GloFAS-Seasonal streamflow forecast to detect the streamflow condition according to selected thresholds.

The rest of the paper is organized as follows: In Section 2, after introducing the study area, we build a cascade reservoir impoundment model for reservoir early refill operation. Section 3 reviews the GloFAS-Seasonal forecasting system and explains different measures to evaluate the skill of the forecast for streamflow conditions. We demonstrate and discuss the results for the streamflow thresholds and performance of the GloFAS-Seasonal forecast in Section 4. Finally, we draw the conclusion in Section 5.

2. Cascade Reservoir Impoundment Model

2.1. Study Area

The Yangtze River, the longest river in Asia, flows 6300 km to the East China Sea with a total drainage area of 1.8 million km^2 and has abundant hydropower resources. A series of cascade reservoirs have been constructed along the upper Yangtze River which provides a wide range of services including flood control, hydropower generation, water supply, as well as navigation. There are five cascade reservoirs in the upper Yangtze River, WDD (Wu-Dong-De), BHT (Bai-He-Tan), XLD (Xi-Luo-Du), XJB (Xiang-Jia-Ba), and TGR (Three Gorges Reservoir). These reservoirs, along with their characteristics, are listed in Table 1. There are no main tributaries between WDD and XJB reservoirs, while there are three main tributaries between XJB and TGR, Min River, Jia-Ling River, and Wu River. The inflow to WDD (Q_{WDD}) and TGR (Q_{TGR}) are derived from gauges at Hua-Tan and Yi-Chang hydrological stations by revivification, respectively. Figure 1 shows the sketch map of the cascade reservoirs, hydrological stations, and tributaries in the upper Yangtze River basin.

Table 1. Characteristics of the five cascade reservoirs in the upper Yangtze River.

Reservoir	Basin Area (Thousand km^2)	Annual Top of Buffer Pool (m)	Top of Conservation Pool (m)	Total Storage Capacity (Billion m^3)	Storage for Flood Control (Billion m^3)	Installed Hydropower Capacity (GW)
WDD	406.1	952	975	3.94	2.44	10.20
BHT	430.3	785	825	20.60	7.50	16.00
XLD	454.4	560	600	12.67	4.65	13.86
XJB	458.8	370	380	5.16	0.90	7.75
TGR	1000	145	175	45.07	22.15	22.50

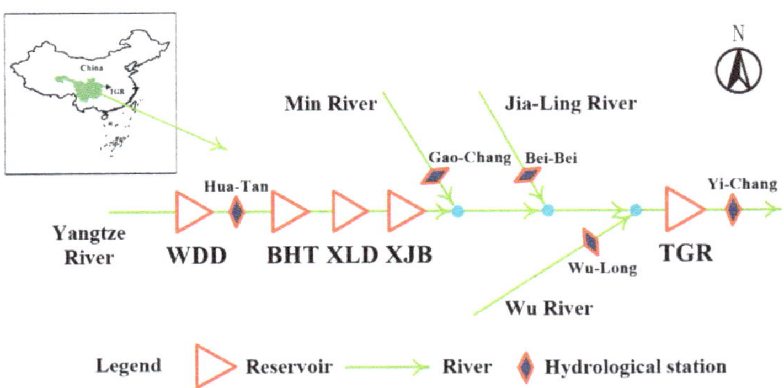

Figure 1. Sketch map of the five cascade reservoirs in the upper Yangtze River.

2.2. Impoundment Operation Rules for Cascade Reservoirs

The impoundment operation rules are employed to refill reservoir storage during the impoundment period. The impoundment operation rules (Figure 2c) delineate trajectories to raise the water level from the annual top of buffer pool at the initial impoundment time to the top of conservation pool by the end of impoundment period. The SIOR derived from historical flow records initiates the impoundment operation at fixed predefined dates. However, the SIOR may fail to refill reservoir during the impoundment period in dry years. Hence, in low flow conditions and dry years, an early impoundment operation is more desirable to refill the storage capacity. Table 2 lists the potential time for employing early initial impoundment in the upper Yangtze River obtained from previous investigations [7,9]. We employ these initial dates along with the inflow conditions for the WDD and TGR reservoirs (Q_{WDD} and Q_{TGR}) to evaluate the possibility of an early impoundment for the cascade reservoirs.

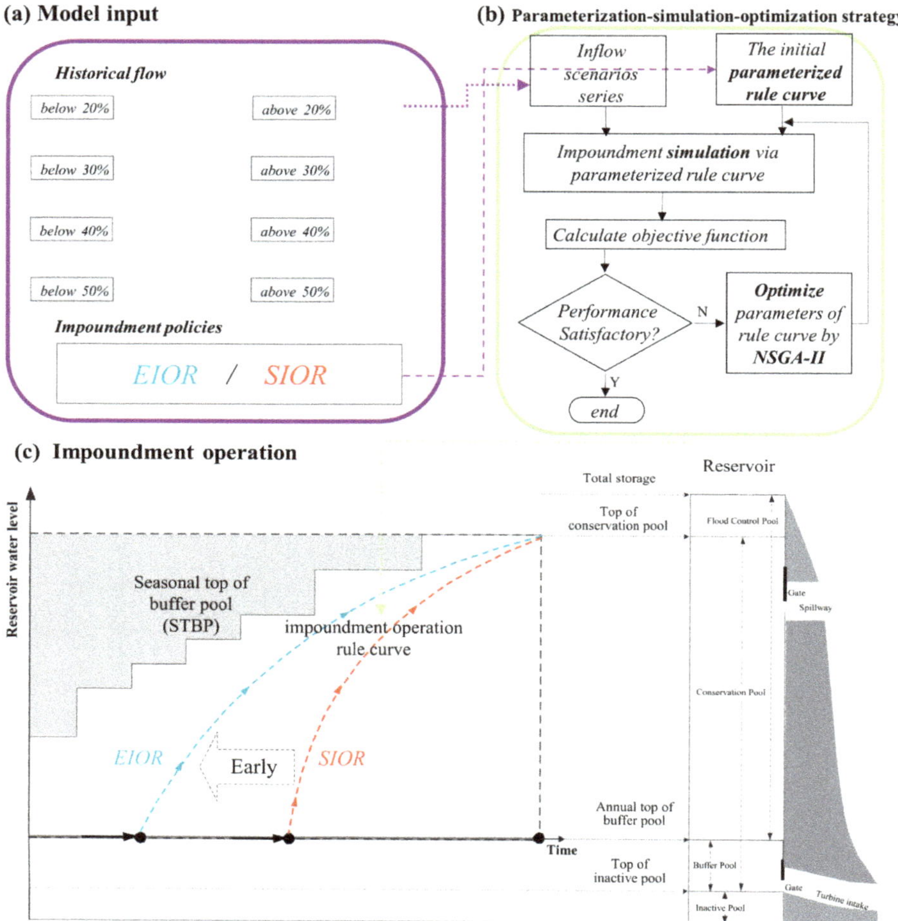

Figure 2. The flowchart of the parameterization–simulation–optimization (PSO) strategy to derive reservoir impoundment operation rule curves (**a**) input data of PSO, (**b**) optimization strategy of PSO, (**c**) concept of the seasonal top of buffer pool (STBP) and impoundment operation rules, similar to a previous study [9].

Table 2. Impounding periods of Standard impoundment operation rules (SIOR) and early impoundment operation rules (EIOR) for the five reservoirs.

Reservoir	Initial Impoundment Time		Final Impoundment Time
	(SIOR)	(EIOR)	(SIOR, EIOR)
WDD	Aug. 10th	Aug.1st	Sep. 10th
BHT	Aug. 10th	Aug.1st	Sep. 30th
XLD	Sep. 10st	Aug. 25th	Sep. 30th
XJB	Sep. 10st	Aug. 25th	Sep. 30th
TGR	Sep. 15th	Aug. 25th	Oct. 31st

Employing EIOR without considering the seasonal top of buffer pool (STBP) could increase the risk of flooding. Figure 2c shows the concept of the STBP and Table 3 lists the value of the STBP for selected reservoir (calculation method obtained from previous studies [7,9]). STBP is employed as the maximum water level to mitigate the risk of flooding for the impoundment period. We follow an iterative process to find the STBP for the reservoirs by evaluating the most extreme event. For further information for this process, please refer to [7,9]. We use the STBP in our model to control and assess the risk of flooding for the selected reservoirs.

Table 3. Seasonal top of buffer pool (STBP) for the five reservoirs in different periods.

Reservoir	Seasonal Top of Buffer Pool (m)					
	Aug.15th	Aug.25th	Sep.1st	Sep.10th	Sep.30th	Oct.31st
WDD	965	965	970	975	975	975
BHT	800	810	810	810	825	825
XLD	560	565	575	575	600	600
XJB	370	372	375	375	380	380
TGR	145	145	152	152	165	175

According to Table 2, the impoundment process for most reservoirs starts before September. Hence, streamflow forecast data with 2-month lead time can be used in early August to evaluate the possibility of using EIOR. For this purpose, we define quantile-based thresholds based on historical September monthly inflow to the Wu-Dong-De and Three Gorges Reservoir (Q_{WDD} and Q_{TGR}). These thresholds are used to determine the streamflow condition to help decision-makers decide to use either EIOR or SIOR. For instance, if the September monthly Q_{WDD} is forecasted to be below the threshold percentile, EIOR is recommended as the suitable impoundment operation rule.

It is worth noting that, employing higher percentiles thresholds for inflow would increase the possibility of using EIOR. However, it also increases the risk of flooding. So, careful consideration should be devoted to the selection of these thresholds. Here, we examine four quantiles of historical monthly inflow in September, including 20, 30, 40, and 50-percentile (Figure 2a), for the Q_{WDD} and Q_{TGR} to select the best thresholds. For instance, considering the 20-percentile threshold of the historical monthly inflow in September, the observed inflow can fall into the above 20-percentile and below 20-percentile category. Then, we evaluate the potential benefit and risk for each of these thresholds by employing a cascade reservoirs impoundment simulation–optimization model under EIOR and SIOR scenarios. Since each of these thresholds divide the historical streamflow observation into two groups, we employ each group to find the impoundment rule curve separately. Hence, there are eight scenarios for the thresholds that need to be evaluated for each impoundment approach by the simulation–optimization model.

The reservoir simulation–optimization model is generally used to construct the rule curves by simulating the reservoir responses to predefined operating rules. Due to a large number of policies and constraints, mathematical optimization techniques can be used to identify the optimal operation rules by evaluating all possible alternatives [37]. The parameterization–simulation–optimization (PSO) approach is a popular and effective way of deriving optimal rule curves for cascade reservoirs [38]. Initially, PSO employs a linear rule curve (impoundment operation rule curve shown in Figure 2c), which connects the annual top of the buffer pool to the top of the conservation pool. It then employs a heuristic strategy to find the optimal rule curve according to predefined objective functions under possible inflow scenarios. Figure 2 shows the scheme of the PSO approach [39]. Finally, the objective function values of the PSO are employed to select the best threshold for reservoir impoundment decision-making.

Here, we employ PSO at a daily timescale to find the optimum rule curve for each threshold and scenario (Figure 2b). By optimizing the parameters of the rule curves for SIOR and EIOR, decision-makers can decide to employ EIOR or SIOR based on the obtained objective function values. The objective functions and the constraints for the impoundment operation employed in the PSO model are discussed in Sections 2.3.1 and 2.3.2. In Section 2.4, we describe the NSGA-II algorithm employed for optimizing impoundment operation rule curve.

2.3. Objective Functions and Constrains

2.3.1. Objective Functions

Decision-makers rely on different criteria to make a comprehensive assessment of operation rules and address trade-offs among different users and services. In the Yangtze River, the goal of reservoir impoundment is to enhance water conservation in order to maximize hydropower generation and fullness storage rate, while minimizing the risk of flooding [7,18]. Hence, we employ objective functions that can measure the degree that these goals are achieved. These objectives are adopted from previous studies [7,18] and can be mathematically expressed as:

(1) Maximum hydropower generation (HG),

$$\max HG = \max\left[\frac{1}{N}\sum_{i=1}^{N}\left(\sum_{k=1}^{M} HG_{i,k}\right)\right]; \tag{1}$$

(2) Maximum fullness storage rate (FSR),

$$\max FSR = \max\left[\frac{1}{N}\sum_{i=1}^{N}\left(\sum_{k=1}^{M} \alpha_k FSR_{i,k}\right)\right], \tag{2}$$

$$FSR_{i,k} = \frac{V_{high,i}^k - V_{min}^k}{V_{max}^k - V_{min}^k} \times 100\%; \tag{3}$$

(3) Minimum flood control risk (R),

$$\min R = \min[\max(R_1, R_2, \cdots, R_k, \cdots, R_M)], \tag{4a}$$

$$R_k = P \text{ (Water level > STBP)} = N_{risk,k}/N; \tag{4b}$$

where

M	the number of reservoirs;
N	the number of years for hydrological time series;
$HG_{i,k}$	hydropower generation of the kth reservoir in the ith simulated year, kW·h
α_k	the weight for fullness storage rate of the kth reservoir, calculated by the ratio about the total storage capacity of M reservoirs;
$FSR_{i,k}$	fullness storage rate of the kth reservoir in the ith simulated year;
V^k_{min}	the storage capacity corresponding to the annual top of buffer pool of the kth reservoir, m³;
V^k_{max}	the storage capacity corresponding to the top of conservation pool of the kth reservoir, m³;
$V^k_{high,i}$	the highest storage during impoundment operation of the kth reservoir in the ith simulated year, m³;
R_k	flood control risk of the kth reservoir;
$N_{risk,k}$	the number of years when the water level exceeds STBP of the kth reservoir.

2.3.2. Operation Constraints

In addition to the objective functions, the constraints of the reservoir system need to be specified for the optimization process. The following equality and inequality operational constraints need to be satisfied in the cascade reservoirs impoundment operation. Adopted from previous studies [7,18,40], the mathematical formulations of these constraints are as follows:

(1) Water balance equation,

$$V^k_{i,j+1} = V^k_{i,j} + (Q^k_{in(i,j)} - Q^k_{out(i,j)})\Delta t \quad i = 1,\ldots,N \; j = 1,\ldots,T; \tag{5}$$

(2) Reservoir capacity,

$$V^k_{min} \leq V^k_{i,j} \leq V^k_{max} \quad i = 1,\ldots,N \; j = 1,\ldots,T; \tag{6}$$

(3) Power generation,

$$P^k_{min} \leq A^k Q^k_{o(i,j)} H^k_{i,j} \leq P^k_{max} \quad i = 1,\ldots,N \; j = 1,\ldots,T; \tag{7}$$

(4) Reservoir discharge,

$$Q^k_{min} \leq Q^k_{out(i,j)} \leq Q^k_{safe} \quad i = 1,\ldots,N \; j = 1,\ldots,T; \tag{8}$$

$$\left| Q^k_{out(i,j+1)} - Q^k_{out(i,j)} \right| \leq \Delta Q^k \quad i = 1,\ldots,N \; j = 1,\ldots,T \tag{9}$$

(5) Navigation,

$$Z^k_{dmin} \leq Z^k_{d(i,j)} \leq Z^k_{dmax} \quad i = 1,\ldots,N \; j = 1,\ldots,T; \tag{10}$$

$$Z^k_{d(i,j)} = f\left(Q^k_{out(i,j)}\right) i = 1,\ldots,N \; j = 1,\ldots,T; \tag{11}$$

where

N	the number of years for hydrological time series;
T	total number of days during impoundment operation in the ith simulated year;
$V_{i,j}^k$	the kth reservoir storage at the beginning of the jth day in the ith simulated year, m^3;
$Q_{in(i,j)}^k$	the kth reservoir inflow on the jth day in the ith simulated year, m^3/s;
$Q_{out(i,j)}^k$	the water discharge of kth reservoir on the jth day in the ith simulated year, m^3/s;
$Q_{o(i,j)}^k$	the water discharge for hydropower generation of the kth reservoir on the jth day in the ith simulated year, m^3/s;
$Q_{w(i,j)}^k$	the spilled water discharge of the kth reservoir on the jth day in the ith simulated year, m^3/s;
A^k	the hydropower generation efficiency of the kth reservoir;
$H_{i,j}^k$	the average hydropower head of the kth reservoir on the jth day in the ith simulated year, m;
P_{min}^k	the minimum power limits of the kth hydropower plant, kW;
P_{max}^k	the maximum power limits of the kth hydropower plant, kW;
Q_{min}^k	the minimum water discharge for downstream of the kth reservoir, m^3/s;
Q_{safe}^k	the maximum water discharge for flood control safety in downstream of the kth reservoir, m^3/s;
ΔQ^k	the maximum water discharge fluctuation of the kth reservoir, m^3/s;
Z_{dmin}^k	the minimum water level at downstream of the kth dam site, m;
Z_{dmax}^k	the maximum water level at downstream of the kth dam site, m;
$Z_{d(i,j)}^k$	the water level at downstream of kth dam site on the jth day in the ith simulated year, m;
$f(\cdot)$	the function provided by reservoir managers expressing the relationship between reservoir discharge and downstream water level.

2.4. NSGA-II Optimization Algorithm

The nonlinearity of the reservoir systems, along with the existing constraints, require an effective optimization algorithm to solve these types of problems [41]. Here, we employ the non-dominated sorting genetic algorithm-II (NSGA-II), which is a robust multi-objective optimization algorithm [36], to derive the parameters of the rule curves. The NSGA-II algorithm has been applied to a wide range of complex multi-objective reservoir optimization and water resources management problems [18,38,42–44].

The NSGA-II algorithm has four parameters, including population size, generation number, crossover rate, and mutation rate, that need to be tuned by the user. Population size and generation number determine the effectiveness and efficiency of the algorithm and control the convergence speed to the optimal non-dominated solutions. Crossover and mutation rates control the ability of the algorithm to perform an effective search over the problem space [38,45]. In this study, the population size and the generation number were set to 50 and 200, respectively. These values are selected based on trial and error to obtain reasonable non-dominated solutions with acceptable simulation time. The crossover and mutation rates were empirically set to 0.9 and 0.1, respectively. The non-dominated solutions are used to evaluate the three objective functions for each threshold and each of the EIOR and SIOR scenarios.

3. Evaluation of GloFAS-Seasonal Forecasts

GloFAS-Seasonal forecasts combine the ECMWF's latest seasonal meteorological forecasting system, SEAS5, and a river routing model, Lisflood, to provide streamflow forecasts at global scale [35]. This dataset provides weekly-averaged river flow with 4-month lead time. The first component of the GloFAS-Seasonal forecast is the meteorological input from SEAS5 which employs a data assimilation system along with a global circulation model. SEAS5 is executed once a month to produce seasonal weather forecasts with 7-month lead time. The second model component is a revised Hydrology Tiled ECMWF Scheme of Surface Exchanges over Land (HTESSEL) which computes the land surface response to atmospheric forcing and simulates the evolution of soil temperature, moisture content, and snowpack conditions through the forecast horizon to produce a corresponding forecast of surface and subsurface run-off [46]. The third model component is Lisflood which simulates the groundwater (subsurface water storage and transport) processes and routing of the water through the river network. While SEAS5 provides forecasts for the 7 months ahead, the GloFAS-Seasonal uses only the first 4 months and produces forecasts of river flow for the next 4 months. For more details on the forecast method, please refer to paper [35].

The GloFAS-Seasonal is a real-time forecast dataset which contains data from January 2018 and updates every month, with a total of 51 ensemble members. In order to evaluate the skill of the dataset, a set of retrospective seasonal forecasts for past dates, which are called reforecasts (also known as hindcasts), are available to compare with the historical observation streamflow. GloFAS-Seasonal reforecasts are available at http://www.globalfloods.eu/ and have 25 ensemble members from January 1981 to December 2017. In this study, GloFAS-Seasonal reforecasts at Hua-Tan and Yi-Chang hydrological stations in the Yangtze River are downloaded and analyzed. Also, the original weekly-averaged reforecasts are converted into monthly products for reservoir impoundment operation. Hence, monthly-averaged streamflow in September is obtained at the beginning of August with 2-month lead time (LM2).

We evaluate the GloFAS-Seasonal reforecasts to measure the capability of the dataset to predict the condition of the streamflow, i.e., the ability of the reforecast to predict that September monthly averaged flow falls below the selected thresholds which is defined in Section 2.1. Since seasonal climate is inherently probabilistic, seasonal forecasts should be evaluated probabilistically [47]. If each of the 25 ensemble members of the GloFAS-Seasonal reforecasts are equally likely, the proportion of ensemble members below each percentile threshold is calculated as the probability of the forecast. In addition, the percentile thresholds are calculated separately for historical observed and reforecast data [48]. This approach takes into account the systematic additive error (bias) of the reforecast data, hence further bias adjustment for the reforecast data is not required [48,49].

The conversion of raw ensemble members to forecast probabilities enables us to validate GloFAS-Seasonal reforecasts by using probabilistic forecasts verification measures. Here, we employ multiple metrics for our evaluation. These metrics include: I) discrimination, ability of the forecast to discriminate among observations; II) skill, the relative accuracy of the forecast over a reference forecast; III) reliability, the agreement between forecast probability and mean observed frequency; IV) resolution, the ability of the forecast to resolve the set of sample events into subsets; and V) sharpness, the tendency to forecast probabilities near 0 or 1. These metrics are briefly discussed here. Interested readers can refer to https://www.cawcr.gov.au/projects/verification/ for further details.

3.1. Discrimination

To assess the potential application of GloFAS-Seasonal forecasts for the prediction of the streamflow condition, the relative operating characteristic (ROC) curve, a measure of discrimination [50], is calculated for the selected thresholds. If the forecasts indicate that flow will be below threshold, which means a dry and unfavorable condition for reservoir impoundment operation, then a warning is issued. The forecasts are converted into a binary (e.g., "yes" or "no") format depending on whether a warning has been issued or not issued. Then the ROC curve is plotted based on hit rate (HR) and false-alarm rate (FAR) of the forecast for streamflow condition. The HR and FAR can be calculated by Equation (12):

$$\mathrm{HR} \ = \ \frac{h}{h \ + \ m} \tag{12a}$$

$$\mathrm{FAR} \ = \ \frac{f}{f \ + \ r} \tag{12b}$$

where h refers to a correct warning (hit), m refers to a missed warning, f refers to a false warning, and r correct no warning detection.

The area under the ROC curve (referred as AUC) is then calculated, which is used to measure whether the forecast is informative for decision-making. Most of the time, the ROC curve does not clearly indicate the accuracy of forecast. As a numerical value, it is more intuitive to use the AUC value as the evaluation standard. The larger the AUC value, the more skillful the forecast is. The value of the AUC ranges from 0 to 1. If the AUC is equal to 0.5, it indicates that forecasts are consistent with the random guess and provides no information. Generally, when the AUC value is greater than 0.6, the seasonal forecast can be regarded as useful [25,35].

3.2. Skill

Skill implies information about the relative accuracy of the forecast according to a reference forecast. The reference forecast is generally an unskilled forecast such as random chance, persistence, or climatology. To assess the skill of GloFAS-Seasonal reforecasts, we compare the reforecasts with climatology [51], an ensemble of observed flows, and use the ROC skill score (ROCSS), which has been used in previous studies for the verification of seasonal forecasts [52]. ROCSS is computed as follows:

$$\text{ROCSS} = \frac{AUC_{fc} - AUC_{cm}}{1 - AUC_{cm}} \tag{13}$$

where AUC_{fc} refers to the AUC value of reforecasts and AUC_{cm} refers to the AUC value of climatological forecasts. ROCSS of one means a perfect forecasting system; ROCSS of zero indicates no improvement over the climatology.

3.3. Reliability, Resolution, and Sharpness

For assessing the reliability of forecasts, the reliability diagram is used here, where X and Y axes represent the forecast probability and the observed frequency of the future below the streamflow threshold, respectively. When the forecast probability and the observed frequency are equal, the reliability of forecasts is perfect. For example, if an event will occur with a forecast probability of 70%, then, on average, the event should occur on 70% of the occasions that this forecast is made. So, reliability is indicated by the proximity of the plotted curve to the diagonal. If the plotted curve lies below the diagonal, this indicates over-estimation (forecast probabilities are too high); curve above the diagonal indicates under-estimation (forecast probabilities are too low).

The climatological average can produce high reliability, but it lacks information for practice. In theory, we are interested in probability forecast systems which give a forecast probability that deviates from the climatological average and approaches 0% or 100% while maintaining a high level of reliability [35]. So, the reliability diagram can also be used to assess the resolution of forecasts. Forecasts that discriminate between events and non-events are said to have a resolution (a forecast of climatological average, a curve lying on or near the horizontal line would have no resolution). For assessing the sharpness of forecasts, the reliability diagram is usually accompanied by a histogram. If the histogram is U-shaped, then the frequency of forecasts approaches 0% and 100% and the forecast system sharpness is well. Forecasts with no or low sharpness will show a peak in the forecast frequency near the climatological average.

4. Results and Discussion

4.1. The Selected Thresholds

As the first step of our evaluation, we select thresholds to evaluate the streamflow condition for impoundment operation. The Changjiang (Yangtze River) Water Resources Commission (CWRC) provides daily inflow and discharge data series for the selected five reservoirs and streamflow for adjacent gauges at hydrological stations, which covers the whole impoundment operation period from 1 August to 31 October (92 days) for 1950–2015 (66 years). We use 20, 30, 40, and 50-percentile of historical inflow as thresholds to determine the inflow condition in September according to the monthly-averaged inflow (Q_{WDD} and Q_{TGR}). For example, the 20-percentile historical average inflow in September divides data into two groups where one group (above 20-percentile in Figure 2a) includes 53 years of data and the other group (below 20-percentile in Figure 2a) has 13 years of data. Therefore, we get two scenarios for these thresholds, one above and one below each threshold. By this approach, we get 16 different flow scenario groups (two Q_{WDD} or Q_{TGR} × four quantiles × two groups for each quantile).

These scenarios are evaluated independently along with the EIOR and SIOR by the PSO approach. Since the population size for the NSGA-II algorithm is set to 50, the algorithm provides 50 Pareto-optimal solutions (non-dominated solutions). Since there are three objective functions for each scenario and considering the multi-purpose nature of these reservoirs, a single value cannot be reported as the best answer from these 50 Pareto-optimal solutions. Therefore, we average the objective function of the 50 Pareto-optimal solutions as a potential benefit and risk in response to this combination of historical flow group and impoundment rules. For 16 different groups and two operation rules, the averaged three objective functions of 50 Pareto-optimal solutions are shown in Table 4. Comparing these 16 different scenarios, we can see that the HG and FSR values are improved by employing larger thresholds. This improvement is due to the increase in streamflow and reservoirs storage in September from the lowest, below 20-percentile, to the highest, above 50-percentile, threshold. It is clearly shown that low flow in September has an adverse impact on impoundment operation.

Table 4. Benefit and risk results of the cascade reservoir system in response to the combination of historical flow group and impoundment rule.

Flow Group		EIOR			SIOR		
		HG (10^8 kW·h)	*FSR* (%)	*R* (%)	*HG* (10^8 kW·h)	*FSR* (%)	*R* (%)
Q_{WDD}	below 20%	813.784	82.68%	0	805.070	79.78%	0
	below 30%	845.857	85.43%	0	834.963	82.45%	0
	below 40%	868.680	89.40%	0	857.848	87.25%	0
	below 50%	895.822	91.72%	0	884.002	89.81%	0
	above 20%	1025.617	99.25%	1.802%	1011.410	98.86%	2.229%
	above 30%	1040.072	99.98%	2.035%	1022.395	99.96%	2.123%
	above 40%	1060.128	100.00%	2.924%	1042.975	100.00%	2.284%
	above 50%	1074.293	100.00%	3.730%	1057.887	100.00%	2.842%
Q_{TGR}	below 20%	816.973	81.86%	0	808.284	79.18%	0
	below 30%	838.596	86.20%	0	829.380	83.76%	0
	below 40%	876.391	89.49%	0	864.720	87.80%	0
	below 50%	895.437	91.56%	2.055%	881.728	90.13%	0
	above 20%	1024.592	99.23%	2.217%	1005.852	98.97%	2.742%
	above 30%	1042.250	99.70%	2.313%	1023.483	99.65%	2.098%
	above 40%	1048.775	99.88%	2.237%	1030.204	99.87%	2.902%
	above 50%	1072.242	99.96%	2.263%	1052.043	99.96%	3.043%

Note: EIOR: early impoundment operation rules; SIOR: standard impoundment operation rules; *HG*: hydropower generation; *FSR*: fullness storage rate; *R*: flood control risk.

Comparing EIOR with SIOR for cascade reservoirs, the EIOR improves the *HG* and *FSR* from the flow group below 20% to below 40% for both of Q_{WDD} and Q_{TGR} without affecting the risk of flooding. We employ these results to select the most suitable threshold among these 16 scenarios for our analysis. Figure 3 shows the relationship between increased benefit ratio and different flow groups of Q_{WDD} and Q_{TGR}. According to Figure 3, *HG* is less affected by the selected thresholds. On the contrary, *FSR* values are decreased by increasing the threshold or inflow. For the group below the 20-percentile and below the 30-percentile, the *FSRs* of the proposed EIOR are increased significantly around or above 3% in comparison to the SIOR, without increasing the risk of flooding. Hence, we select the 20-percentile and 30-percentile as the thresholds for our study, as their performance is superior to others. In early August, we use these thresholds to evaluate the performance of the GloFAS-Seasonal in predicting the streamflow condition for Q_{WDD} and Q_{TGR} next month.

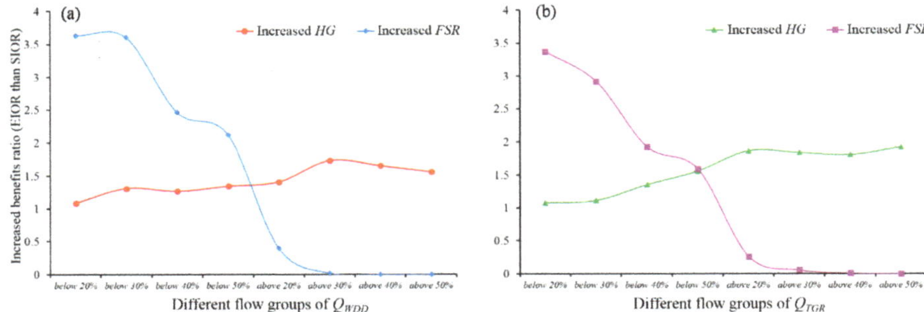

Figure 3. The relationship between increased benefit ratio of EIOR and different flow groups of (a) Q_{WDD} and (b) Q_{TGR}.

4.2. Evaluation of GloFAS-Seasonal Reforecasts

GloFAS-Seasonal reforecasts are evaluated using adjusted historical river flow data at the Hua-Tan and Yi-Chang hydrologic stations in the Yangtze River. GloFAS-Seasonal reforecasts represent natural flow and do not consider any reservoir routing. The CWRC provides monthly averaged historical flow records which have been adjusted to represent the natural flow. These adjusted historical natural streamflow timeseries span over thirty years (1981–2013). So, GloFAS-Seasonal reforecasts are evaluated over the same 33-year period. Since the impoundment operation starts before September, we investigate the GloFAS-Seasonal reforecasts on 1 August (2-month lead, LM2) to evaluate the potential for employing EIOR. We also investigate the 1-month lead, LM1, on 1 September to evaluate the performance of GloFAS-Seasonal for different lead times.

4.2.1. AUC Values

In order to compare AUC values for different stations, lead times, and thresholds, we employ the Nightingale's Rose chart. This chart is suitable to visually evaluate the evident differences between various categorical data. The results are shown in Figure 4, and it is clearly shown that all AUC values are greater than 0.6, which means that the forecasts can be regarded as informative and have the ability to predict the streamflow condition (whether streamflow is below the threshold or not). Besides, the AUC values exhibit a decline from the LM1 (around 0.9) to the LM2 (below 0.8) as expected.

For different stations and thresholds, AUC values of forecasts vary more significantly with lead times. So, the discrimination of GloFAS-Seasonal reforecasts is relatively stable over space in the upper Yangtze River. Moreover, an interesting finding is that the performance of thresholds varies for hydrological stations. For Hua-Tan, the 20-percentile has the best performance, whereas the 30-percentile for the Yi-Chang station. This emphasizes that a spatial evaluation of thresholds is necessary for the Yangtze River to find the best thresholds for employing the GloFAS-Seasonal forecast at the basin.

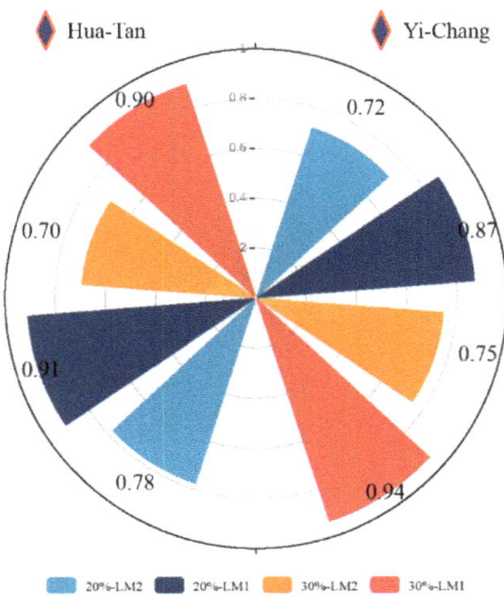

Figure 4. The area under the ROC curve (AUC) values for GloFAS-Seasonal reforecast at Hua-Tan (left) and Yi-Chang (right) hydrologic stations.

4.2.2. ROCSS Values

Tercile plots are designed to show the performance of a forecast system at different periods [53]. Here, we employ these plots to compare reforecast probabilities (color coded from light to dark color for lower to higher probability) for different threshold events with the observed condition (white dots). We defined three different categories of threshold events for our comparison. Since low flow condition leads to employing the EIOR, we only employ 0–20%, 20–40%, and 40–60% quantiles of the streamflow data to evaluate the performance of GloFAS-Seasonal reforecasts for predicting the correct flow condition. However, the evaluation can be done for other flow ranges based on the selected thresholds. ROCSS values for each quantile is shown on the right axis for comparison. Significant values of ROCSS with a 95% confidence are marked with an asterisk for statistical evaluation.

Results show that for GloFAS-Seasonal reforecasts below 20%, the ROCSS exhibit a decline in skill from the LM1 (0.8 and 0.76) to the LM2 (0.46 and 0.42) for both Hua-Tan and Yi-Chang hydrological stations. However, skills (ROCSS greater than 0) still prevail in the LM2 and are marked with asterisks, which means that forecasts of LM2 are better than climatology. Furthermore, forecast sharpness is also evident in this tercile plot. The darker the color of the square, the better the sharpness of that probabilistic forecasts is. Forecasts for both LM1 and LM2 exhibit sharpness, although the sharpness is higher for LM1, which is indicated by the colors of the squares in Figure 5.

According to Figure 5, the streamflow condition for the Hua-Tan hydrological station is below the 20-percentile in seven years, among which the forecast predicted the highest probability for five and three of these years by LM1 and LM2, respectively. For the Yi-Chang hydrological station, the number of years with streamflow below the 20-percentile is seven, out of which GloFAS-Seasonal reforecasts with LM1 and LM2 predicted the highest probability for five and three of these years, respectively. Consistent with the results of AUC, although LM1 shows better performance with shorter lead time, aiming at reservoir impoundment operation, GloFAS-Seasonal reforecasts with 2-month lead time (LM2) are still informative. Further, compared with the LM1, LM2 still has a lot of potential improvement in the future, which depends on developing the seasonal climate prediction. A similar analysis can be performed for the 30-percentile threshold with other ranges.

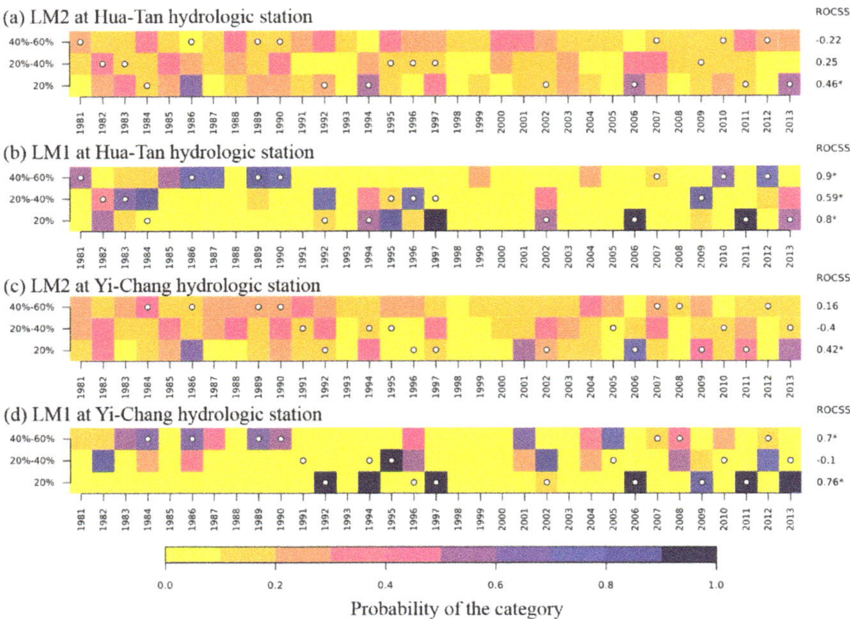

Figure 5. Tercile plots and ROC skill score (ROCSS) for GloFAS-Seasonal reforecasts (**a**) 2-month lead at the Hua-Tan, (**b**) 1-month lead at the Hua-Tan, (**c**) 1-month lead at the Yi-Chang, and (**d**) 2-month lead at the Yi-Chang hydrologic station.

4.2.3. Reliability Diagram

Similar to ROC calculations, the reliability is assessed for both the 20-percentile and 30-percentile threshold. Due to the limit number of samples, the range of forecast probabilities is divided into five bins (for every 20% from 0% to 100%) rather than ten bins in order to avoid sparseness of the probability categories. Since GloFAS-Seasonal reforecasts have similar performance at hydrological stations, reliability diagrams are only presented for Yi-Chang hydrologic station here. Figure 6 shows the effect of (a) the lead time (LM1 and LM2) and (b) the threshold (20-percentile and 30-percentile) on reliability by combining the contingency table for thresholds and lead times, respectively.

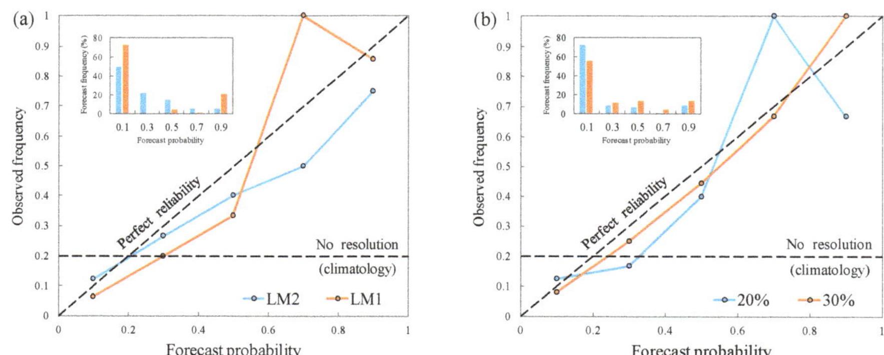

Figure 6. Reliability diagrams of GloFAS-Seasonal reforecasts for comparing (**a**) two lead times and (**b**) thresholds.

Figure 6a shows that forecasts have more reliability than climatology, regardless of the lead time. It is worth noting that the observed frequency is unrealistically equal to 1 for 60–80% and the LM1 due to sampling limitations rather than necessarily true deviations from reliability [54]. Overall, the reliability appears to be slightly better for forecasts of LM2 than LM1. The forecast data for both LM2 and LM1 exhibit sharpness, which means that forecast probabilities are more informative than climatology. Similar behavior is observed for the 60–80% bin in Figure 6b, because of the limited number of samples. Figure 6b shows that the reliability of the 30-percentile threshold is better than the 20-percentile. In contrast to reliability, sharpness is better for forecasts of the 20-percentile rather than the 30-percentile threshold. Differences in reliability and sharpness can be explained by the limited number of samples. So, the performance of the two selected thresholds is close and hard to distinguish.

Due to most dots laying below the diagonal, Figure 6 suggests that in general, GloFAS-Seasonal reforecasts have a tendency to over-estimate the likelihood of a below percentile streamflow condition, which is a common situation for seasonal forecasting [55]. This conclusion is consistent with the reliability diagram of GloFAS-Seasonal reforcasts aggregated across all observation stations globally [35], and reflects the characteristics of the GloFAS-Seasonal forecasting system. However, with respect to the impoundment of the reservoirs, it is more favorable to over-estimate the below threshold conditions rather than under-estimating. The reservoir operators could employ GloFAS-Seasonal forecasts for decision-making for the early impoundment operation, while control the risk of flooding through short-term hydrological forecasting in real-time operation.

4.3. Specific Analysis and Benefits of the EIOR

The above results demonstrate that GloFAS-Seasonal forecasts have the potential to give water managers the flexibility to employ early impoundment in the upper Yangtze River. Here, we try to analyze the EIOR and find its benefits. As an example, Figure 7 shows the Pareto-optimal solutions of EIOR (plot a) and SIOR (plot b) for the Q_{TGR} below the 20-percentile threshold. These Pareto-optimal solutions are averaged for derving parts of Table 4. We are employing three objective functions. Therefore, three subplots are needed for Pareto-optimal solutions to show three objectives in pairs. However, the flood control risk (R) of almost all Pareto-optimal solutions are equal to zero. Hence, we only show two objective functions (*FSR* and *HG)* in Figure 7. Each one of the 50 Pareto-optimal solutions obtained from the NSGA-II algorithm represents impoundment rule curve for each of the five cascade reservoirs. Figure 7 also shows EIOR and SIOR rule curves of WDD and TGR reservoirs focusing on the extreme solution of the *FSR* objective function. The figure shows that the average water level of EIOR is higher than SIOR for the selected 20-percentile threshold, while the risk of flooding is zero.

To illustrate the potential maximum benefits of EIOR, Figure 7 also shows a linear operation rule (LOR), which connects the initial water level to the top of STBP at the end of impoundment period. As a benchmark, LOR could present the maximum benefit of the EIOR from two parts where one is the earlier initial impoundment time and the other is the optimized rule curve. For curves of EIOR and LOR in Figure 7, Table 5 shows the benefit and risk of EIOR compared to the LOR. The proposed EIOR improves the *FSR* by 5.63% and increases *HG* by 4.02%. In conclusion, during dry years, our proposed methodology could significantly increase the hydropower generation and water utilization by employing GloFAS-Seasonal forecasts and early reservoir impoundment.

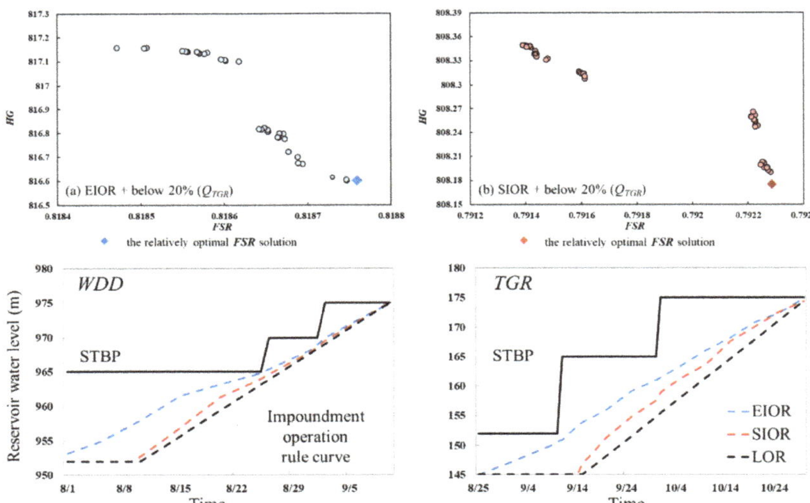

Figure 7. Optimized EIOR and SIOR rule curves of WDD and TGR reservoirs focusing on fullness storage rate (*FSR*) for flow scenario group (below 20%) of Q_{TGR}.

Table 5. Comparison of the benefit and risk of EIOR and linear operation rule (LOR) rule curves for the Q_{TGR} below the 20-percentile threshold.

Flow Group		Rule Curve in Figure 7	Benefit and Risk		
			HG (10^8 kW·h)	FSR (%)	R (%)
Q_{TGR}	below 20%	LOR	785.042	77.512%	0
		EIOR	816.601	81.876%	0
		Increased ratio	4.02%	5.63%	0

Note: Q_{TGR}: inflow to the Three Gorges Reservoir; LOR: linear operation rule; EIOR: early impoundment operation rules; *HG*: hydropower generation; *FSR*: fullness storage rate; *R*: flood control risk.

5. Conclusions

In this study, we evaluated the potential application of GloFAS-Seasonal forecasts for early reservoir impoundment in the upper Yangtze River. A cascade reservoirs impoundment simulation–optimization model was employed to select suitable low flow thresholds for decision-making for EIOR or SIOR. These thresholds were selected by analyzing the historical inflow data of WDD and TGR reservoirs, which were derived from Hua-Tan and Yi-Chang hydrologic stations. The performance of GloFAS-Seasonal reforecasts to predict the streamflow condition at these two hydrological stations was evaluated using AUC, ROCSS, and reliability diagram for two different lead times (LM1 and LM2) and selected thresholds. The main findings of our study can be summarized as follows:

(1) The low flow condition in September has a very significant impact on reservoir impoundment operation in the upper Yangtze River. The 20-percentile and 30-percentile selected thresholds of inflow at WDD and TGR are suitable for evaluating the possibility of early impoundment. These two selected thresholds can be used as a measure for flow condition and decision-making for early impoundment operation.

(2) All AUC values of reforecasts are greater than 0.6 which shows that GloFAS-Seasonal forecasts can be used to predict the streamflow condition according to the selected thresholds. However, AUC decreases from the LM1 (around 0.9) to the LM2 (below 0.8) as expected. The ROCSS reveals that both

LM1 and LM2 are significantly better than climatology. The reliability diagrams also show that both LM1 and LM2 forecasts have more reliability and sharpness than climatology. Furthermore, results also indicate a tendency of the two lead time forecasts to over-estimate, which is more favorable for water managers.

(3) GloFAS-Seasonal forecasts with 2-month lead time (LM2) are valuable for reservoir impoundment operation. During dry years, the proposed EIOR improves the fullness storage rate by 5.63% and the annual average hydropower generation by 4.02% without increasing the risk of flooding.

This paper demonstrates that GloFAS-Seasonal forecasts has the potential to improve the standard impoundment operation rules in the upper Yangtze River and give water managers the flexibility to employ early impoundment.

Author Contributions: Conceptualization and software. K.C. and S.G.; Data Curation, J.W. and P.Q.; Formal Analysis, K.C., S.H. and S.S.; Writing—Original Draft Preparation, K.C. and M.R.N.; Writing—Review and Editing, S.G.

Funding: This study was funded by the National Key Research and Development Project (Grant NO. 2016YFC0402206) of China, the National Natural Science Foundation of China (Grant NO. 51879192), the "111 Project" Fund of China (B18037), the joint U.S.-China Clean Energy Research Center for Water-Energy Technologies (CERC-WET) project (2018YFE0196000) and the U.S. Department of Energy (DOE Prime Award # DE-IA0000018).

Acknowledgments: The authors would like to express their gratitude to anonymous reviewers for their insightful and constructive comments.

Conflicts of Interest: The authors declare no conflict of interest.

References

1. Labadie, J.W. Optimal operation of multireservoir systems: State-of-the-art review. *J. Water Res. Plan. Man.* **2004**, *130*, 93–111. [CrossRef]
2. Guo, S.L.; Zhang, H.G.; Chen, H.; Peng, D.Z.; Liu, P.; Pang, B. A reservoir flood forecasting and control system for China. *Hydrolog. Sci. J.* **2004**, *49*, 959–972. [CrossRef]
3. Mehran, A.; Mazdiyasni, O.; AghaKouchak, A. A hybrid framework for assessing socioeconomic drought: Linking climate variability, local resilience, and demand. *J. Geophys. Res. Atmos.* **2015**, *120*, 7520–7533. [CrossRef]
4. Rahnamay Naeini, M. A Framework for Optimization and Simulation of Reservoir Systems Using Advanced Optimization and Data Mining Tools. Ph.D. Dissertation, University of California Irvine, Irvine, CA, USA, 2019.
5. Zhou, C.; Sun, N.; Chen, L.; Ding, Y.; Zhou, J.Z.; Zha, G.; Luo, G.L.; Dai, L.; Yang, X. Optimal Operation of Cascade Reservoirs for Flood Control of Multiple Areas Downstream: A Case Study in the Upper Yangtze River Basin. *Water* **2018**, *10*, 1250. [CrossRef]
6. Wang, Q.S.; Zhou, J.Z.; Huang, K.D.; Dai, L.; Zha, G.; Chen, L.; Qin, H. Risk Assessment and Decision-Making Based on Mean-CVaR-Entropy for Flood Control Operation of Large Scale Reservoirs. *Water* **2019**, *11*, 649. [CrossRef]
7. Zhou, Y.L.; Guo, S.L.; Xu, C.Y.; Liu, P.; Qin, H. Deriving joint optimal refill rules for cascade reservoirs with multi-objective evaluation. *J. Hydrol.* **2015**, *524*, 166–181. [CrossRef]
8. Li, H.; Liu, P.; Guo, S.L.; Ming, B.; Cheng, L.; Zhou, Y.L. Hybrid Two-Stage Stochastic Methods Using Scenario-Based Forecasts for Reservoir Refill Operations. *J. Water Res. Plan. Man.* **2018**, *144*. [CrossRef]
9. He, S.K.; Guo, S.L.; Chen, K.B.; Deng, L.L.; Liao, Z.; Xiong, F.; Yin, J.B. Optimal impoundment operation for cascade reservoirs coupling parallel dynamic programming with importance sampling and successive approximation. *Adv. Water Resour.* **2019**, *131*, 103375. [CrossRef]
10. Batalla, R.J.; Gomez, C.M.; Kondolf, G.M. Reservoir-induced hydrological changes in the Ebro River basin (NE Spain). *J. Hydrol.* **2004**, *290*, 117–136. [CrossRef]
11. Clark, E.J. Impounding reservoirs. *J. Am. Water Works Assoc.* **1956**, *48*, 349–354.
12. Lund, J.R. *Developing Seasonal and Long-Term Reservoir System Operation Plans Using HEC-PRM*; Hydrologic Engineering Center: Davis, CA, USA, 1996.

13. Lund, J.R.; Guzman, J. Derived operating rules for reservoirs in series or in parallel. *J. Water Res. Plan. Man.* **1999**, *125*, 143–153. [CrossRef]

14. Liu, P.; Guo, S.L.; Xiong, L.H.; Li, W.; Zhang, H.G. Deriving reservoir refill operating rules by using the proposed DPNS model. *Water Resour. Manag.* **2006**, *20*, 337–357. [CrossRef]

15. Liu, X.Y.; Guo, S.L.; Liu, P.; Chen, L.; Li, X.A. Deriving Optimal Refill Rules for Multi-Purpose Reservoir Operation. *Water Resour. Manag.* **2011**, *25*, 431–448. [CrossRef]

16. Li, Y.; Guo, S.L.; Quo, J.L.; Wang, Y.; Li, T.Y.; Chen, J.H. Deriving the optimal refill rule for multi-purpose reservoir considering flood control risk. *J. Hydro Environ. Res.* **2014**, *8*, 248–259. [CrossRef]

17. Wang, X.M.; Zhou, J.Z.; Ouyang, S.; Li, C.L. Research on Joint Impoundment Dispatching Model for Cascade Reservoir. *Water Resour. Manag.* **2014**, *28*, 5527–5542. [CrossRef]

18. Zhou, Y.L.; Guo, S.L.; Chang, F.J.; Xu, C.Y. Boosting hydropower output of mega cascade reservoirs using an evolutionary algorithm with successive approximation. *Appl. Energ.* **2018**, *228*, 1726–1739. [CrossRef]

19. Jasperse, J.; Ralph, M.; Anderson, M.; Brekke, L.D.; Dillabough, M.; Dettinger, M.D.; Haynes, A.; Hartman, R.; Jones, C.; Forbis, J. *Preliminary Viability Assessment of Lake Mendocino Forecast Informed Reservoir Operations;* Center For Western Weather and Water Extremes: San Diego, CA, USA, 2017.

20. Liu, L.L.; Xiao, C.; Du, L.M.; Zhang, P.Q.; Wang, G.F. Extended-Range Runoff Forecasting Using a One-Way Coupled Climate-Hydrological Model: Case Studies of the Yiluo and Beijiang Rivers in China. *Water* **2019**, *11*, 1150. [CrossRef]

21. Panagoulia, D. Impacts of Giss-Modeled Climate Changes on Catchment Hydrology. *Hydrolog. Sci. J.* **1992**, *37*, 141–163. [CrossRef]

22. Yuan, X.; Wood, E.F.; Ma, Z.G. A review on climate-model-based seasonal hydrologic forecasting: Physical understanding and system development. *Wires Water* **2015**, *2*, 523–536. [CrossRef]

23. Panagoulia, D.; Dimou, G. Sensitivity of flood events to global climate change. *J. Hydrol.* **1997**, *191*, 208–222. [CrossRef]

24. Yuan, X.; Zhang, M.; Wang, L.Y.; Zhou, T. Understanding and seasonal forecasting of hydrological drought in the Anthropocene. *Hydrol. Earth Syst. Sci.* **2017**, *21*, 5477–5492. [CrossRef]

25. Arnal, L.; Cloke, H.L.; Stephens, E.; Wetterhall, F.; Prudhomme, C.; Neumann, J.; Krzeminski, B.; Pappenberger, F. Skilful seasonal forecasts of streamflow over Europe? *Hydrol. Earth Syst. Sci.* **2018**, *22*, 2057–2072. [CrossRef]

26. Bennett, J.C.; Wang, Q.J.; Robertson, D.E.; Schepen, A.; Li, M.; Michael, K. Assessment of an ensemble seasonal streamflow forecasting system for Australia. *Hydrol. Earth Syst. Sci.* **2017**, *21*, 6007–6030. [CrossRef]

27. Demargne, J.; Wu, L.M.; Regonda, S.K.; Brown, J.D.; Lee, H.; He, M.X.; Seo, D.J.; Hartman, R.; Herr, H.D.; Fresch, M.; et al. The Science of NOAA's Operational Hydrologic Ensemble Forecast Service. *Bull. Am. Meteorol. Soc.* **2014**, *95*, 79–98. [CrossRef]

28. Block, P. Tailoring seasonal climate forecasts for hydropower operations. *Hydrol. Earth Syst. Sci.* **2011**, *15*, 1355–1368. [CrossRef]

29. Delorit, J.; Ortuya, E.C.G.; Block, P. Evaluation of model-based seasonal streamflow and water allocation forecasts for the Elqui Valley, Chile. *Hydrol. Earth Syst. Sci.* **2017**, *21*, 4711–4725. [CrossRef]

30. Baker, S.A.; Wood, A.W.; Rajagopalan, B. Developing Subseasonal to Seasonal Climate Forecast Products for Hydrology and Water Management. *J. Am. Water Resour. Assoc.* **2019**, *55*, 1024–1037. [CrossRef]

31. Chen, K.B.; Guo, S.L.; He, S.K.; Xu, T.; Zhong, Y.X.; Sun, S.R. The Value of Hydrologic Information in Reservoir Outflow Decision-Making. *Water* **2018**, *10*, 1372. [CrossRef]

32. Zhu, S.; Zhou, J.Z.; Ye, L.; Meng, C.Q. Streamflow estimation by support vector machine coupled with different methods of time series decomposition in the upper reaches of Yangtze River, China. *Environ. Earth* **2016**, *75*. [CrossRef]

33. He, R.R.; Chen, Y.F.; Huang, Q.; Yu, S.N.; Hou, Y. Evaluation of ocean-atmospheric indices as predictors for summer streamflow of the Yangtze River based on ROC analysis. *Stoch. Env. Res. Risk Assess.* **2018**, *32*, 1903–1918. [CrossRef]

34. Tan, Q.F.; Lei, X.H.; Wang, X.; Wang, H.; Wen, X.; Ji, Y.; Kang, A.Q. An adaptive middle and long-term runoff forecast model using EEMD-ANN hybrid approach. *J. Hydrol.* **2018**, *567*, 767–780. [CrossRef]

35. Emerton, R.; Zsoter, E.; Arnal, L.; Cloke, H.L.; Muraro, D.; Prudhomme, C.; Stephens, E.M.; Salamon, P.; Pappenberger, F. Developing a global operational seasonal hydro-meteorological forecasting system: GloFAS-Seasonal v1.0. *Geosci. Model Dev.* **2018**, *11*, 3327–3346. [CrossRef]

36. Deb, K.; Pratap, A.; Agarwal, S.; Meyarivan, T. A fast and elitist multiobjective genetic algorithm: NSGA-II. *IEEE Trans. Evolut. Comput.* **2002**, *6*, 182–197. [CrossRef]
37. Yeh, W.W.G. Reservoir Management and Operations Models - a State-of-the-Art Review. *Water Resour. Res.* **1985**, *21*, 1797–1818. [CrossRef]
38. Lei, X.H.; Zhang, J.W.; Wang, H.; Wang, M.N.; Khu, S.T.; Li, Z.J.; Tan, Q.F. Deriving mixed reservoir operating rules for flood control based on weighted non-dominated sorting genetic algorithm II. *J. Hydrol.* **2018**, *564*, 967–983. [CrossRef]
39. Celeste, A.B.; Billib, M. Evaluation of stochastic reservoir operation optimization models. *Adv. Water Resour.* **2009**, *32*, 1429–1443. [CrossRef]
40. Zhao, T.T.G.; Zhao, J.S. Optimizing Operation of Water Supply Reservoir: The Role of Constraints. *Math. Probl. Eng.* **2014**. [CrossRef]
41. Naeini, M.R.; Yang, T.T.; Sadegh, M.; AghaKouchak, A.; Hsu, K.L.; Sorooshian, S.; Duan, Q.Y.; Lei, X.H. Shuffled Complex-Self Adaptive Hybrid EvoLution (SC-SAHEL) optimization framework. *Environ. Modell. Softw.* **2018**, *104*, 215–235. [CrossRef]
42. Chang, L.C.; Chang, F.J. Multi-objective evolutionary algorithm for operating parallel reservoir system. *J. Hydrol.* **2009**, *377*, 12–20. [CrossRef]
43. Tsai, W.P.; Chang, F.J.; Chang, L.C.; Herricks, E.E. AI techniques for optimizing multi-objective reservoir operation upon human and riverine ecosystem demands. *J. Hydrol.* **2015**, *530*, 634–644. [CrossRef]
44. Uen, T.S.; Chang, F.J.; Zhou, Y.L.; Tsai, W.P. Exploring synergistic benefits of Water-Food-Energy Nexus through multi-objective reservoir optimization schemes. *Sci. Total Environ.* **2018**, *633*, 341–351. [CrossRef] [PubMed]
45. Chen, Y.H.; Chang, F.J. Evolutionary artificial neural networks for hydrological systems forecasting. *J. Hydrol.* **2009**, *367*, 125–137. [CrossRef]
46. Balsamo, G.; Pappenberger, F.; Dutra, E.; Viterbo, P.; van den Hurk, B. A revised land hydrology in the ECMWF model: A step towards daily water flux prediction in a fully-closed water cycle. *Hydrol. Process.* **2011**, *25*, 1046–1054. [CrossRef]
47. Landman, W.A.; Beraki, A. Multi-model forecast skill for mid-summer rainfall over southern Africa. *Int. J. Climatol.* **2012**, *32*, 303–314. [CrossRef]
48. Fundel, F.; Jorg-Hess, S.; Zappa, M. Monthly hydrometeorological ensemble prediction of streamflow droughts and corresponding drought indices. *Hydrol. Earth Syst. Sci.* **2013**, *17*, 395–407. [CrossRef]
49. Manzanas, R.; Gutierrez, J.M.; Bhend, J.; Hemri, S.; Doblas-Reyes, F.J.; Torralba, V.; Penabad, E.; Brookshaw, A. Bias adjustment and ensemble recalibration methods for seasonal forecasting: A comprehensive intercomparison using the C3S dataset. *Clim. Dynam.* **2019**, *53*, 1287–1305. [CrossRef]
50. Mason, S.J.; Graham, N.E. Conditional probabilities, relative operating characteristics, and relative operating levels. *Weather Forecast.* **1999**, *14*, 713–725. [CrossRef]
51. Pappenberger, F.; Ramos, M.H.; Cloke, H.L.; Wetterhall, F.; Alfieri, L.; Bogner, K.; Mueller, A.; Salamon, P. How do I know if my forecasts are better? Using benchmarks in hydrological ensemble prediction. *J. Hydrol.* **2015**, *522*, 697–713. [CrossRef]
52. Ogutu, G.E.O.; Franssen, W.H.P.; Supit, I.; Omondi, P.; Hutjes, R.W.A. Skill of ECMWF system-4 ensemble seasonal climate forecasts for East Africa. *Int. J. Climatol.* **2017**, *37*, 2734–2756. [CrossRef]
53. Diez, E.; Orfila, B.; Frias, M.D.; Fernandez, J.; Cofino, A.S.; Gutierrez, J.M. Downscaling ECMWF seasonal precipitation forecasts in Europe using the RCA model. *Tellus A* **2011**, *63*, 757–762. [CrossRef]
54. Jolliffe, I.T.; Stephenson, D.B. *Forecast Verification: A Practitioner's Guide in Atmospheric Science*; John Wiley & Sons: Chichester, UK, 2012.
55. Mason, S.J.; Stephenson, D.B. How do we know whether seasonal climate forecasts are any good? In *Seasonal Climate: Forecasting and Managing Risk*; Springer: Dordrecht, The Netherlands, 2008; pp. 259–289.

Article

Multi-Objective Operation of Cascade Hydropower Reservoirs Using TOPSIS and Gravitational Search Algorithm with Opposition Learning and Mutation

Zhong-kai Feng [1], Shuai Liu [1], Wen-jing Niu [2,*], Zhi-qiang Jiang [1], Bin Luo [3] and Shu-min Miao [4]

[1] School of Hydropower and Information Engineering, Huazhong University of Science and Technology, Wuhan 430074, China; myfellow@163.com (Z.-k.F.); m201873748@hust.edu.cn (S.L.); zqjzq@hust.edu.cn (Z.-q.J.)

[2] Bureau of Hydrology, ChangJiang Water Resources Commission, Wuhan 430010, China

[3] Tsinghua Sichuan Energy Internet Research Institute, Chengdu 610200, China; luobin1590@163.com

[4] State Grid Sichuan Electric Power Research Institute, Chengdu 610072, China; hao410miao@163.com

*   Correspondence: dgniuwenjing@163.com or niuwj@cjh.com.cn

Received: 27 August 2019; Accepted: 28 September 2019; Published: 29 September 2019

Abstract: In this research, a novel enhanced gravitational search algorithm (EGSA) is proposed to resolve the multi-objective optimization model, considering the power generation of a hydropower enterprise and the peak operation requirement of a power system. In the proposed method, the standard gravity search algorithm (GSA) was chosen as the fundamental execution framework; the opposition learning strategy was adopted to increase the convergence speed of the swarm; the mutation search strategy was chosen to enhance the individual diversity; the elastic-ball modification strategy was used to promote the solution feasibility. Additionally, a practical constraint handling technique was introduced to improve the quality of the obtained agents, while the technique for order preference by similarity to an ideal solution method (TOPSIS) was used for the multi-objective decision. The numerical tests of twelve benchmark functions showed that the EGSA method could produce better results than several existing evolutionary algorithms. Then, the hydropower system located on the Wu River of China was chosen to test the engineering practicality of the proposed method. The results showed that the EGSA method could obtain satisfying scheduling schemes in different cases. Hence, an effective optimization method was provided for the multi-objective operation of hydropower system.

Keywords: cascade hydropower reservoirs; multi-objective optimization; TOPSIS; gravitational search algorithm; opposition learning; partial mutation; elastic-ball modification

---

1. Introduction

With the merits of low pollution emission and high work efficiency, hydropower is seen as one of the most important renewable energy sources and usually shoulders many different kinds of management requirements, in practice [1–3]. For hydropower enterprises, it is preferred to make full use of the water resources to obtain the maximum economic benefit [4]; for power systems, the flexible hydropower generators are often asked to respond to the peak loads and smooth the energy demand curves [5]. Under such a background, the operation optimization of cascade hydropower reservoirs balancing a power generation and peak operation has become one of the most important tasks in both water resource systems and modern power systems [6]. However, as far as we know, there are few reports that have addressed this engineering problem, and therefore, the goal of this research was to refill this huge gap between theoretical research and engineering requirement.

From a mathematical point of view, the target problem was a typical multi-objective constrained optimization problem with a set of complex equality or inequality constraints [7], and many classical methods have been successfully developed by scholars, during the past decades [8–13], like linear programming [14], quadratic programming [15], dynamic programming [16–18], Lagrange relaxation, and network optimization [19–21]. However, the hydropower operation problem is usually modeled with nonlinear characteristic curves, physical constraints, or objective functions [22–24]. The above-mentioned traditional methods might fail to address the complexity due to various defects, like dimensionality problem [25], high computational burden [26], duality gap [27], or parameter tuning [28–30]. In recent years, with the booming development of computer technology, many evolutionary algorithms have been proposed to resolve these kind of problems [31–33], like genetic algorithm (GA) [34], differential evolution (DE) [35,36], particle swarm algorithm (PSO) [37–40], cuckoo search (CS) [41], Covariance Matrix Adaptation Evolution Strategy with a Directed Target to Best Perturbation (CMA-ES-DTBP) [42], and a clustered adaptive teaching–learning-based optimization (CATLBO) [43]. Compared with traditional methods, the evolutionary algorithms can produce satisfying solutions in most cases, regardless of the problem features (like continuity or nonconvexity) [44–46]. However, due to the premature convergence, evolutionary algorithms often fall into local optima, which have limited their widespread applications in practical engineering [47–50]. Hence, it is necessary to find some effective modified strategies to enhance the performances of the evolutionary algorithms in the hydropower operation problems.

Gravitational search algorithm (GSA) is a famous evolutionary algorithm based on the laws of gravity and mass interactions in Newtonian mechanics [51–53]. Although GSA outperforms the standard PSO and GA methods in many problems, certain defects (like exploration–exploitation imbalance and local convergence) still exist. For improving the performance of the standard GSA method, this study proposes an enhanced GSA algorithm (EGSA) where three modified strategies (opposition learning strategy, mutation search strategy, and elastic-ball modification strategy) are integrated to alleviate the premature convergence problem. Several famous test functions are used to verify the feasibility of the developed method in solving the global optimization problems, and the results demonstrate that the modified strategies could effectively enhance the convergence speed and the global search ability of the population at the same time. Then, two specific strategies were embedded into the EGSA method to help address the multi-objective operation problem of a hydropower system—(i) the practical constraint handling strategy for solution feasibility and (ii) the Technique for Order Preference by Similarity to an Ideal Solution (TOPSIS) for multi-objective decision. Finally, the EGSA method was applied to address the operation optimization of cascade hydropower system in the Wu River of China. The simulations showed that the proposed method could produce better scheduling schemes in different application scenarios.

The rest of this paper is organized as below—the EGSA method is presented in Section 2; several benchmark functions are used to verify the performance of the EGSA method in Section 3; Section 4 gives the mathematical model for the multi-objective operation of cascade hydropower reservoirs; Section 5 compares the results of different methods; and the conclusions are drawn in the end.

2. Enhanced Gravitational Search Algorithm (EGSA)

2.1. Gravitational Search Algorithm (GSA)

GSA is a novel swarm-based evolutionary algorithm inspired by the universal gravitation and mass interactions in classical Newton's Mechanics [54]. In GSA, the potential solution for the target optimization problems is considered to be an agent with a certain quality, and all the agents are attracted by other agents via the so-called gravity force, which can effectively help the swarm share the acquired beneficial information about the surrounding [55]. Figure 1 draws the sketch map of the GSA method. It can be seen that during the evolutionary process, the gravity force of the two agents is directly and inversely proportional to their qualities and distances, respectively; while the agents with better performances usually have larger attraction forces and slower velocity than those agents with smaller quality. By these means, all agents gradually finish the transition from the global exploration to local exploitation in the decision space, which help the swarm obtain the global optima.

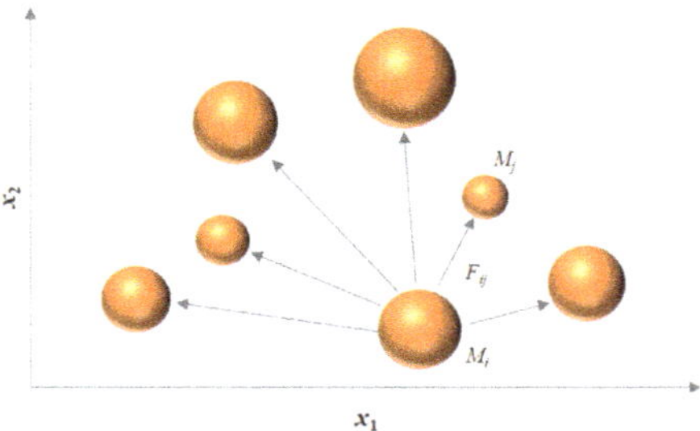

Figure 1. Sketch map of the gravitational search algorithm (GSA) method for a two-variable problem.

Without a loss of generality, it is assumed that there are N agents in the evolutionary swarm to determine the optimal solution for the minimization–optimization problem with D variables, and then the position of the ith agent at the kth iteration ($X_i(k)$ for short) can be expressed as follows:

$$X_i(k) = (x_i^1(k), \cdots, x_i^d(k), \cdots, x_i^D(k)) \tag{1}$$

where $x_i^d(k)$ is the position of the ith agent in the dth dimension at the kth iteration.

In the evolutionary process, the gravitational force of the nth agent acting on the ith agent in the dth dimension at the kth iteration can be expressed as follows:

$$F_{in}^d(k) = G(k) \frac{M_i(k) \times M_n(k)}{R_{in}(k) + \varphi} (x_i^d(k) - x_n^d(k)) \tag{2}$$

where $M_i(k)$ and $M_n(k)$ are the gravitational masses of the ith agent and the nth agent at the kth iteration, respectively. $R_{in}(k) = \|X_i(k), X_n(k)\|_2$ is the Euclidian distance between the ith agent and the nth agent at the kth iteration. φ is a small constant used to prevent the denominator being 0. $G(k)$ is the gravitational constant of the swarm at the kth iteration, which is dynamically updated by:

$$G(k) = G_0 \times \exp(-\alpha \frac{k}{\bar{k}}) \tag{3}$$

where G_0 is the initial gravitational constant. α is the attenuation coefficient. $\bar{k}$ is the maximum number of iterations.

After obtaining the latest positions of all agents, the mass of the ith agent at the kth iteration ($m_i(k)$ for short) can be expressed as follows:

$$\begin{cases} m_i(k) = \dfrac{fit_i(k) - worst(k)}{best(k) - worst(k)} \\ M_i(k) = m_i(k) \Big/ \sum\limits_{i=1}^{N} m_i(k) \end{cases} \tag{4}$$

where $fit_i(k)$ is the fitness of the ith agent at the kth iteration. $best(k)$ and $worst(k)$ are the best and worst fitness values of the swarm at the kth iteration, respectively.

In order to reduce the negative influence of agents that are not conducive to the evolution of the swarm, the resultant force of the ith agent in the dth dimension at the kth iteration ($F_i^d(k)$ for short) is obtained by:

$$F_i^d(k) = \sum_{j=1, j \neq i}^{Kbest} rand_j F_{ij}^d(k) \tag{5}$$

where $rand_j$ is the random number uniformly distributed in the range of [0,1]. *Kbest* is the number of agents with better fitness values.

Based on the motion law in the classical Newtonian mechanics, the acceleration of the ith agent in the dth dimension at the kth iteration ($acc_i^d(k)$ for short) can be expressed as follows:

$$acc_i^d(k) = \frac{F_i^d(k)}{M_i(k)} \tag{6}$$

The velocity and position values of the ith agent in the dth dimension at the $k+1$th iteration ($x_i^d(k+1)$ and $v_i^d(k+1)$ for short) can be updated as below:

$$\begin{cases} v_i^d(k+1) = rand_i \times v_i^d(k) + acc_i^d(k) \\ x_i^d(k+1) = x_i^d(k) + v_i^d(k+1) \end{cases} \tag{7}$$

where $rand_i$ is the random number uniformly distributed in the range of [0,1].

2.2. Opposition Learning Strategy to Improve the Convergence Speed of the Swarm

In practice, a large number of situations can be expressed by the opposition concept, like west–east, south–north, up–down, and left–right. Inspired by this case, opposition learning was proposed for intelligent computing. For the point $x \in [a, b]$, the corresponding opposite point (x^1 for short) of x could be easily obtained by $x^1 = a + b - x$. Based on previous references, when the optimal objective function value was unknown, the opposite directions of the current solution could increase the probability of finding better solutions [56–58]. The opposition learning strategy could enlarge the search region in the three-dimensional space. Thus, the modified opposition learning strategy based on the social learning mechanism in PSO is proposed to improve the convergence speed of the swarm, which could be expressed as below:

$$\begin{cases} re_i^d(k) = Ub^d + Lb^d - a_i^d(k) \\ a_i^d(k) = c_1 \times x_i^d(k) - c_2 \times rand \times \left[gBest^d(k) - x_i^d(k)\right] \end{cases} \tag{8}$$

where $re_i^d(k)$ is the ith opposite agent in the dth dimension at the kth iteration. Ub^d and Lb^d are the upper and lower limits of the dth dimension. c_1 and c_2 are two learning factors. $rand$ is the random number uniformly distributed in the range of [0,1]. $gBest^d(k)$ is the dth dimension of the global optimal agent at the kth iteration.

2.3. Partial Mutation Strategy to Enhance the Individual Diversity

Generally, most agents at a later evolutionary stage are often attracted by the heavier individuals, leading to the loss of swarm diversity [47]. As a result, the GSA method easily falls into the local optima, and it is necessary to find some ways to help the agents search in different regions of the problem space. In some evolutionary algorithms represented by DE and GA, the elitism strategy is often employed to conserve better solutions, while the mutation strategy is used to promote the swarm diversity [59]. Inspired by this, a new partial mutation strategy is introduced to achieve the above goal. Specifically, the agents produced by the opposition learning strategy are first merged with the original parent swarm to form the offspring swarm v; second, the first *cbest* agents in v directly enter the next cycle, while the left agents in v are used to generate the mutated individuals using Equation (9). In this way, some elite agents are maintained, while the search directions of other agents are promoted, which effectively enhance the individual diversity.

$$B_i^d(k) = pBest_\lambda^d(k) + r_1 \times (pBest_i^d(k) - gBest^d(k)) \tag{9}$$

where $B_i^d(k)$ is the position of the ith mutated agent in the dth dimension at the kth iteration. $pBest_i^d(k)$ is the best-known position of the ith agent in the dth dimension at the kth iteration. λ is the number randomly selected from the index set $\{1, 2, \cdots, N\}$. r_1 is the random number uniformly distributed in the range of $[-0.5, 0.5]$.

2.4. Elastic-Ball Modification Strategy to Promote Solution Feasibility

During the random search process of the algorithm, the newly-obtained agent falls out of the feasible space with a certain probability. In the conventional method, the infeasible agents are often forced to be equal to the boundary value. As a result, the number of the agents with boundary values increase gradually with the increasing number of iterations, which might produce a certain negative influence on the evolution of the swarm. To alleviate this problem, an improved elastic-ball strategy was proposed to modify the infeasible agents [60]. Specifically, the infeasible agents were first modified to the feasible position using Equations (10) and (11), and if the newly-obtained agent was still infeasible, the corresponding position would be randomly initialized in the feasible space.

$$x_i^d(k) = \begin{cases} Ub^d - r_2 \times \Delta_i^d(k) \ \text{if}\!\left(x_i^d(k) > Ub^d\right) \\ Lb^d + r_2 \times \Delta_i^d(k) \ \text{if}\!\left(x_i^d(k) < Lb^d\right) \end{cases} \tag{10}$$

$$\Delta_i^d(k) = \begin{cases} x_i^d(k) - Ub^d \ \text{if}\!\left(x_i^d(k) > Ub^d\right) \\ Lb^d - x_i^d(k) \ \text{if}\!\left(x_i^d(k) < Lb^d\right) \end{cases} \tag{11}$$

where r_2 is the random number uniformly distributed in the range of $[0,1]$.

2.5. Execution Procedure of the Proposed EGSA Method

In the EGSA method, three modified strategies (the opposition learning strategy, mutation search strategy, and the elastic-ball modification strategy) were introduced to enhance the global search ability of the standard GSA method in the complex nonlinear constrained optimization problems. Specially, the GSA module could well guarantee the search performance of the swarm; the opposition learning strategy could increase social interaction ability of the agents by exploring the reverse areas; the mutation search strategy was used to balance the convergence speed and the swarm diversity; while the elastic-ball modification strategy could effectively guarantee the feasibility of the newly-generated agents. By combining the advantages of the above strategies, the EGSA method had a stronger ability to enhance the local exploration and global search, in comparison with the conventional GSA method, which would help alleviate the premature convergence problem. Then, the brief execution procedure of the EGSA method was summarized as below:

Step 1: Set the values of the computational parameters and then randomly generate the initial swarm in the problem space.
Step 2: Calculate the fitness values of all the agents in the current population, and then update the personal best-known of each agent and the global best-known agent of the swarm.
Step 3: Calculate the correlated variables (like the gravitational coefficient, mass, and acceleration) to update the velocity and position values of all the agents.
Step 4: Execute the opposition learning strategy to increase the convergence speed of the swarm.
Step 5: Execute the partial mutation search strategy to enhance the individual diversity.
Step 6: Execute the elastic-ball modification strategy to promote the solution feasibility.
Step 7: Repeat Step 2–6 until the stopping criterion is met, and then the global optimal position is regarded as the final solution of the optimization problem.

3. Numerical Experiments to Verify the Performance of the EGSA Method

3.1. Benchmark Functions

To verify the performance of the EGSA algorithm, 12 classic benchmark functions are chosen for numerical experiments. Table 1 shows the details of the selected functions, where D is the number of variables; range is the search scope for each variable; and f_{min} is the optimal objective value in theory. For all test functions, the optimal value of F_8, $-418.9 \times D$, varies with the dimension; while the optimal values for other functions are 0. Meanwhile, the benchmark functions are divided into two categories—unimodal functions (F_1–F_8) with one global optimum, and multimodal functions (F_9–F_{12}) with multiple local optimums. The unimodal functions are used to test the convergence speed of the algorithm, while the multimodal function can test the ability of jumping out of the local optimum, which can fully verify the performance of the evolutionary methods.

Table 1. Detailed information of the twelve benchmark functions.

Function	D	Range	f_{min}
$F_1(x) = \sum_{i=1}^{n} x_i^2$	30	$[-100,100]$	0
$F_2(x) = \sum_{i=1}^{n} |x_i| + \prod_{i=1}^{n} |x_i|$	30	$[-10,10]$	0
$F_3(x) = \sum_{i=1}^{n} (\sum_{j=1}^{i} x_j)^2$	30	$[-100,100]$	0
$F_4(x) = \max\{|x_i|, 1 \le i \le n\}$	30	$[-100,100]$	0
$F_5(x) = \sum_{i=1}^{n-1} [100(x_{i+1} - x_i^2)^2 + (x_i - 1)^2]$	30	$[-30,30]$	0
$F_6(x) = \sum_{i=1}^{n} (x_i + 0.5)^2$	30	$[-100,100]$	0
$F_7(x) = \sum_{i=1}^{n} ix_i^4 + random[0,1)$	30	$[-1.28,1.28]$	0
$F_8(x) = \sum_{i=1}^{n} -x_i \sin(\sqrt{|x_i|})$	30	$[-500,500]$	$-418.9 \times D$
$F_9(x) = \sum_{i=1}^{n} [x_i^2 - 10\cos(2\pi x_i) + 10]$	30	$[-5.12,5.12]$	0
$F_{10}(x) = -20\exp(-0.2\sqrt{\frac{1}{n}\sum_{i=1}^{n} x_i^2}) - \exp(\frac{1}{n}\sum_{i=1}^{n} \cos(2\pi x_i))$ $+20 + e$	30	$[-32,32]$	0
$F_{11}(x) = \frac{1}{4000}\sum_{i=1}^{n} x_i^2 - \prod_{i=1}^{n} \cos(\frac{x_i}{\sqrt{i}}) + 1$	30	$[-600,600]$	0
$F_{12}(x) = \frac{\pi}{n}\{10\sin^2(\pi y_1) + \sum_{i=1}^{n-1} (y_i - 1)^2[1 + 10\sin^2(\pi y_{i+1})]$ $+(y_n - 1)^2\} + \sum_{i=1}^{n} u(x_i, 10, 100, 4)$ $y_i = 1 + \frac{x_i+1}{4}, u(x_i, a, k, m) = \begin{cases} k(x_i - a)^m & x_i > a \\ 0 & a \le x_i < a \\ k(-x_i - a)^m & x_i < -a \end{cases}$	30	$[-50,50]$	0

3.2. Parameters Settings

For the sake of fair comparison, the results of 11 famous evolutionary algorithms are introduced in this section, including differential evolution (DE), particle swarm optimization (PSO), sine cosine algorithm (SCA), gravitational search algorithm (GSA), ant lion optimizer (ALO) [61], cuckoo search algorithm (CS) [62], modified cuckoo search algorithm (MCS) [63], lightning search algorithm (LSA) [64], grey wolf optimizer (GWO) [65], firefly algorithm (FA) [66], and whale optimization algorithm (WOA) [67]. The results of these five methods (DE, PSO, SCA, GSA, and EGSA) were developed in the JAVA procedures, while the results of the other methods were taken from the corresponding literature. For the five developed methods, the swarm size and maximum iteration were set as 50 and 1000; while the other parameters were set as:

DE: The scaling factor was set as 0.5 and the crossover probability was set as 0.6.

PSO: The inertia weight w was linearly decremented from 0.9 to 0.3; while two learning factors (c_1 and c_2) were set as 2.0, respectively.

SCA: The computational constant a was set as 2.0.

GSA: The attenuation coefficient was set as 20; the initial gravitational constant was set as 100.

EGSA: The attenuation coefficient was set to 20; the initial gravitational constant was set to 100; while the value of *cbest* was set to 0.7.

3.3. Comparison with Other Evolutionary Algorithms in Small-Scale Problems

3.3.1. Result Comparison in Multiple Runs

For the sake of alleviating the adverse influence of random initial seeds, all approaches were independently executed 30 times. Table 2 shows the average (Ave.) and standard deviation (Std.) of 12 algorithms for all the test functions, where the number of variables was set as 30. It could be seen that for all the functions, EGSA was always superior to the GSA method in terms of both the average and standard deviation; as compared with other methods, the EGSA method could obtain satisfying results in most functions. For instance, EGSA outperformed ALO in the average of all test functions except for F_5, the WOA performance was tied with EGSA in F_9 and defeated by EGSA in the other functions. To sum up, the proposed method could obtain better results than the other evolutionary algorithms in the 12 test functions.

3.3.2. Box and Whisker Test

In this section, the box and whisker test was employed to demonstrate the objective distribution of the solutions since it can show the abundant information (including minimum, second and third quartile, median, and maximum) on the studied data. Figure 2 shows the detailed data obtained by the three methods in 30 independent runs. It could be clearly observed that compared to the other methods, the proposed method had a smaller distribution and better values in all indices. For instance, the standard GSA algorithm had different degrees of outliers on F_4–F_5, which indicated that this method easily fell into the state of premature convergence in the search process; the SCA algorithm exhibited outliers in most functions but had a distribution of relatively discrete solutions in F_3, which indicated that SCA was not ideal in terms of robustness. Additionally, the EGSA method had a more concentrated solution distribution and fewer outliers in all functions, demonstrating its satisfying robustness and search ability. Thus, the performances of the EGSA method were superior to the comparative methods in the 12 functions.

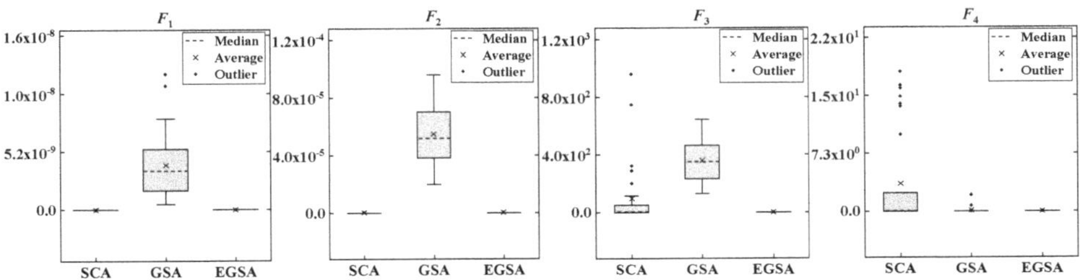

Figure 2. *Cont.*

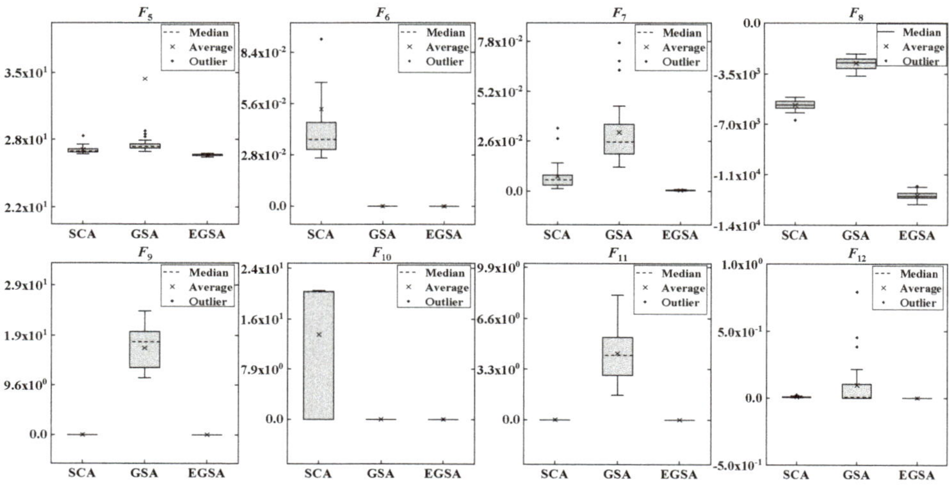

Figure 2. Box and whisker test of three algorithms for 12 functions with 30 variables.

3.3.3. Wilcoxon Nonparametric Test

To achieve statistically meaningful conclusions, the Wilcoxon nonparametric test was employed to test the significance level of various methods. Table 3 shows the statistical results of various methods. If the results of the EGSA was better than the control method, it was recorded as a win; if two methods were tied, it was recorded as a tie; otherwise, it was recorded as a loss. From Table 3, it can be seen that the proposed method could produce better solutions as compared to the other methods, due to the larger number of the 'Win' symbol. Then, the multiple-problem statistical results obtained by the Wilcoxon nonparametric test with a significance level of 0.05 are given in Table 4, where the mean objective value obtained by each method is chosen as the data sample. From Table 4, it can be found that the proposed method could achieve higher R^+ than R^- in all comparisons, while all values of p were smaller than 0.05. This case proved that the EGSA method outperformed the other methods in a statistical manner.

Table 2. Statistical indexes of 12 algorithms for all the test functions.

Function	Item	CS	MCS	LSA	GWO	FA	WOA	ALO	DE	PSO	SCA	GSA	EGSA
F_1	Ave.	9.06×10^1	1.01×10^0	4.81×10^{-8}	6.59×10^{-28}	1.20×10^{-2}	1.41×10^{-30}	2.59×10^{-10}	7.80×10^{-6}	4.58×10^{-7}	2.87×10^{-35}	4.00×10^{-9}	6.96×10^{-134}
	Std.	2.62×10^1	2.72×10^{-1}	3.40×10^{-7}	6.34×10^{-5}	4.30×10^{-3}	4.91×10^{-30}	1.65×10^{-10}	3.14×10^{-6}	3.51×10^{-7}	1.26×10^{-34}	2.92×10^{-9}	3.70×10^{-133}
F_2	Ave.	9.70×10^0	1.81×10^{-1}	3.68×10^{-2}	7.18×10^{-17}	3.73×10^{-1}	1.06×10^{-21}	1.84×10^{-6}	3.73×10^{-4}	1.19×10^{-4}	2.21×10^{-27}	5.48×10^{-5}	5.21×10^{-69}
	Std.	1.98×10^0	3.31×10^{-2}	1.56×10^{-1}	2.90×10^{-2}	1.01×10^{-1}	2.39×10^{-21}	6.58×10^{-7}	9.05×10^{-5}	8.03×10^{-5}	6.99×10^{-27}	2.24×10^{-5}	2.09×10^{-68}
F_3	Ave.	3.84×10^3	4.62×10^2	4.32×10^1	3.29×10^{-6}	1.81×10^3	5.39×10^{-7}	6.06×10^{-10}	2.94×10^4	2.33×10^0	9.74×10^1	3.61×10^2	4.30×10^{-119}
	Std.	7.24×10^2	1.23×10^2	2.99×10^1	7.91×10^1	6.60×10^2	2.93×10^{-6}	6.34×10^{-10}	4.13×10^3	1.14×10^0	2.24×10^2	1.45×10^2	2.29×10^{-118}
F_4	Ave.	7.23×10^0	1.73×10^0	1.49×10^0	5.61×10^{-7}	7.67×10^{-2}	7.26×10^{-2}	1.36×10^{-8}	1.43×10^0	5.35×10^{-1}	2.66×10^0	9.48×10^{-2}	5.58×10^{-70}
	Std.	6.76×10^{-1}	5.12×10^{-1}	1.30×10^0	1.32×10^0	1.46×10^{-2}	3.97×10^{-1}	1.81×10^{-9}	3.21×10^{-1}	1.33×10^{-1}	4.16×10^0	3.92×10^{-1}	2.12×10^{-69}
F_5	Ave.	4.98×10^0	5.81×10^1	6.43×10^1	2.68×10^1	1.28×10^2	2.79×10^1	3.47×10^{-1}	4.45×10^2	1.64×10^2	2.74×10^1	3.45×10^1	2.69×10^1
	Std.	1.75×10^0	3.31×10^1	4.38×10^1	6.99×10^1	2.79×10^2	7.64×10^{-1}	1.10×10^{-1}	1.55×10^2	1.55×10^2	3.85×10^{-1}	3.65×10^1	9.72×10^{-2}
F_6	Ave.	4.31×10^4	4.44×10^3	3.34×10^0	8.17×10^{-1}	0	3.12×10^0	2.56×10^{-10}	7.94×10^{-6}	9.29×10^{-7}	5.30×10^{-2}	4.99×10^{-9}	8.23×10^{-15}
	Std.	7.21×10^3	7.52×10^2	2.09×10^0	1.26×10^{-4}	0	5.32×10^{-1}	1.09×10^{-10}	3.39×10^{-6}	1.93×10^{-6}	4.98×10^{-2}	3.74×10^{-9}	4.61×10^{-15}
F_7	Ave.	2.46×10^{-2}	9.10×10^{-3}	2.41×10^{-2}	2.21×10^{-3}	3.52×10^{-2}	1.43×10^{-3}	4.29×10^{-3}	1.35×10^{-1}	3.55×10^0	7.48×10^{-3}	3.05×10^{-2}	4.76×10^{-4}
	Std.	7.90×10^{-3}	2.20×10^{-3}	5.72×10^{-3}	1.00×10^{-1}	2.40×10^{-2}	1.15×10^{-3}	5.09×10^{-3}	2.93×10^{-2}	4.79×10^0	6.96×10^{-3}	1.59×10^{-2}	2.32×10^{-4}
F_8	Ave.	-8.98×10^3	-9.80×10^3	-8.00×10^3	-6.12×10^3	-5.90×10^3	-5.08×10^3	-1.61×10^3	-7.95×10^3	-6.39×10^3	-5.66×10^3	-2.75×10^3	-1.19×10^4
	Std.	1.98×10^2	5.31×10^2	6.69×10^2	4.09×10^3	6.56×10^2	6.96×10^2	3.14×10^2	3.33×10^2	1.29×10^3	3.52×10^2	4.00×10^2	3.13×10^2
F_9	Ave.	2.94×10^2	1.35×10^2	6.28×10^1	3.11×10^{-1}	2.63×10^1	0	7.71×10^{-6}	1.32×10^2	7.87×10^1	1.87×10^{-9}	1.67×10^1	0
	Std.	1.43×10^1	2.16×10^1	1.49×10^1	4.74×10^1	9.15×10^0	0	8.45×10^{-6}	8.78×10^0	2.91×10^1	1.02×10^{-8}	3.82×10^0	0
F_{10}	Ave.	1.93×10^1	1.21×10^1	2.69×10^0	1.06×10^{-13}	5.12×10^{-2}	7.40×10^0	3.73×10^{-15}	1.37×10^{-3}	5.99×10^{-4}	1.34×10^1	3.25×10^{-5}	3.64×10^{-15}
	Std.	3.50×10^{-1}	7.52×10^{-1}	9.11×10^{-1}	7.78×10^{-2}	1.37×10^{-2}	9.90×10^0	1.50×10^{-15}	8.85×10^{-4}	5.43×10^{-4}	9.64×10^0	1.10×10^{-5}	1.08×10^{-15}
F_{11}	Ave.	2.12×10^2	8.32×10^0	7.24×10^{-3}	4.49×10^{-3}	5.84×10^{-3}	2.89×10^{-4}	1.86×10^{-2}	3.36×10^{-3}	6.98×10^{-3}	1.86×10^{-5}	4.34×10^0	0
	Std.	3.97×10^1	1.54×10^0	6.70×10^{-3}	6.66×10^{-3}	1.43×10^{-3}	1.59×10^{-3}	9.55×10^{-3}	1.06×10^{-2}	8.16×10^{-3}	1.02×10^{-4}	1.66×10^0	0
F_{12}	Ave.	1.47×10^0	1.38×10^{-1}	3.58×10^{-1}	5.34×10^{-2}	2.40×10^{-4}	3.40×10^{-1}	9.74×10^{-12}	6.87×10^{-1}	2.01×10^{-2}	8.52×10^{-3}	9.59×10^{-2}	5.30×10^{-17}
	Std.	3.61×10^{-1}	2.86×10^{-1}	7.44×10^{-1}	2.07×10^{-2}	1.00×10^{-4}	2.15×10^{-1}	9.33×10^{-12}	3.05×10^{-1}	2.26×10^{-2}	4.69×10^{-3}	1.75×10^{-1}	1.92×10^{-17}

Note: Ave. = average; Std. = standard deviation.

Table 3. The results of Wilcoxon test for a single problem for statistical significance level at alpha = 0.05.

Function	EGSA–CS	EGSA–MCS	EGSA–LSA	EGSA–GWO	EGSA–FA	EGSA–WOA	EGSA–ALO	EGSA–DE	EGSA–PSO	EGSA–SCA	EGSA–GSA
F_1	EGSA	EGSA	EGSA	EGSA	EGSA	EGSA	EGSA	EGSA	EGSA	EGSA	EGSA
F_2	EGSA	EGSA	EGSA	EGSA	EGSA	EGSA	EGSA	EGSA	EGSA	EGSA	EGSA
F_3	EGSA	EGSA	EGSA	EGSA	EGSA	EGSA	EGSA	EGSA	EGSA	EGSA	EGSA
F_4	EGSA	EGSA	EGSA	EGSA	EGSA	EGSA	EGSA	EGSA	EGSA	EGSA	EGSA
F_5	CS	EGSA	EGSA	GWO	EGSA	EGSA	ALO	EGSA	EGSA	EGSA	EGSA
F_6	EGSA	EGSA	EGSA	EGSA	FA	EGSA	EGSA	EGSA	EGSA	EGSA	EGSA
F_7	EGSA	EGSA	EGSA	EGSA	EGSA	EGSA	EGSA	EGSA	EGSA	EGSA	EGSA
F_8	EGSA	EGSA	EGSA	EGSA	EGSA	EGSA	EGSA	EGSA	EGSA	EGSA	EGSA
F_9	EGSA	EGSA	EGSA	EGSA	EGSA	Tie	EGSA	EGSA	EGSA	EGSA	EGSA
F_{10}	EGSA	EGSA	EGSA	EGSA	EGSA	EGSA	EGSA	EGSA	EGSA	EGSA	EGSA
F_{11}	EGSA	EGSA	EGSA	EGSA	EGSA	EGSA	EGSA	EGSA	EGSA	EGSA	EGSA
F_{12}	EGSA	EGSA	EGSA	EGSA	EGSA	EGSA	EGSA	EGSA	EGSA	EGSA	EGSA
W/T/L	11/0/1	12/0/0	12/0/0	11/0/1	11/0/1	11/1/0	11/0/1	12/0/0	12/0/0	12/0/0	12/0/0

Table 4. The results of Wilcoxon signed-rank test for statistically significance level at alpha = 0.05.

Item	EGSA–CS	EGSA–MCS	EGSA–LSA	EGSA–GWO	EGSA–FA	EGSA–WOA	EGSA–ALO	EGSA–DE	EGSA–PSO	EGSA–SCA	EGSA–GSA
R^+	72	78	78	70	77	77.5	67	78	78	78	78
R^-	6	0	0	8	1	0.5	11	0	0	0	0
p-value	6.84×10^{-3}	4.88×10^{-4}	4.88×10^{-4}	1.22×10^{-2}	9.77×10^{-4}	9.77×10^{-4}	2.69×10^{-2}	4.88×10^{-4}	4.88×10^{-4}	4.88×10^{-4}	4.88×10^{-4}
Significant	√	√	√	√	√	√	√	√	√	√	√

3.3.4. Convergence Analysis

Figure 3 shows the convergence trajectories of five algorithms for 12 functions with 30 variables. It can be seen that the proposed method could constantly improve the solution's quality from the beginning to the end, and could find the best solutions in the end. Taking the functions F_1–F_4 as examples, three methods (GSA, PSO, and DE) quickly fail into local optima because their best-so-far solutions change slightly in the evolutionary process; both EGSA and SCA converge to a smaller objective at the initial stage, but EGSA can quickly find better solutions at iteration 800, while SCA fails to make it. Additionally, for multimodal functions, EGSAs achieve the global optimal solution in both F_9 and F_{11}, and better results than other methods in F_8, F_{10}, and F_{12}. Thus, the above analysis fully proves that the presented method has a superior convergence speed and global search ability.

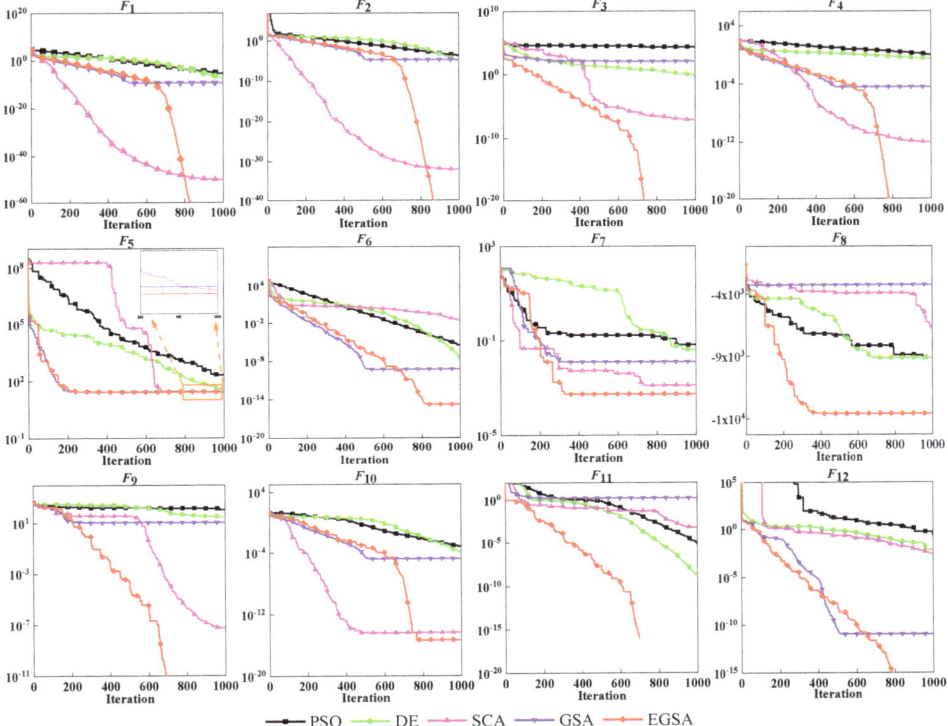

Figure 3. Convergence trajectories of the different methods for 12 benchmark functions with 30 variables.

4. EGSA for the Multi-Objective Operation of Cascade Hydropower Reservoirs

4.1. Mathematical Model

4.1.1. Objective Functions

With the advantages of low pollutant emission and high operational efficiency, hydropower now plays an increasingly important role in the energy system throughout the world [68–70]. As a result, the cascade hydropower reservoirs are often responsible for a variety of operational requirements. Considering the practical requirements of generation enterprises and electrical power systems in China, the generation benefit and peak operations were chosen as the focus of this paper.

(1) Power Generation

For almost all hydropower enterprises, the economic benefit is an important indicator of performance check under the market environment [71–73]. Generally, the same amount of available water resource can produce numerous scheduling schemes with different generation benefits. Hence, the first objective was chosen to find the best scheme that could maximize the total power generation of all hydropower reservoirs, which could be expressed as below:

$$E = \max \sum_{t=1}^{T} \sum_{n=1}^{N} P_{n,t} \times \Delta_t \tag{12}$$

where E is the power generation of all hydropower reservoirs. T is the number of periods. N is the number of hydropower reservoirs. $P_{n,t}$ is the output of the nth hydroplant at the tth period. Δ_t is the total hours of the tth period.

(2)   Peak Operation

In modern power systems, the hydropower system is often asked to respond to the peak loads of power grid to produce a relatively smooth load series for other kinds of energy systems (like thermal, solar, and wind) [74–76]. In this way, the total operational cost of the electrical power system in the long run can be sharply reduced [77–79]. Hence, the second objective function was chosen to minimize the mean square deviation of the residual load curve obtained by subtracting all power outputs of hydropower reservoirs from the original energy load curve, which could be expressed as:

$$F = \min \sqrt{\frac{1}{2} \sum_{t=1}^{T} \left[Load_t - \sum_{n=1}^{N} P_{n,t} \right]^2} \tag{13}$$

where $Load_t$ is the load demand of the power system at the tth period.

4.1.2. Physical Constraints

(1)   Water Balance Equation:

$$V_{n,t+1} = V_{n,t} + 3600\Delta_t \times \left[I_{n,t} + \sum_{i=1}^{Num} (Q_{i,t} + S_{i,t}) - Q_{n,t} - S_{n,t} \right] \tag{14}$$

where $V_{n,t}$, $I_{n,t}$, $Q_{n,t}$, $S_{n,t}$ are the storage volume, local inflow, turbine discharge and spillage of the nth hydroplant at the tth period, respectively. Num is the number of upstream hydroplants directly connected to the nth hydroplant.

(2)   Reservoir Storage Constraint:

$$\underline{V}_{n,t} \leq V_{n,t} \leq \overline{V}_{n,t} \tag{15}$$

where $\overline{V}_{n,t}$ and $\underline{V}_{n,t}$ denote the maximum and minimum storage volumes of the nth hydroplant at the tth period, respectively.

(3)   Turbine Discharge Constraint:

$$\underline{Q}_{n,t} \leq Q_{n,t} \leq \overline{Q}_{n,t} \tag{16}$$

where $\overline{Q}_{n,t}$ and $\underline{Q}_{n,t}$ denote the maximum and minimum turbine discharges of the nth hydroplant at the tth period, respectively.

(4)   Total Discharge Constraint:

$$\underline{q}_{n,t} \leq q_{n,t} \leq \overline{q}_{n,t} \tag{17}$$

where $\bar{q}_{n,t}$ and $\underline{q}_{n,t}$ denote the maximum and minimum total discharges of the nth hydroplant at the tth period, respectively.

(5) Power Output Constraint:

$$\underline{P}_{n,t} \leq P_{n,t} \leq \bar{P}_{n,t} \tag{18}$$

where $\bar{P}_{n,t}$ and $\underline{P}_{n,t}$ denote the maximum and minimum power outputs of the nth hydroplant at the tth period, respectively.

(6) Initial and Final Water Levels Constraint:

$$\begin{cases} Z_{n,0} = Z_n^{beg} \\ Z_{n,T} = Z_n^{end} \end{cases} \tag{19}$$

where Z_n^{beg} and Z_n^{end} represent the initial and final water level of the nth hydroplant, respectively.

4.2. Technique for Order Preference by Similarity to an Ideal Solution (TOPSIS)

As mentioned above, the multi-objective operation of cascade hydropower reservoirs is a typical multiple-criteria decision-making problem. Thus, it was necessary to find some tools to transform the original multi-objective problem into the single-objective optimization problem that can be addressed by EGSA. After a comprehensive comparison, TOPSIS was introduced to achieve this goal. In TOPSIS, the best scheme had the smallest distance from the positive ideal scheme and the largest distance from the negative ideal scheme, respectively. Then, the procedure of TOPSIS was give as below:

Step 1: Create an initial decision evaluation matrix with m schemes and J attributes. For the target problem, each scheme is composed of two attributes—the power generation E and the peak operation F. Then, the decision-making matrix A can be expressed as follows:

$$A = \left(a_{i,j}\right)_{m \times J} = \begin{bmatrix} a_{11} & a_{12} & \cdots & a_{1J} \\ a_{21} & a_{22} & \cdots & a_{2J} \\ \vdots & \vdots & a_{i,j} & \vdots \\ a_{m1} & a_{m2} & \cdots & a_{mJ} \end{bmatrix} \tag{20}$$

where $a_{i,j}$ is the original value of the jth attribute in the ith scheme.

Step 2: Obtain the modified decision evaluation matrix to reduce the potential impact of various attribute properties (like measurement unit and tendency); given below:

$$B = \left(b_{i,j}\right)_{m \times J} = \begin{bmatrix} b_{11} & b_{12} & \cdots & b_{1J} \\ b_{21} & b_{22} & \cdots & b_{2J} \\ \vdots & \vdots & b_{i,j} & \vdots \\ b_{m1} & b_{m2} & \cdots & b_{mJ} \end{bmatrix} \tag{21}$$

$$b_{ij} = \begin{cases} \dfrac{a_j^{max} - a_{ij}}{a_j^{max} - a_j^{min}} & \text{minimization attribute} \\[2ex] \dfrac{a_{ij} - a_j^{min}}{a_j^{max} - a_j^{min}} & \text{maximization attribute} \end{cases}, i = 1, 2, \cdots, m; j = 1, 2, \cdots, J \tag{22}$$

where a_j^{max} and a_j^{min} are the maximum and minimum values of the jth attribute.

Step 3: Obtain the normalized decision-making matrix to reduce the possible numerical problem caused by the feature differences (like units or magnitude), which could be expressed as below:

$$R = \left(r_{i,j}\right)_{m \times J} = \begin{bmatrix} r_{11} & r_{12} & \cdots & r_{1J} \\ r_{21} & r_{22} & \cdots & r_{2J} \\ \vdots & \vdots & r_{i,j} & \vdots \\ r_{m1} & r_{m2} & \cdots & r_{mJ} \end{bmatrix} \tag{23}$$

$$r_{ij} = \frac{b_{ij}}{\sqrt{\sum\limits_{i=1}^{m} b_{ij}^2}}, i = 1, 2, \cdots, m; \ j = 1, 2, \cdots, J \tag{24}$$

where r_{ij} is the normalized value of the jth attribute in the ith scheme.

Step 4: Set the weight vector $w = \left[w_1, \cdots, w_j, \cdots, w_J\right]$, where w_j is the weight of the jth attribute and there is $\sum\limits_{j=1}^{J} w_j = 1$. Then, calculate the weighted normalized decision-making matrix V by:

$$V = \left(v_{i,j}\right)_{m \times J} = \left(w_j \times r_{i,j}\right)_{m \times J} \tag{25}$$

Step 5: Determine the positive ideal scheme C^+ and negative ideal scheme C^- by:

$$\begin{cases} C^+ = \left[c_1^+, c_2^+, \ldots, c_J^+\right] = \max\{v_{i,j} | i = 1, 2, \cdots, m\} \\ C^- = \left[c_1^-, c_2^-, \ldots, c_J^-\right] = \min\{v_{i,j} | i = 1, 2, \cdots, m\} \end{cases} j = 1, 2, \ldots, J \tag{26}$$

Step 6: Calculate the Euclidean distances between the decision scheme and the positive (negative) ideal scheme, which could be expressed as below:

$$\begin{cases} D_i^+ = \sqrt{\sum\limits_{j=1}^{J} (v_{ij} - c_j^+)^2} \\ D_i^- = \sqrt{\sum\limits_{j=1}^{J} (v_{ij} - c_j^-)^2} \end{cases}, i = 1, 2, \ldots m \tag{27}$$

where D_i^+ or D_i^- is the distance between the ith scheme and the position (or negative) ideal scheme.

Step 7: Calculate the closeness of each decision plan to the negative ideal scheme by

$$f_i = \frac{D_i^-}{D_i^- + D_i^+}, i = 1, 2, \ldots m \tag{28}$$

Step 8: Use the obtained closeness values to sort all the schemes, and the scheme possessing the maximum closeness will be treated as the optimal decision scheme.

$$f_{best} = \max\{f_1, f_2, \ldots, f_m\} \tag{29}$$

4.3. Details of EGSA for Multi-Objective Operation of Cascade Hydropower Reservoirs

4.3.1. Individual Structure and Swarm Initialization

To enhance the execution efficiency of reservoir operation, the total discharge of the hydropower reservoir is selected as the decision variable for encoding, and then the ith agent at the kth iteration $X_i(k)$ can be expressed as below:

$$X_i(k) = \begin{bmatrix} q_{1,1} & q_{1,2} & \cdots & q_{1,T} \\ q_{2,1} & q_{2,2} & \cdots & q_{2,T} \\ \vdots & \vdots & q_{n,t} & \vdots \\ q_{N,1} & q_{N,2} & \cdots & q_{N,T} \end{bmatrix} \tag{30}$$

During the initialization phase, the elements of the solution $X_i(k)$ are randomly generated in the preset of the total discharge scopes of all hydroplants, which could be expressed as follows:

$$q_{n,t} = \underline{q}_{n,t} + random \times \left(\bar{q}_{n,t} - \underline{q}_{n,t} \right) \tag{31}$$

where *random* is the random number uniformly distributed in the range of [0,1].

4.3.2. Heuristic Constraint Handling Method

During the evolutionary process, it is possible that some agents violate the equality or inequality constraints imposed on the hydropower system. In order to effectively modify the infeasible agents, this section presents a heuristic constraint handling method that tries to equally allocate the violated final storage volume to the turbine discharges of all adjusting periods. The detailed procedures for the solution $X_i(k)$ are given as below:

Step 1: Set $n = 1$ and determine the necessary parameters for the constraint processing.

Step 2: Set the counter $L = 1$, and then use the water balance equation to calculate the storage of the nth hydroplant at the tth period by

$$V_{n,t} = \max\left\{ \underline{V}_{n,t}, \min\left\{ V_{n,t-1} + 3600\Delta_{t-1} \times \left[I_{n,t-1} + \sum_{i=1}^{Num} (Q_{i,t-1} + S_{i,t-1}) - Q_{n,t-1} - S_{n,t-1} \right], \bar{V}_{n,t} \right\} \right\} \tag{32}$$

Step 3: Set $L = L + 1$, and then calculate the violation value η_n of the final storage by Equation (33). Then, if $|\eta_n|$ is smaller than the accuracy ε or L is larger than the maximum iteration, go to Step 5; otherwise, turn to Step 4.

$$\eta_n = f_n(Z_{n,T}) - f_n\left(Z_n^{end}\right) \tag{33}$$

where $f_n(\cdot)$ is the nonlinear stage-storage curve of the nth hydropower reservoir.

Step 4: Use Equation (34) to obtain the modified total discharges of the target reservoir, and then calculate the corresponding reservoir storage volume by Equation (32) before turning to Step 3.

$$q_{n,t} = \max\left\{ \underline{q}_{n,t}, \min\left\{ q_{n,t} + \frac{\eta_n}{T \times 3600\Delta_t}, \bar{q}_{n,t} \right\} \right\} \tag{34}$$

Step 5: Set $n = n+1$. If $n \leq N$, turn to Step 2 for the new hydroplant; otherwise, stop the iteration.

4.3.3. Calculation of the Modified Objective Values

Generally, the agents modified by the above heuristic constraint handling procedures would satisfy most of the physical constraints. Nevertheless, the agents might still violate some constraints due to reasons like unreasonable operation trajectory and limited adjusted times. To eliminate the negative effect of the infeasible agents, the value of the constraint violation involved in the solution $X_i(k)$ ($Viol[X_i(k)]$ for short) was obtained by Equation (35), and then merged into two original objective values (E and F) by Equation (36) to obtain the modified objective values (E' and F').

$$Viol[X_i(k)] = \sum_{l_1=1}^{L_g} c_{l_1} \cdot \max\left\{ g_{l_1}(X_i(k)), 0 \right\} + \sum_{l_2=1}^{L_e} c_{l_2} \cdot \left[e_{l_2}(X_i(k)) \right]^2 \tag{35}$$

$$\begin{cases} E'[X_i(k)] = \sum_{t=1}^{T} \sum_{n=1}^{N} P_{n,t} \times \Delta t - Viol[X_i(k)] \\ F'[X_i(k)] = \sqrt{\frac{1}{2} \sum_{t=1}^{T} \left[Load_t - \sum_{n=1}^{N} P_{n,t} \right]^2} + Viol[X_i(k)] \end{cases} \tag{36}$$

where $g_{l_1}(\cdot) \leq 0$ is the l_1th inequality constraint; $e_{l_2}(\cdot) \leq 0$ is the l_2th equality constraint; c_{l_1} and c_{l_2} are the penalty coefficients for the relevant inequality or equality constraint. L_g and L_e denote the number of inequality and equality constraints, respectively.

4.3.4. Execution Procedures of the EGSA Method for the Target Problem

The flowchart of the developed EGSA method for solving multi-objective operation of cascade hydropower reservoirs is given in Figure 4.

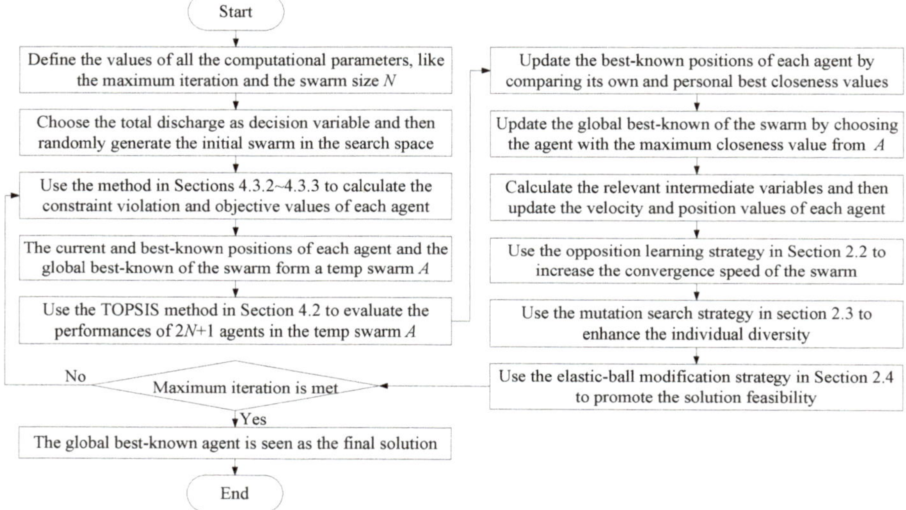

Figure 4. The flowchart of enhanced gravitational search algorithm (EGSA) for multi-objective operation of cascade hydropower reservoirs.

5. Case Studies

5.1. Engineering Background

In this section, five reservoirs of the Wu hydropower system in Southwest China were chosen to test the performance of the proposed method, including Hongjiadu (HJD), Dongfeng (DF), Suofengying (SFY), Wujiangdu (WJD), and Goupitan (GPT). Since being put into operation, the Wu hydropower system has played a vitally important role in the environment protection and economic development of the Guizhou province. Figure 5 shows the topological structure of the selected hydropower reservoirs, while the typical load demand curves of the four seasons are shown in Figure 6. From Figures 5 and 6, it could be observed that in the Wu hydropower system, there were tight hydraulic and electrical connections from upstream to downstream reservoirs, while four load curves with different features further increased the optimization difficulty.

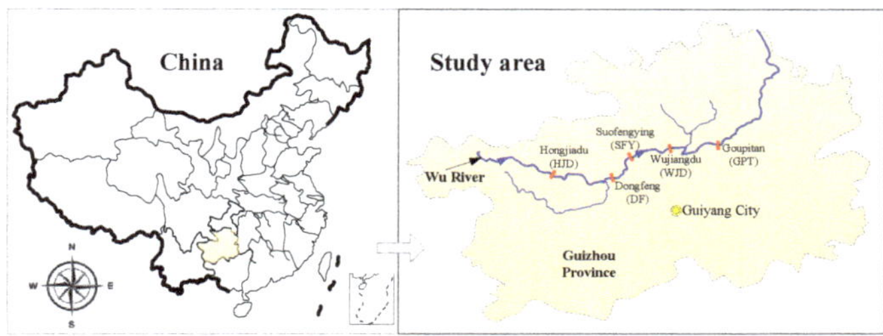

Figure 5. The topological structure of the studied hydropower reservoirs.

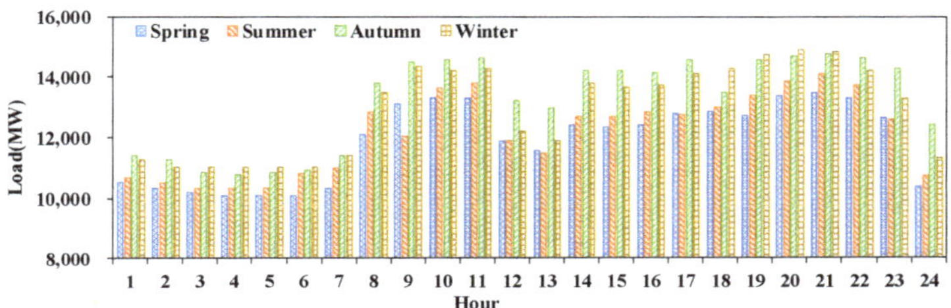

Figure 6. Typical load curves in different season.

5.2. Case Study 1: Power Generation of Cascade Hydropower Reservoirs

5.2.1. Robustness Testing of Different Evolutionary Algorithms

In this section, to verify the robustness of the developed method, four famous methods (DE, PSO, SCA, and GSA) were introduced for comparative analysis. Table 5 shows the detailed statistical results of the 5 methods in 20 independent runs for 4 cases, where the swarm size and maximum iteration were set as 50 and 500, respectively. It can be clearly seen that the solutions obtained by the EGSA method were more stable than the other methods. For instance, in Case 1, the EGSA method could make about 96.4%, 96.5%, 97.8%, and 97.7% reductions in the standard deviations of the DE, PSO, SCA, and GSA. Hence, the feasibility of the EGSA method was fully demonstrated in this case.

Table 5. Statistical results of the 5 methods in 20 independent runs for 4 cases of power generation (10^4 kW·h).

Case	Method	Best	Worst	Average	Standard Deviation	Range
	DE	5388.10	5386.25	5387.19	0.56	1.85
	PSO	5393.67	5391.27	5392.49	0.57	2.40
Case 1	SCA	5388.26	5384.34	5385.65	0.89	3.92
	GSA	5388.37	5385.34	5386.76	0.88	3.03
	EGSA	5396.87	5396.81	5396.86	0.02	0.06
	DE	6439.12	6434.22	6436.14	1.43	4.90
	PSO	6445.48	6442.35	6444.12	0.79	3.13
Case 2	SCA	6439.06	6435.25	6437.07	1.03	3.81
	GSA	6440.17	6437.00	6438.55	0.79	3.17
	EGSA	6450.34	6450.33	6450.33	0.01	0.01
	DE	4977.94	4976.38	4977.17	0.40	1.56
	PSO	4981.62	4979.71	4980.58	0.55	1.91
Case3	SCA	4975.51	4973.58	4974.54	0.58	1.93
	GSA	4976.74	4973.83	4975.25	0.71	2.91
	EGSA	4984.56	4984.54	4984.55	0.01	0.02
	DE	4433.08	4431.68	4432.32	0.44	1.40
	PSO	4435.89	4434.23	4435.11	0.49	1.66
Case4	SCA	4431.51	4428.62	4429.58	0.70	2.89
	GSA	4431.61	4429.07	4430.37	0.64	2.54
	EGSA	4438.33	4438.32	4438.33	0.01	0.01

Figure 7 draws the box and whisker test of the best-so-far solutions obtained by 5 algorithms in 4 cases. It can be observed that the performances of four control methods in the power generation of hydropower system are relatively stable but still inferior to the EGSA method possessing the smallest variation ranges in the final solutions. Therefore, the proposed method proves to be an effective tool to deal with the power generation of a hydropower system.

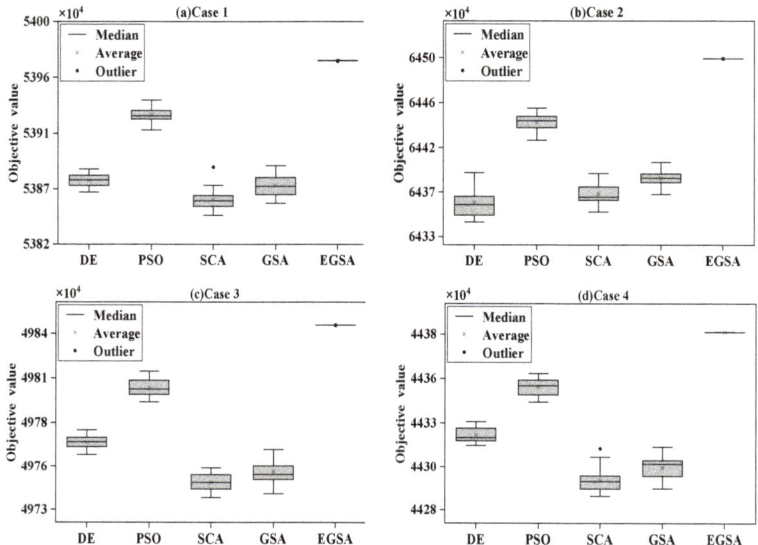

Figure 7. Box and whisker test of different algorithms in different cases for power generation.

5.2.2. Comparison of the Optimal Results Obtained by Different Evolutionary Algorithms

Table 6 shows the detailed results in the best solutions obtained by different methods. It can be seen that in all cases, EGSA can always find the best results among all algorithms. For instance, compared with DE, PSO, SCA, and GSA, the EGSA method could make about 8.77×10^4 kW·h, 3.2×10^4 kW·h, 8.61×10^4 kW·h, and 8.5×10^4 kW·h improvements in power generation in Case 1, respectively. Additionally, among all hydroplants, the GPT hydroplant with a huge installed capacity produced the largest proportion of power generation in all solutions, which was in line with the actual operation situation of the hydropower system [80–83]. Thus, the above analysis demonstrated the effectiveness and rationality of the scheduling process obtained by the proposed method.

Table 6. Detailed results in the best solutions by different methods (10^4 kW·h).

Runoff	Method	HJD	DF	SFY	WJD	GPT	Sum
	DE	726.73	646.19	527.41	1106.65	2381.12	5388.10
	PSO	728.66	647.76	526.51	1107.99	2382.75	5393.67
Case 1	SCA	727.92	644.60	527.68	1106.77	2381.29	5388.26
	GSA	728.35	647.35	526.77	1107.15	2378.75	5388.37
	EGSA	728.84	646.94	530.33	1108.10	2382.66	5396.87
	DE	869.46	770.94	632.25	1322.47	2844.00	6439.12
	PSO	871.70	774.09	627.59	1325.38	2846.72	6445.48
Case 2	SCA	871.15	771.70	628.34	1322.76	2845.11	6439.06
	GSA	871.48	774.01	627.42	1324.03	2843.23	6440.17
	EGSA	871.58	773.17	633.84	1324.91	2846.84	6450.34

Table 6. *Cont.*

Runoff	Method	HJD	DF	SFY	WJD	GPT	Sum
	DE	234.57	400.46	461.27	1052.33	2829.31	4977.94
	PSO	235.21	402.56	459.19	1054.13	2830.53	4981.62
Case 3	SCA	233.99	401.01	459.94	1052.01	2828.56	4975.51
	GSA	235.02	402.45	457.71	1053.20	2828.36	4976.74
	EGSA	235.50	401.67	462.62	1054.23	2830.54	4984.56
	DE	208.46	355.81	411.02	936.99	2520.8	4433.08
	PSO	208.58	358.21	409.24	938.46	2521.4	4435.89
Case 4	SCA	208.26	356.74	409.21	937.08	2520.22	4431.51
	GSA	208.82	357.66	408.43	937.47	2519.23	4431.61
	EGSA	209.13	357.35	411.88	938.51	2521.46	4438.33

5.2.3. Convergence Analysis of Different Evolutionary Algorithms

Figure 8 shows the convergence trajectories of 5 algorithms in different cases. It could be observed that in four cases, SCA failed into a local optima at the early stage, and then started to improve the quality of solutions as the iteration number reached 300, but still failed to find out the satisfying solutions at the end; due to the loss of individual diversity [84], three other methods (PSO, GSA, and DE) outperformed the SCA method but were still inferior to the proposed method from the beginning to end. Additionally, the proposed method could quickly converge to the scheduling schemes that were close to the optimal solution at the early stage, and then change slightly with an increasing number of iterations. Thus, it could be concluded that three modified strategies could effectively enhance the performance of the standard GSA method.

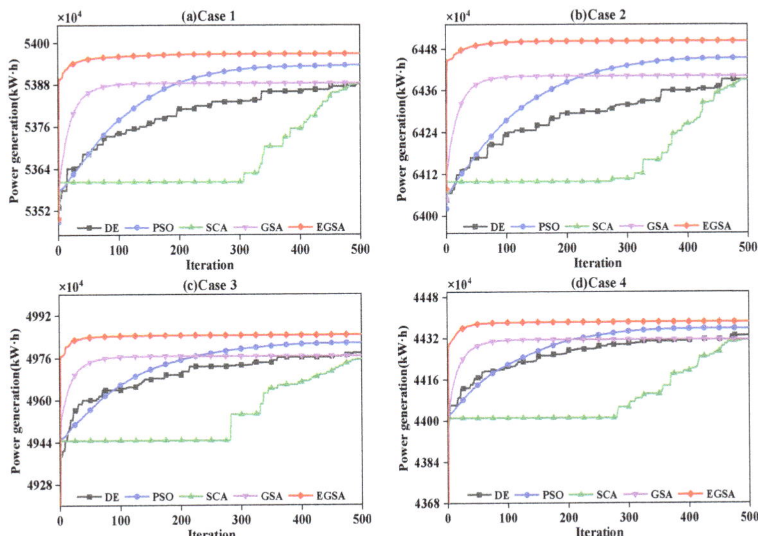

Figure 8. Convergence trajectories of 5 methods in different cases.

5.3. Case Study 2: Peak Operation of Cascade Hydropower Reservoirs

5.3.1. Robustness Testing of Different Evolutionary Algorithms

In this section, the proposed method was applied to deal with the peak operation of a hydropower system. Table 7 shows the statistical results of the 5 methods in 20 independent runs for 4 cases. It was observed that all of the methods could achieve the goal of reducing the peak loads of a power system due to the reductions in the statistical indices; while the proposed method had the performances among all methods. For instance, the average improvements in the range of the objective values of EGSA was about 99.7%, compared to the three other methods in the 4 cases. Hence, the feasibility of the EGSA method in addressing the peak operation of the hydropower system was proved in this case.

Table 7. Statistical results of the 5 methods in 20 independent runs for 4 cases of peak operation.

Season	Item	Best	Worst	Average	Standard Deviation	Range
Spring	DE	32,130.69	32,183.80	32,153.90	16.22	53.11
	PSO	32,052.91	32,127.22	32,084.17	20.24	74.31
	SCA	32,160.67	32,230.89	32,194.00	22.32	70.22
	GSA	32,173.70	32,224.22	32,199.76	14.41	50.52
	EGSA	31,986.47	31,986.62	31,986.53	0.04	0.15
Summer	DE	33,006.63	33,093.05	33,051.56	23.51	86.42
	PSO	32,962.41	33,057.03	32,993.63	22.72	94.62
	SCA	33,032.40	33,129.77	33,100.08	24.05	97.37
	GSA	33,069.24	33,134.56	33,097.29	19.83	65.32
	EGSA	32,888.05	32,888.16	32,888.10	0.03	0.11
Autumn	DE	36,681.47	36,752.01	36,719.49	17.51	70.54
	PSO	36,595.43	36,714.71	36,667.14	30.32	119.28
	SCA	36,745.91	36,840.43	36,781.14	24.73	94.52
	GSA	36,738.72	36,835.34	36,802.57	27.25	96.62
	EGSA	36,536.95	36,537.06	36,537.00	0.03	0.11
Winter	DE	35,825.89	35,890.14	35,857.52	14.79	64.25
	PSO	35,762.63	35,850.26	35,811.97	23.97	87.63
	SCA	35,822.72	35,950.51	35,896.25	32.32	127.79
	GSA	35,879.31	35,965.39	35,931.63	22.63	86.08
	EGSA	35,671.70	35,671.80	35,671.74	0.03	0.10

Figure 9 shows the box and whisker test of 5 algorithms in different cases. It can be seen that the distribution ranges of the solutions obtained by the proposed method were much smaller than the other methods, demonstrating the effectiveness of the constraint handling method and the modified strategies. Thus, it can be concluded that the EGSA method could produce stable scheduling schemes for the complex hydropower system peak operation problem.

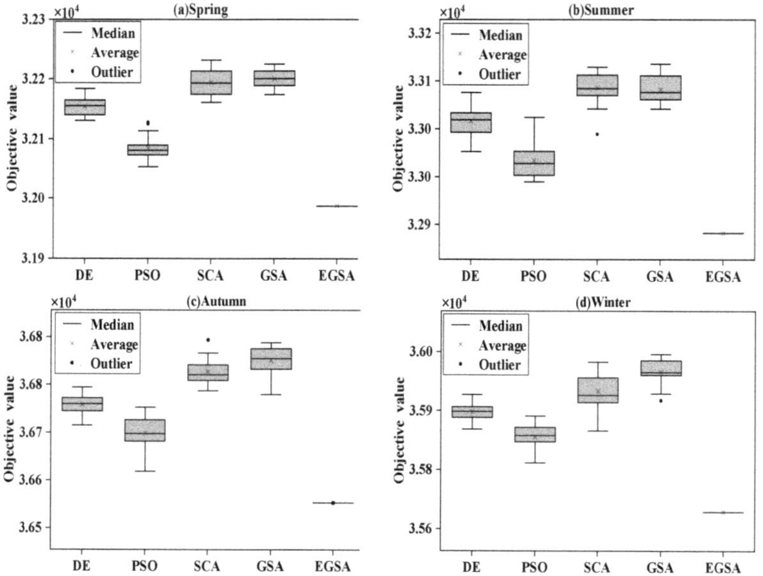

Figure 9. Box and whisker test of 5 algorithms in different cases for peak operation.

5.3.2. Comparison of the Optimal Results Obtained by Different Evolutionary Algorithms

To clearly show the feasibility of the EGSA method, the detailed results in the best solutions as obtained by different methods are given in Table 8. It was found that the EGSA method could achieve satisfactory results. For instance, the proposed method could bring about a 30.9% reduction in the peak values in Summer, which was obviously better than 21.0% of DE and 24.9% of PSO, 23.9% of SCA and 19.4% of GSA. Thus, this simulation clearly proved the superiority of the proposed method in responding to the peak operation requirement of the power system.

Table 8. Detailed results in the best solutions as obtained by the different methods (MW).

Season	Method	Item	Peak	Valley	Peak-valley	Average	Standard Deviation
		Original	13,477.93	10,101.60	3376.33	11,910.93	1281.95
	DE	Optimization	10,952.90	7716.77	3236.13	9242.92	789.32
		Reduction	2525.03	2384.83	140.20	2668.01	492.63
	PSO	Optimization	10,050.98	8169.90	1881.08	9231.64	640.11
		Reduction	3426.95	1931.7	1495.25	2679.29	641.84
Spring	SCA	Optimization	10,752.18	7809.30	2942.88	9239.61	926.13
		Reduction	2725.75	2292.3	433.45	2671.32	355.82
	GSA	Optimization	10,737.30	7592.29	3145.01	9229.75	1058.60
		Reduction	2740.63	2509.31	231.32	2681.18	223.35
	EGSA	Optimization	9428.01	8987.22	440.79	9232.28	165.13
		Reduction	4049.92	1114.38	2935.54	2678.65	1116.82
		Original	14,119.78	10,342.98	3776.80	12,170.94	1302.21
	DE	Optimization	11,151.13	8535.00	2616.13	9505.54	670.54
		Reduction	2968.65	1807.98	1160.67	2665.40	631.67
	PSO	Optimization	10,610.47	8392.24	2218.23	9492.57	673.27
		Reduction	3509.31	1950.74	1558.57	2678.37	628.94
Summer	SCA	Optimization	10,749.92	7591.09	3158.83	9500.12	839.87
		Reduction	3369.86	2751.89	617.97	2670.82	462.34
	GSA	Optimization	11,379.10	7774.71	3604.39	9493.04	1028.32
		Reduction	2740.68	2568.27	172.41	2677.90	273.89
	EGSA	Optimization	9759.71	9234.99	524.72	9492.46	172.66
		Reduction	4360.07	1107.99	3252.08	2678.48	1129.55
		Original	14,773.84	10,786.00	3987.84	13,220.16	1528.83
	DE	Optimization	12,615.24	8928.53	3686.71	10,553.86	880.28
		Reduction	2158.6	1857.5	301.1	2666.3	648.6
	PSO	Optimization	11,234.24	9366.35	1867.89	10,547.10	613.62
		Reduction	3539.6	1419.65	2119.95	2673.06	915.21
Autumn	SCA	Optimization	11,891.57	8142.23	3749.34	10,552.90	1099.32
		Reduction	2882.27	2643.77	238.50	2667.26	429.51
	GSA	Optimization	12,649.33	8347.14	4302.19	10,541.04	1193.12
		Reduction	2124.51	2438.86	−314.35	2679.12	335.71
	EGSA	Optimization	10,760.16	10,182.25	577.91	10,545.07	221.99
		Reduction	4013.68	603.75	3409.93	2675.09	1306.84
		Original	14,913.26	11,028.24	3885.02	12,971.28	1489.76
	DE	Optimization	11,709.53	8772.72	2936.81	10,303.38	955.49
		Reduction	3203.73	2255.52	948.21	2667.90	534.27
	PSO	Optimization	11,884.52	8637.32	3247.20	10,293.13	970.91
		Reduction	3028.74	2390.92	637.82	2678.15	518.85
Winter	SCA	Optimization	12,391.09	8667.77	3723.32	10,301.54	1052.86
		Reduction	2522.17	2360.47	161.70	2669.74	436.90
	GSA	Optimization	11,946.42	8414.17	3532.25	10,293.50	1174.01
		Reduction	2966.84	2614.07	352.77	2677.78	315.75
	EGSA	Optimization	10,568.93	10,012.57	556.36	10,295.50	208.98
		Reduction	4344.33	1015.67	3328.66	2675.78	1280.78

Note: Reduction = original–optimization of the target method.

5.3.3. Rationality Analysis of the Best Results Obtained by the Different Evolutionary Algorithms

Figure 10 shows the detailed output process obtained by the different methods in four cases. It was found that there were often two peak periods (morning and evening) in the original load curves, and obvious differences in the load features (like peak times or numbers) of the four load curves, which would increase the optimization difficulty of the peak operation. The two methods (GSA and EGSA) exhibited excellent responses to the load changes by collecting the hydropower generation at the peak periods and storing the energy at valley periods. The EGSA method was superior to the GSA method due to its smoother residual load curves; besides, the outputs of all the hydroplants varied in the feasible range between the minimum output limit and the installed capacity. Thus, the EGSA method could produce feasible solutions for the peak operation of a hydropower system.

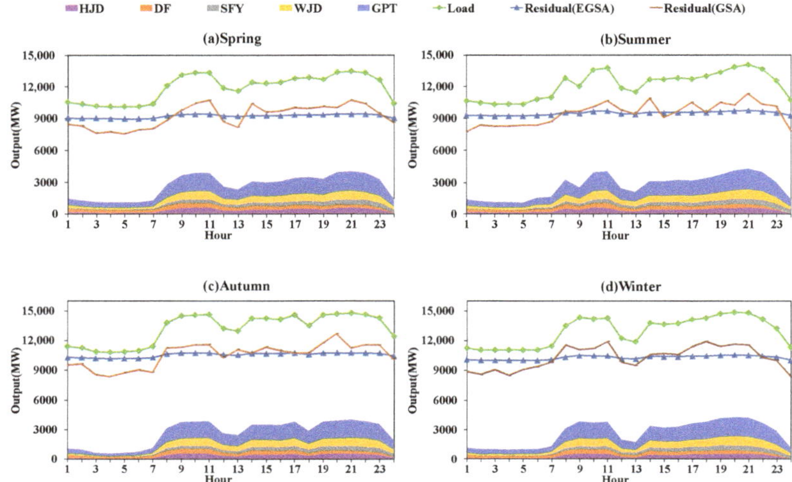

Figure 10. Detailed output process obtained by the different methods in the four cases.

5.4. Case Study 3: Mutli-Objective Operation of Cascade Hydropower Reservoirs

5.4.1. Comparative Analysis of the Optimal Results Obtained by the Different Methods with 100 Weights

The focus of this section is on the demonstration of the feasibility of the proposed method in addressing the multi-objective optimization problems of a hydropower system. Figure 11 draws the distributions of the objective functions obtained by different methods, where each method was executed in 100 different weight combinations. Specifically, the weight w_1 for power generation was increased from 0 to 1.0 at the same interval of 0.01, while the weight w_2 for peak operation was set as $1.0 - w_1$. From Figure 11, it was observed that the total generation was gradually decreasing with the increasing objective value of peak operation, which implied that there was an obvious conflict between power generation and peak operation. Additionally, the solutions of PSO and GSA were dominated by that of the EGSA method, which meant that the EGSA method had a higher probability to obtain the Pareto optimal solutions than the PSO and GSA method. Thus, the proposed method can generate the near-optimal Pareto solutions to approximate the relationship between power generation and peak shaving in practice.

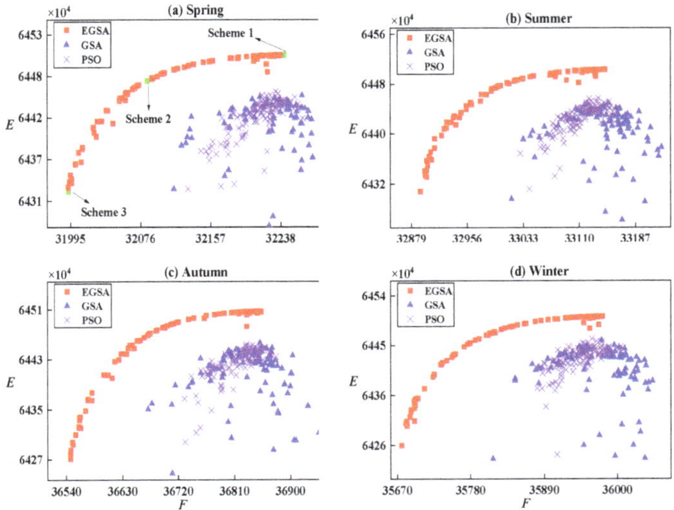

Figure 11. Distributions of the objective functions obtained by different methods with 100 weights.

5.4.2. Rationality Analysis of the Best Results Obtained by the Different Evolutionary Algorithms

Figure 12 shows the detailed results of three typical schemes obtained by the EGSA method in Spring, where Scheme 1 showed the maximum power generation, Scheme 3 had the best peak operation performance, while Scheme 2 could achieve a balance between power generation and peak operation. It could be seen that for all the hydropower reservoirs, the water levels varied in the preset range between a dead water level and a normal water level while the power outputs were smaller than the installed capacity, demonstrating the effectiveness of the heuristic constraint handling method in guaranteeing the feasibility of the solution. Meanwhile, there were obvious differences in the peak operation and power generations of the three typical schemes, which indicated that it was necessary to choose a suitable scheme based on the actual requirement. Thus, this case again proved the practicability of the EGSA method in solving the multi-objective operation problems.

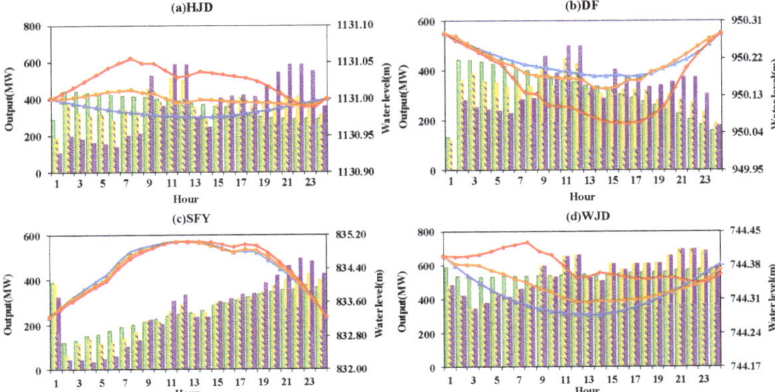

Figure 12. *Cont.*

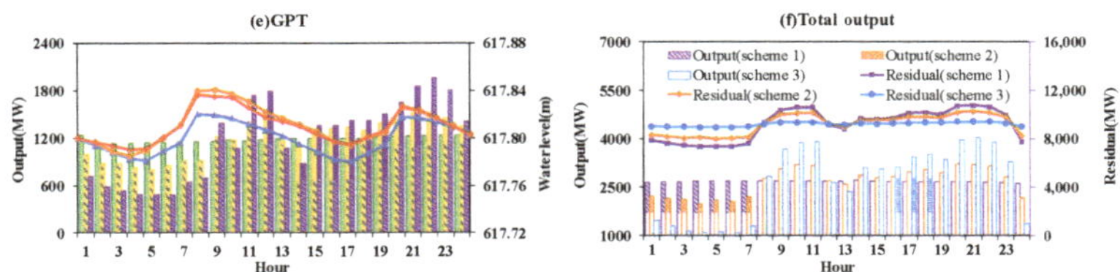

Figure 12. Detailed scheduling results of three typical schemes obtained by EGSA in Spring.

6. Conclusions

In this research, an enhanced gravitational search algorithm coupled with the Technique for Order Preference by Similarity to an Ideal Solution (TOPSIS) was developed to deal with the complex multi-objective operation problem of cascade hydropower reservoirs balancing the power generation and peak operation benefits. The proposed method and several famous evolutionary methods were used to solve the 12 famous benchmark functions and the optimal operation of the Wu hydropower system in China. The critical findings obtained from the experimental results are given as below:

(1) Due to the loss of the population diversity, the conventional GSA method suffered from severe premature convergence shortcomings. The proposed method based on the three modified strategies (opposite learning strategy, partial mutation strategy and elastic-ball modification strategy) could effectively improve the convergence speed, swarm diversity, and solution feasibility of the standard GSA method, respectively.

(2) For the original complex multi-objective optimization problem, balancing power generation and peak operation requirements, the famous TOPSIS method was used to transform it into the relatively simple single-objective problem, which could help make an obvious reduction in the modeling difficulty of the multi-objective decision.

(3) There was a competitive relationship between the generation benefit of the hydropower enterprise and the peak operation of the power system. In other words, an increasing value of one objective would obviously reduce another objective value. Thus, it was necessary for operators to carefully determine the scheduling schemes so as to effectively balance the practical requirements of power generation enterprises and power grid companies.

Author Contributions: All authors contributed extensively to the work presented in this paper. Z.-k.F. and W.-j.N. contributed to the finalized the manuscripts. S.L. contributed to the modeling and programming. Z.-q.J. contributed to the literature review. M.-s.M. and B.L. contributed to the data analyses.

Funding: This research was supported by the Natural Science Foundation of Hubei Province (2018CFB573), National Natural Science Foundation of China (51709119 and 51909133), and the Fundamental Research Funds for the Central Universities (HUST: 2017KFYXJJ193).

Acknowledgments: The writers would like to express appreciation to both editors and reviewers for their valuable comments and suggestions.

Conflicts of Interest: The authors declare no conflict of interest.

References

1. Chang, J.; Li, Y.; Yuan, M.; Wang, Y. Efficiency evaluation of hydropower station operation: A case study of Longyangxia station in the Yellow River, China. *Energy* **2017**, *135*, 23–31. [CrossRef]
2. Feng, M.; Liu, P.; Guo, S.; Gui, Z.; Zhang, X.; Zhang, W.; Xiong, L. Identifying changing patterns of reservoir operating rules under various inflow alteration scenarios. *Adv. Water Resour.* **2017**, *104*, 23–36. [CrossRef]
3. Liu, P.; Li, L.; Chen, G.; Rheinheimer, D.E. Parameter uncertainty analysis of reservoir operating rules based on implicit stochastic optimization. *J. Hydrol.* **2014**, *514*, 102–113. [CrossRef]
4. Zhao, T.; Zhao, J.; Yang, D. Improved dynamic programming for hydropower reservoir operation. *J. Water Res. Plan. Manag.* **2014**, *140*, 365–374. [CrossRef]

5. Liu, S.; Feng, Z.K.; Niu, W.J.; Zhang, H.R.; Song, Z.G. Peak operation problem solving for hydropower reservoirs by elite-guide sine cosine algorithm with Gaussian local search and random mutation. *Energies* **2019**, *12*, 2189. [CrossRef]

6. Feng, Z.K.; Niu, W.J.; Cheng, C.T. China's large-scale hydropower system: Operation characteristics, modeling challenge and dimensionality reduction possibilities. *Renew. Energy* **2019**, *136*, 805–818. [CrossRef]

7. Bai, T.; Wu, L.; Chang, J.X.; Huang, Q. Multi-Objective Optimal Operation Model of Cascade Reservoirs and Its Application on Water and Sediment Regulation. *Water Resour. Manag.* **2015**, *29*, 2751–2770. [CrossRef]

8. Guo, X.Y.; Ma, C.; Tang, Z.B. Multi-timescale joint optimal dispatch model considering environmental flow requirements for a dewatered river channel: Case study of Jinping cascade hydropower stations. *J. Water Res. Plan. Manag.* **2018**, *144*, 05018014. [CrossRef]

9. Liu, P.; Nguyen, T.D.; Cai, X.; Jiang, X. Finding multiple optimal solutions to optimal load distribution problem in hydropower plant. *Energies* **2012**, *5*, 1413–1432. [CrossRef]

10. Chang, J.; Wang, X.; Li, Y.; Wang, Y.; Zhang, H. Hydropower plant operation rules optimization response to climate change. *Energy* **2018**, *160*, 886–897. [CrossRef]

11. Guo, S.; Chen, J.; Li, Y.; Liu, P.; Li, T. Joint operation of the multi-reservoir system of the Three Gorges and the Qingjiang cascade reservoirs. *Energies* **2011**, *4*, 1036–1050. [CrossRef]

12. Zhang, Y.; Jiang, Z.; Ji, C.; Sun, P. Contrastive analysis of three parallel modes in multi-dimensional dynamic programming and its application in cascade reservoirs operation. *J. Hydrol.* **2015**, *529*, 22–34. [CrossRef]

13. Reddy, S.; Abhyankar, A.R.; Bijwe, P.R. Reactive power price clearing using multi-objective optimization. *Energy* **2011**, *36*, 3579–3589. [CrossRef]

14. Feng, Z.K.; Niu, W.J.; Wang, S.; Cheng, C.T.; Jiang, Z.Q.; Qin, H.; Liu, Y. Developing a successive linear programming model for head-sensitive hydropower system operation considering power shortage aspect. *Energy* **2018**, *155*, 252–261. [CrossRef]

15. Catalao, J.P.S.; Mariano, S.J.P.S.; Mendes, V.M.F.; Ferreira, L.A.F.M. Scheduling of head-sensitive cascaded hydro systems: A nonlinear approach. *IEEE Trans. Power Syst.* **2009**, *24*, 337–346. [CrossRef]

16. Jiang, Z.; Wang, C.; Liu, Y.; Feng, Z.; Ji, C.; Zhang, H. Study on the raw water allocation and optimization in shenzhen city, China. *Water* **2019**, *11*, 1426. [CrossRef]

17. Jiang, Z.; Ji, C.; Qin, H.; Feng, Z. Multi-stage progressive optimality algorithm and its application in energy storage operation chart optimization of cascade reservoirs. *Energy* **2018**, *148*, 309–323. [CrossRef]

18. Huang, G.H.; He, L.; Zeng, G.M.; Lu, H.W. Identification of optimal urban solid waste flow schemes under impacts of energy prices. *Environ. Eng. Sci.* **2008**, *25*, 685–695. [CrossRef]

19. Ren, G.; Cao, Y.; Wen, S.; Huang, T.; Zeng, Z. A modified Elman neural network with a new learning rate scheme. *Neurocomputing* **2018**, *286*, 11–18. [CrossRef]

20. Xie, X.; Wen, S.; Zeng, Z.; Huang, T. Memristor-based circuit implementation of pulse-coupled neural network with dynamical threshold generators. *Neurocomputing* **2018**, *284*, 10–16. [CrossRef]

21. Wen, S.; Xie, X.; Yan, Z.; Huang, T.; Zeng, Z. General memristor with applications in multilayer neural networks. *Neural Netw.* **2018**, *103*, 142–149. [CrossRef] [PubMed]

22. Yang, T.; Gao, X.; Sellars, S.L.; Sorooshian, S. Improving the multi-objective evolutionary optimization algorithm for hydropower reservoir operations in the California Oroville-Thermalito complex. *Environ. Model. Softw.* **2015**, *69*, 262–279. [CrossRef]

23. Xu, B.; Zhong, P.A.; Du, B.; Chen, J.; Liu, W.; Li, J.; Guo, L.; Zhao, Y. Analysis of a stochastic programming model for optimal hydropower system operation under a deregulated electricity market by considering forecasting uncertainty. *Water* **2018**, *10*, 885. [CrossRef]

24. Zhao, T.; Zhao, J. Improved multiple-objective dynamic programming model for reservoir operation optimization. *J. Hydroinform.* **2014**, *16*, 1142–1157. [CrossRef]

25. Feng, Z.K.; Niu, W.J.; Cheng, C.T. Optimizing electrical power production of hydropower system by uniform progressive optimality algorithm based on two-stage search mechanism and uniform design. *J. Clean. Prod.* **2018**, *190*, 432–442. [CrossRef]

26. Catalão, J.P.S.; Mariano, S.J.P.S.; Mendes, V.M.F.; Ferreira, L.A.F.M. Nonlinear optimization method for short-term hydro scheduling considering head-dependency. *Eur. Trans. Electr. Power* **2010**, *20*, 172–183. [CrossRef]

27. Jia, B.; Zhong, P.; Wan, X.; Xu, B.; Chen, J. Decomposition-coordination model of reservoir group and flood storage basin for real-time flood control operation. *Hydrol. Res.* **2015**, *46*, 11–25. [CrossRef]

28. Dong, M.; Wen, S.; Zeng, Z.; Yan, Z.; Huang, T. Sparse fully convolutional network for face labeling. *Neurocomputing* **2019**, *331*, 465–472. [CrossRef]
29. Li, Z.; Dong, M.; Wen, S.; Hu, X.; Zhou, P.; Zeng, Z. CLU-CNNs: Object detection for medical images. *Neurocomputing* **2019**, *350*, 53–59. [CrossRef]
30. Cao, Y.; Wang, S.; Guo, Z.; Huang, T.; Wen, S. Synchronization of memristive neural networks with leakage delay and parameters mismatch via event-triggered control. *Neural Netw.* **2019**, *119*, 178–189. [CrossRef]
31. Zheng, F. Comparing the real-time searching behavior of four differential-evolution variants applied to water-distribution-network design optimization. *J. Water Res. Plan. Manag.* **2015**, *141*, 04015016. [CrossRef]
32. Yang, T.; Asanjan, A.A.; Faridzad, M.; Hayatbini, N.; Gao, X.; Sorooshian, S. An enhanced artificial neural network with a shuffled complex evolutionary global optimization with principal component analysis. *Inf. Sci.* **2017**, *418–419*, 302–316. [CrossRef]
33. Wan, W.; Guo, X.; Lei, X.; Jiang, Y.; Wang, H. A Novel Optimization Method for Multi-Reservoir Operation Policy Derivation in Complex Inter-Basin Water Transfer System. *Water Resour. Manag.* **2018**, *32*, 31–51. [CrossRef]
34. Feng, Z.K.; Niu, W.J.; Cheng, C.T. Optimization of hydropower reservoirs operation balancing generation benefit and ecological requirement with parallel multi-objective genetic algorithm. *Energy* **2018**, *153*, 706–718. [CrossRef]
35. Zheng, F.; Zecchin, A.C.; Maier, H.R.; Simpson, A.R. Comparison of the searching behavior of NSGA-II, SAMODE, and borg MOEAs applied to water distribution system design problems. *J. Water Res. Plan. Manag.* **2016**, *142*, 04016017. [CrossRef]
36. Salkuti, S.R. Short-term electrical load forecasting using hybrid ANN–DE and wavelet transforms approach. *Electr. Eng.* **2018**, *100*, 2755–2763. [CrossRef]
37. Chang, J.X.; Bai, T.; Huang, Q.; Yang, D.W. Optimization of Water Resources Utilization by PSO-GA. *Water Resour. Manag.* **2013**, *27*, 3525–3540. [CrossRef]
38. Ji, Y.; Lei, X.; Cai, S.; Wang, X. Hedging rules for water supply reservoir based on the model of simulation and optimization. *Water* **2016**, *8*, 249. [CrossRef]
39. Zhang, J.; Li, Z.; Wang, X.; Lei, X.; Liu, P.; Feng, M.; Khu, S.T.; Wang, H. A novel method for deriving reservoir operating rules based on flood classification-aggregation-decomposition. *J. Hydrol.* **2019**, *568*, 722–734. [CrossRef]
40. Niu, W.J.; Feng, Z.K.; Cheng, C.T.; Wu, X.Y. A parallel multi-objective particle swarm optimization for cascade hydropower reservoir operation in southwest China. *Appl. Soft Comput. J.* **2018**, *70*, 562–575. [CrossRef]
41. Ming, B.; Chang, J.X.; Huang, Q.; Wang, Y.M.; Huang, S.Z. Optimal Operation of Multi-Reservoir System Based-On Cuckoo Search Algorithm. *Water Resour. Manag.* **2015**, *29*, 5671–5687. [CrossRef]
42. Reddy, S.S.; Panigrahi, B.K.; Debchoudhury, S.; Kundu, R.; Mukherjee, R. Short-term hydro-thermal scheduling using CMA-ES with directed target to best perturbation scheme. *Int. J. Bio-Inspired Comput.* **2015**, *7*, 195–208. [CrossRef]
43. Salkuti, S.R. Short-term optimal hydro-thermal scheduling using clustered adaptive teaching learning based optimization. *Int. J. Electr. Comput. Eng.* **2019**, *9*, 3359–3365.
44. Niu, W.; Feng, Z.; Cheng, C.; Zhou, J. Forecasting daily runoff by extreme learning machine based on quantum-behaved particle swarm optimization. *J. Hydrol. Eng.* **2018**, *23*, 04018002. [CrossRef]
45. Zheng, F.; Zecchin, A.C.; Simpson, A.R. Self-adaptive differential evolution algorithm applied to water distribution system optimization. *J. Comput. Civ. Eng.* **2013**, *27*, 148–158. [CrossRef]
46. Zheng, F.; Simpson, A.R.; Zecchin, A.C. A combined NLP-differential evolution algorithm approach for the optimization of looped water distribution systems. *Water Resour. Res.* **2011**, *47*. [CrossRef]
47. Feng, Z.K.; Niu, W.J.; Cheng, C.T. Multi-objective quantum-behaved particle swarm optimization for economic environmental hydrothermal energy system scheduling. *Energy* **2017**, *131*, 165–178. [CrossRef]
48. Bai, T.; Kan, Y.B.; Chang, J.X.; Huang, Q.; Chang, F.J. Fusing feasible search space into PSO for multi-objective cascade reservoir optimization. *Appl. Soft Comput. J.* **2017**, *51*, 328–340. [CrossRef]
49. Rezaei, F.; Safavi, H.R.; Mirchi, A.; Madani, K. f-MOPSO: An alternative multi-objective PSO algorithm for conjunctive water use management. *J. Hydro-Environ. Res.* **2017**, *14*, 1–18. [CrossRef]
50. Reddy, S.S.; Bijwe, P.R. Multi-Objective Optimal Power Flow Using Efficient Evolutionary Algorithm. *Int. J. Emerg. Electr. Power Syst.* **2017**, *18*. [CrossRef]

51. Ji, B.; Yuan, X.; Yuan, Y.; Lei, X.; Fernando, T.; Iu, H.H.C. Exact and heuristic methods for optimizing lock-quay system in inland waterway. *Eur. J. Oper. Res.* **2019**, *277*, 740–755. [CrossRef]

52. Mirjalili, S.; Lewis, A. Adaptive gbest-guided gravitational search algorithm. *Neural Comput. Appl.* **2014**, *25*, 1569–1584. [CrossRef]

53. Rashedi, E.; Nezamabadi-Pour, H.; Saryazdi, S. GSA: A gravitational search algorithm. *Inf. Sci.* **2009**, *179*, 2232–2248. [CrossRef]

54. Niu, W.J.; Feng, Z.K.; Zeng, M.; Feng, B.F.; Min, Y.W.; Cheng, C.T.; Zhou, J.Z. Forecasting reservoir monthly runoff via ensemble empirical mode decomposition and extreme learning machine optimized by an improved gravitational search algorithm. *Appl. Soft Comput. J.* **2019**, *82*, 105589. [CrossRef]

55. Tian, H.; Yuan, X.; Ji, B.; Chen, Z. Multi-objective optimization of short-term hydrothermal scheduling using non-dominated sorting gravitational search algorithm with chaotic mutation. *Energy Convers. Manag.* **2014**, *81*, 504–519. [CrossRef]

56. Mahootchi, M.; Tizhoosh, H.R.; Ponnambalam, K. Oppositional extension of reinforcement learning techniques. *Inf. Sci.* **2014**, *275*, 101–114. [CrossRef]

57. Mahdavi, S.; Rahnamayan, S.; Deb, K. Opposition based learning: A literature review. *Swarm Evol. Comput.* **2018**, *39*, 1–23. [CrossRef]

58. Ewees, A.A.; Abd Elaziz, M.; Houssein, E.H. Improved grasshopper optimization algorithm using opposition-based learning. *Expert Syst. Appl.* **2018**, *112*, 156–172. [CrossRef]

59. Xia, Y.; Feng, Z.; Niu, W.; Qin, H.; Jiang, Z.; Zhou, J. Simplex quantum-behaved particle swarm optimization algorithm with application to ecological operation of cascade hydropower reservoirs. *Appl. Soft Comput.* **2019**, *84*, 105715. [CrossRef]

60. Chen, Z.; Yuan, X.; Tian, H.; Ji, B. Improved gravitational search algorithm for parameter identification of water turbine regulation system. *Energy Convers. Manag.* **2014**, *78*, 306–315. [CrossRef]

61. Mirjalili, S. The ant lion optimizer. *Adv. Eng. Softw.* **2015**, *83*, 80–98. [CrossRef]

62. Yang, X.; Deb, S. *Cuckoo Search via Lévy Flights*; IEEE: Piscataway, NJ, USA, 2009; pp. 210–214.

63. Walton, S.; Hassan, O.; Morgan, K.; Brown, M.R. Modified cuckoo search: A new gradient free optimisation algorithm. *Chaos Solitons Fractals* **2011**, *44*, 710–718. [CrossRef]

64. Shareef, H.; Ibrahim, A.A.; Mutlag, A.H. Lightning search algorithm. *Appl. Soft Comput.* **2015**, *36*, 315–333. [CrossRef]

65. Mirjalili, S.; Mirjalili, S.M.; Lewis, A. Grey wolf optimizer. *Adv. Eng. Softw.* **2014**, *69*, 46–61. [CrossRef]

66. Yang, X.S. *Firefly Algorithm in Engineering Optimization*; John Wiley & Sons Inc.: Hoboken, NJ, USA, 2010.

67. Mirjalili, S.; Lewis, A. The whale optimization algorithm. *Adv. Eng. Softw.* **2016**, *95*, 51–67. [CrossRef]

68. Hua, Z.; Ma, C.; Lian, J.; Pang, X.; Yang, W. Optimal capacity allocation of multiple solar trackers and storage capacity for utility-scale photovoltaic plants considering output characteristics and complementary demand. *Appl. Energy* **2019**, *238*, 721–733. [CrossRef]

69. Reddy, S.S.; Bijwe, P.R.; Abhyankar, A.R. Optimal dynamic emergency reserve activation using spinning, hydro and demand-side reserves. *Front. Energy* **2016**, *10*, 409–423. [CrossRef]

70. Reddy, S.S.; Bijwe, P.R. Real time economic dispatch considering renewable energy resources. *Renew. Energy* **2015**, *83*, 1215–1226. [CrossRef]

71. Feng, Z.K.; Niu, W.J.; Zhang, R.; Wang, S.; Cheng, C.T. Operation rule derivation of hydropower reservoir by k-means clustering method and extreme learning machine based on particle swarm optimization. *J. Hydrol.* **2019**, *576*, 229–238. [CrossRef]

72. Feng, Z.K.; Niu, W.J.; Cheng, C.T.; Wu, X.Y. Optimization of hydropower system operation by uniform dynamic programming for dimensionality reduction. *Energy* **2017**, *134*, 718–730. [CrossRef]

73. Zhang, C.; Ding, W.; Li, Y.; Meng, F.; Fu, G. Cost-Benefit Framework for Optimal Design of Water Transfer Systems. *J. Water Res. Plan. Manag.* **2019**, *145*, 04019007. [CrossRef]

74. Li, H.; Chen, D.; Arzaghi, E.; Abbassi, R.; Xu, B.; Patelli, E.; Tolo, S. Safety assessment of hydro-generating units using experiments and grey-entropy correlation analysis. *Energy* **2018**, *165*, 222–234. [CrossRef]

75. Chen, D.; Liu, S.; Ma, X. Modeling, nonlinear dynamical analysis of a novel power system with random wind power and it's control. *Energy* **2013**, *53*, 139–146. [CrossRef]

76. Zhang, R.; Chen, D.; Ma, X. Nonlinear predictive control of a hydropower system model. *Entropy-Switz* **2015**, *17*, 6129–6149. [CrossRef]

77. Feng, Z.; Niu, W.; Wang, W.; Zhou, J.; Cheng, C. A mixed integer linear programming model for unit commitment of thermal plants with peak shaving operation aspect in regional power grid lack of flexible hydropower energy. *Energy* **2019**, *175*, 618–629. [CrossRef]
78. Feng, Z.; Niu, W.; Wang, S.; Cheng, C.; Song, Z. Mixed integer linear programming model for peak operation of gas-fired generating units with disjoint-prohibited operating zones. *Energies* **2019**, *12*, 2179. [CrossRef]
79. Feng, Z.K.; Niu, W.J.; Cheng, C.T.; Zhou, J.Z. Peak shaving operation of hydro-thermal-nuclear plants serving multiple power grids by linear programming. *Energy* **2017**, *135*, 210–219. [CrossRef]
80. Niu, W.J.; Feng, Z.K.; Feng, B.F.; Min, Y.W.; Cheng, C.T.; Zhou, J.Z. Comparison of multiple linear regression, artificial neural network, extreme learning machine, and support vector machine in deriving operation rule of hydropower reservoir. *Water* **2019**, *11*, 88. [CrossRef]
81. Ma, C.; Lian, J.; Wang, J. Short-term optimal operation of Three-gorge and Gezhouba cascade hydropower stations in non-flood season with operation rules from data mining. *Energy Convers. Manag.* **2013**, *65*, 616–627. [CrossRef]
82. Jiang, Z.; Liu, P.; Ji, C.; Zhang, H.; Chen, Y. Ecological flow considered multi-objective sorage energy operation chart optimization of large-scale mixed reservoirs. *J. Hydrol.* **2019**, *577*, 123949. [CrossRef]
83. Lian, J.; Yao, Y.; Ma, C.; Guo, Q. Reservoir operation rules for controlling algal blooms in a tributary to the impoundment of three Gorges Dam. *Water* **2014**, *6*, 3200–3223. [CrossRef]
84. Fu, W.; Wang, K.; Zhang, C.; Tan, J. A hybrid approach for measuring the vibrational trend of hydroelectric unit with enhanced multi-scale chaotic series analysis and optimized least squares support vector machine. *Trans. Inst. Meas. Control* **2019**, *41*, 4436–4449. [CrossRef]

Article

Small and Medium-Scale River Flood Controls in Highly Urbanized Areas: A Whole Region Perspective

Zengmei Liu [1], Yuting Cai [1], Shangwei Wang [1], Fupeng Lan [1] and Xushu Wu [1,2,*]

[1]   School of Civil Engineering and Transportation, South China University of Technology, Guangzhou 510641, China; liuzm@scut.edu.cn (Z.L.); cytxiao99@163.com (Y.C.); wsw961202@163.com (S.W.); fjshlfp@163.com (F.L.)

[2]   Guangdong Engineering Technology Research Center of Safety and Greenization for Water Conservancy Project, Guangzhou 510641, China

*   Correspondence: xshwu@scut.edu.cn

Received: 20 November 2019; Accepted: 6 January 2020; Published: 9 January 2020

Abstract: While rapid urbanization promotes social and economic development, it poses a serious threat to the health of rivers, especially the small and medium-scale rivers. Flood control for small and medium-scale rivers in highly urbanized areas is particularly important. The purpose of this study is to explore the most effective flood control strategy for small and medium-scale rivers in highly urbanized areas. MIKE 11 and MIKE 21 were coupled with MIKE FLOOD model to simulate flooding with the flood control standard, after which the best flooding control scheme was determined from a whole region perspective (both the mainstream and tributary conditions were considered). The SheGong River basin located near the Guangzhou Baiyun international airport in Guangzhou city over south China was selected for the case study. The results showed that the flooding area in the basin of interest accounts for 42% of the total, with maximum inundation depth up to 0.93 m under the 20-year return period of the designed flood. The flood-prone areas are the midstream and downstream where urbanization is high; however the downstream of the adjacent TieShan River is still able to bear more flooding. Therefore, the probable cost-effective flood control scheme is to construct two new tributaries transferring floodwater in the mid- and downstream of the SheGong River into the downstream of the TieShan River. This infers that flood control for small and medium-scale rivers in highly urbanized areas should not simply consider tributary flood regimes but, rather, involve both tributary and mainstream flood characters from a whole region perspective.

Keywords: small and medium-scale rivers; highly urbanized area; flood control; whole region perspective; coupled models

---

1. Introduction

It is undisputed that frequent rainstorms driven by climate change and land-use change caused by high urbanization have resulted in urban flooding in many countries and regions [1–18]. As one of the most serious natural hazards, urban flooding, especially in highly urbanized areas, threatens lives and hinders society's sustainable development nowadays [19–24]. See, for example, the major flooding that occurred throughout most of the Brisbane River's catchment in January 2011, with more than 20 deaths and an economic loss of $2.55 billion [25]. During the 2011 rainy season, Thailand encountered a large flood of a 50-year return period and a total of 65 provinces were flooded; more than 700 people died and the economic loss reached up to $41.2 billion [26]. In China, an extreme storm attacked Beijing on 21 July 2012 and a flash flood was triggered in the urban area, causing 79 deaths and a direct economic loss of $1.86 billion, damaging infrastructure like roads with trapped cars and buses, bridges, and collapsed buildings [27–29]. Therefore, urban flooding is a hot topic in the field of disaster research at present and has attracted much attention in both developed and developing countries [30–37].

The middle and lower reaches of a river are usually more prosperous than the upper reach, and consequently, flooding problems are more serious. Therefore, more effective measures of flooding should be carried out to deal with increasing flooding risk with high urbanization [38]. In general, there are two ways of flood control: engineering and non-engineering measures. Engineering measures include building reservoirs, dikes, detention basins, pumping stations, and spillways. Non-engineering measures usually include flood forecasting or simulation, land-use management, land acquisition and relocation plans, flood emergency planning and response, and post-flood recovery [39]. Among these, flood forecasting or simulation analysis is essential for flood control. The state-of-the-art method for flood simulation includes a one-dimensional river flow model coupled with a two-dimensional surface flow model [40]. For example, Yazdi et al. coupled MIKE 11 with the NSGA-II model for a small watershed in the central part of Iran, showing that optimal designs of multi-reservoir systems can efficiently reduce construction costs, flood peaks and their corresponding damage costs at the downstream reaches of the basin [41]. For flood control measures, there are usually scientific urban plans, sufficient funds, effective laws and regulations in developed countries to put the flood control system on the right track, while the blind and disordered development in developing or undeveloped countries usually cause river flooding problems [42–44]. Rapid urban growth in developing countries usually results in the proliferation of informal settlements. The housing within informal settlements is virtually always built without the consent of the official planning authorities and rarely conforms to official planning guidelines, building regulations, and construction standards [45]. In most cases, there are many buildings along rivers in highly urbanized areas and there is not enough space for the construction of flood control. Meanwhile, the height of river levees differ from one place to another. To solve the problem of river flooding, many developing countries have paid much attention to flooding control measures. Ali Reza Shokoohi has studied the effect of constructing feasible detention dams in urban areas and found that they had good operational effects [46]. Marcelo et al. (2009) focused upon the use of a wide range of different flood control measures in the Joana River watershed, located at the northern region of Rio de Janeiro City, Brazil, and pointed out that distributed detention reservoirs in upstream reaches, parks, public squares, or at urban sites, are very important flood control alternatives [47]. In Turkey, to cope with floods and decrease any further damage, local authorities have designed a set of measures aiming to improve stream conveyance capacity by straightening reaches, lining channels, and building hydraulic structures such as new dams [48]. While in Vietnam, the Red River Delta, the central part of the country, and the Mekong Delta all have completed master plans for flood defenses including upgrading dikes [49]. However, these conventional flood control measures such as widening rivers and increasing embankment height are usually not the best flood control measure for small and medium-scale rivers in highly urbanized areas, nor do they control floods in such a way that other rivers within the same region are considered and fully used to reduce the flood hazards of the target river [50]. When widening the river, due to a large number of residential buildings around rivers, the surrounding residents need to be compensated with land compensation, resettlement compensation and compensation for attachments and young crops, which is a huge financial burden for the government. In the process of land acquisition and compensation allocation, there are often some distribution disputes. If the levee is raised, it will also not be coordinate with the surrounding buildings. Accordingly, the traditional measures are time-consuming and expensive due to the large population and dense buildings, making it very difficult to build a flood control system for small and medium-scale rivers in highly urbanized areas. Unlike large rivers, it is impossible to carry out unified planning for small rivers. Decision making and optimization should be carried out according to the local and actual situation. From a whole region perspective, river networks in the same region are usually connected to each other, and some might suffer floods while the others may not during a storm event; it is possible to utilize the rivers having enough flood bearing capacity to reduce flood hazards of the flooded rivers. However, such idea is seldom proposed in either the research community or the engineering field; how to achieve this idea remains unsolved at present and requires further study.

Therefore, this paper aims to explore a new flood control measure by fully considering the flood bearing capacities of all rivers within the same region. We specifically looked to use the surplus flood bearing capacities of the rivers free from flooding to reduce flood hazards of the flooded rivers. To demonstrate such an idea, we chose a tributary of TieShan River, i.e., the SheGong River located in the comprehensive development zone of Guangzhou Baiyun international airport in China as the case study, and integrated MIKE 11 with MIKE 21 and MIKE FLOOD [47] to simulate flooding situations of all the rivers in the case region. We expect to provide a new idea of flood control for highly urbanized areas around the world.

2. Study Area and Data

2.1. Study Area

In this study, we chose the river SheGong River located in GuangZhou BaiYun airport development zone in China as the study case. It is a tributary of the TieShan River originated from TieShan New Village in Huashan Town, flowing into the Tieshan River from north to south. The total length of the SheGong River is 7.98 km with a drainage area of 9.58 km^2, and the average slope of the river is 3.4%. The Tieshan River Basin covers an area of 58.3 km^2, and the topographic map of the basin is shown in Figure 1. The SheGong River, located in the central part of Huashan Town, constitutes the main channel for irrigation and flood channel. Many small reservoirs were built along the river channel and these reservoirs could store massive amounts of water. With the rapid development and utilization of land in the river basin, the irrigation function has been gradually weakened, and the request for ecological landscape water use is prevailing. At present, although the TieShan River has been regulated based on the standard of 20-year return period, the SheGong River is basically in a natural state. The twisted, narrow, and seriously silted channel and the reservoirs frequently lead to flooding after rainstorms; the accumulation of water in the plains of the two sides has caused serious impacts on local production and life. The map of land-use type of the study area in the year 2000 and 2019 are showed in Figure 2.

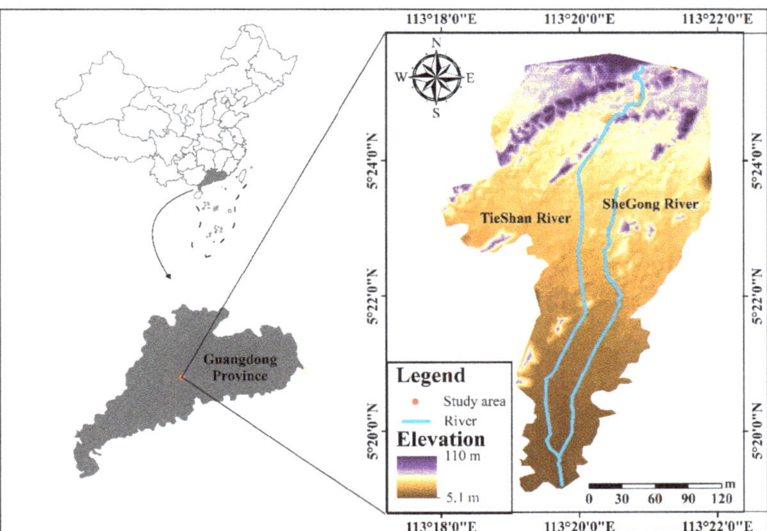

Figure 1. Topographic and water system in the study area.

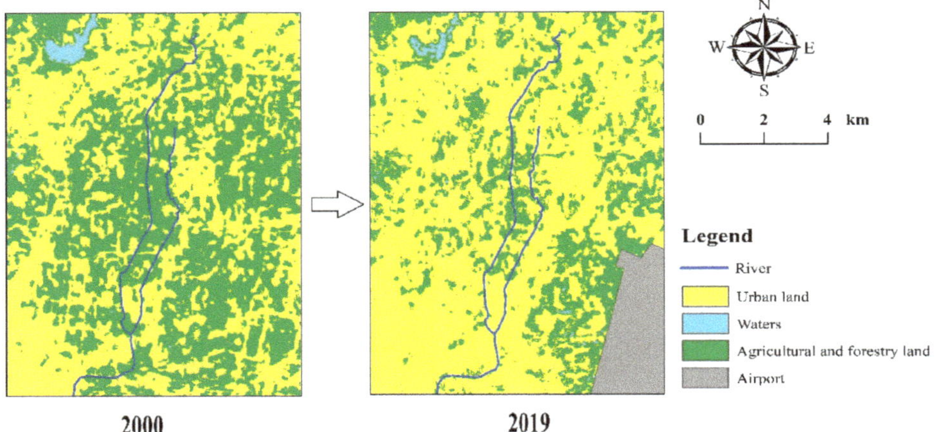

Figure 2. Land-use type in 2000 and 2019.

According to the topographical conditions and possible submergence range, we finally determined the boundary of the study area: the north boundary reaches the Sanjia sluice of the TieShan River and the source of SheGong River and the south boundary reaches the SheGong River estuary, forming a relatively complete river basin (Figure 3). In the study area, the section of the SheGong River is determined as 7.94 km long and that of the TieShan River is 13.4 km.

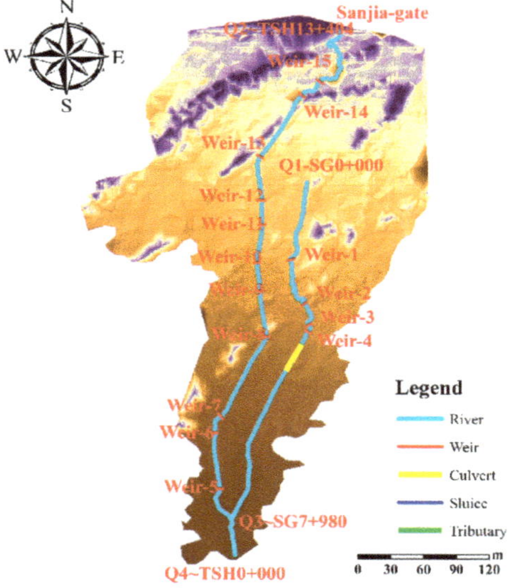

Figure 3. River generalization map of the study area.

2.2. Data

The land-use type of the study area was obtained from satellite remote sensing image maps in the years 2000 and 2019. The data was acquired by combining and correcting the remote sensing data of the study area, and by comprehensive classification methods such as supervised classification and decision tree classification.

The water system planning of Huadu District in Guangzhou indicates that the flood control standard of the SheGong River is a 20-year return period in the near future and a 50-year return period in the long term. Therefore, according to the flood control standard of China (GB50201-2014), the flood control (tide) standard of Guangdong Province (Trial), and the Huadu District water system planning of Guangzhou, the flood standard of this project is a 20-year return period, with the protection level of 4 in both levee and the main building, and level 5 in both the secondary building and the temporary building.

According to the requirements of the code for design of levee engineering of China (GB50286-98), the safety heightening value of level 4 in the levee engineering is as follows: 0.6 m for the levee engineering that is not allowed to cross the waves, and 0.3 m for the levee engineering that is allowed to cross the waves. The elevation of dike top of this project is designed according to the principle that no overtopping is allowed; in this case, the safe heightening is set to 0.6 m. The calculation height induced by run-up wave is 0.2 m and the super elevation of the dike top is 0.8 m. Therefore, the design flood level plus 0.8 m super elevation of the dike top is the elevation value of the dike top.

The design river bottom elevation at 0 + 000 section at the beginning of the regulation in the SheGong River is 22.2 m, and that of the 5 + 600 section at the key point of regulation is 9.1 m. According to the measured data in 2015, the current river bottom elevation of the 0 + 000 section is 22.54 m, and that of the 5 + 600 section is 9.38 m.

A hydrological atlas of Guangdong Province published by Guangdong Hydrological Bureau was used in the study. The lower boundary condition in the simulation model, i.e., river stages, came from the Comprehensive Planning Report of River Basin in Huadu District of Guangzhou City provided by the Guangzhou Water Conservancy Bureau. The topographic map of the Huadu District was provided by the Guangzhou Surveying and Mapping Institute, while that of SheGong River with a measuring scale of 1:2000 and the cross- and vertical section of the river were given by a commissioned surveying and mapping company.

3. Methodology

3.1. Design Flood Calculation

According to the manual regarding the use of rainfall-runoff curves in Guangdong Province and the standard of design flood calculation for water conservancy and hydropower engineering (SL44-2006), the design peak flow of small watershed is as follows [48,49]:

$$Q_m = \frac{0.278\psi S_p}{\tau^n}F \tag{1}$$

$$\tau = \frac{0.278L}{mJ^{\frac{1}{3}}Q^{\frac{1}{4}}} \tag{2}$$

where Q_m is design peak flow (m^3/s); ψ is runoff coefficient of flood peak ; S_p is design rainfall intensity; F is catchment area of the basin (km^2); τ is routing duration; m is routing parameter, L is river length (km); J is average slope of the river basin. These parameters are further given as

$$\psi = 1 - \frac{\mu}{S_p}\tau^n (t_c \geq \tau) \tag{3}$$

$$\psi = n\left(\frac{t_c}{\tau}\right)^{1-n} \tag{4}$$

$$\mu = RS(rl)p \tag{5}$$

$$S_p = H_{24p}t^{n-1} \tag{6}$$

$$t_c = \left[\frac{(1-n)S_p}{\mu} \right]^{\frac{1}{n}} \tag{7}$$

where μ is the soil infiltration rate; R is the loss coefficient; rl is the loss index (for the calculation of rainstorm peak in a small watershed); H_{24p} is the maximum 24-h rainfall; n is the rainstorm decay index (by referring to the Hydrographic Atlas of Guangdong Province).

The cross-section inputs are set up according to the observed data in the SheGong River and the calculated data in the TieShan River channel. The cross sections are set at an average spatial distance of 50 m, and there are 427 sections in total. Since there are no smaller sub-basins in the river channel and the river channel is long enough, the inflow between two sections is regarded as uniform. The results of design peak flow for each section are shown in Table 1.

Table 1. Discharge of each calculated section of river.

Section	Station Number	Discharge Number	Discharge for Different Return Period (m³/s)			
			2a	5a	10a	20a
Sanjia sluice	TieShan River TSH13 + 404	Q1	108.5	156.3	188.1	281.0
7#weir	TieShan River TSH3 + 600	Q2	149.6	218.8	259.6	348.7
The source of SheGong River	SheGong River SG0 + 000	Q3	11.9	18.0	22.4	26.5
SheGong sluice	SheGong River SG5 + 600	Q4	26.7	40.5	50.4	59.2
TieShan River estuary	TieShan River TSH0 + 000	Q5	201.2	297.1	354.7	453.9

3.2. Flood Simulation Model

MIKE FLOOD is used to couple the one-dimensional MIKE 11 model and the two-dimensional MIKE 21 model, for simulating the flood prone area with different return periods.

3.2.1. MIKE 11

MIKE 11 is an implicit finite difference model for one dimensional unsteady flow computation, the basic equations of which are Saint–Venant equations that are solved by the Abbott–Ionescu six-point implicit difference method and are used to simulate flood processes in a river channel [50]. The general steps of model setup include river network generalization, river section setting, boundary conditions setting, and determination of hydraulic parameters (Figure 4). In this study, we generalized the river network based on geographic data of the study area (Figure 3). Based on observed data, a total of 427 river sections were set up, and for boundary conditions, the upstream boundary was selected as the flow boundary, whilst the downstream boundary was selected as the water level boundary; through analysis and demonstration, we selected the design flood level with a 5-year return period as the boundary water level. The setting of hydraulic parameters mainly included the determination of a river roughness coefficient based on the current status of the river and the related data.

3.2.2. MIKE 21

MIKE 21 belongs to the free two-dimensional surface flow model, and uses the finite volume method to solve the planar two-dimensional shallow water equation. The general steps of model setup include grid division, elevation interpolation, boundary conditions setting, and determination of hydraulic parameters (Figure 4). In this study, the study area was divided into 28,800 grids with grid size of 50 × 50 m. We also performed elevation interpolation based on the measured terrain data. For the boundaries connected to rivers in MIKE 21, we set them as open boundaries, and other boundaries were set as closed boundaries. The setting of hydraulic parameters was to determine the surface roughness coefficient based on the type of land-use [51].

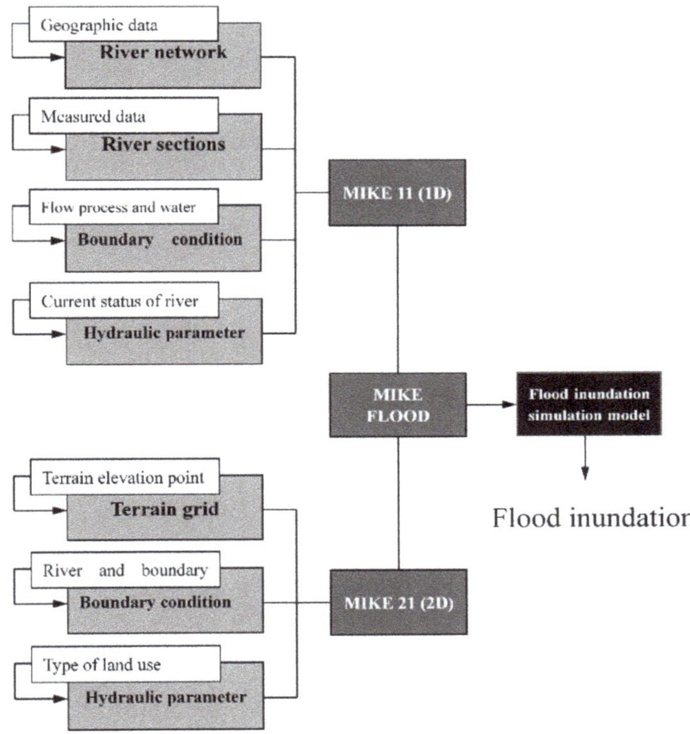

Figure 4. Flow chart of flooding simulation model.

3.2.3. MIKE FLOOD

MIKE FLOOD is used to couple the one-dimensional MIKE 11 model and the two-dimensional MIKE 21 model. By simulating the momentum transfer between the 1D river network and the 2D surface, the flood situation in the study area can be simulated. In this paper, a lateral coupling method was adopted; water above the river bank is exchanged with the two-dimensional surface model along the flow direction perpendicular to the river, and the exchange flow is approximately calculated by the weir flow formula [52]:

$$q = WC(H_{us} - H_{ds})^k \left[1 - \left(\frac{H_{ds} - H_w}{H_{us} - H_w} \right) \right]^{0.385} \tag{8}$$

where q is the exchange flow (m³/s); W is the width of the connection part (m); C is the coefficient of weir flow taken as 1.838; k is the weir index; H_{us} and H_{ds} are the water levels in the upstream and downstream sections of weir (m), respectively; H_w is the elevation at the top of weir (m).

3.2.4. Model Verification

On 7 May 2017, a flood event with a 20-year return period occurred in TieShan River basin, and was selected to verify the flooding simulation model.

(1) One-dimensional model verification

Choosing the water level in the upstream section of overflow weirs 7#–15# in TieShan River to compare with the observed water level, it is found that the differences between simulated and observed values are between 0.01 and 0.18 m (Table 2). Therefore, the one-dimensional model can be considered as accurate enough for flood simulation.

Table 2. Comparison of simulated and observed water levels.

Position	Station Number	Observed Values/m	Simulated Values/m	Bias/m
7#weir	TSH3 + 720	13.02	13.03	−0.01
8#weir	TSH5 + 640	16.77	16.95	−0.18
9#weir	TSH6 + 570	19.31	19.26	0.05
10#weir	TSH7 + 200	21.21	21.27	−0.06
11#weir	TSH8 + 180	23.6	23.69	−0.09
12#weir	TSH8 + 700	26.16	26.34	−0.18
13#weir	TSH9 + 640	28.81	28.69	0.12
14#weir	TSH11 + 210	31.69	31.74	−0.05
15#weir	TSH11 + 800	33.68	33.72	−0.04

(2) Two-dimensional model verification

The survey result shows that the culvert of SheGong River and the urban area in the south are the main flooding areas; meanwhile, the TieShan River is in a safe state. Comparing the actual flooding area and depth with the simulated ones, we found that the simulation results are basically consistent with the actual situation. Therefore, it can be concluded that the parameters in the 2D model are reasonable, and the model can be used to reflect the actual flooding situations in the study area [53].

4. Results

4.1. Flooding Analysis

We applied the flooding simulation model to simulate the situation of the current river channel when it encounters floods with different return periods. For a 2-year return period, the maximum flooding depth increases during the time interval of 2–6 h, and decreases from 6 to 17 h, reaching maximum value of 0.32 m at the time of 6 h; the flooding depth tends to be relatively stable after 17 h. In correspondence, the maximum flooding area increases from 2 to 9 h and decreases during 9–11 h, reaching the maximum value of 0.93 km^2 at 9 h, and remains unchanged after 11 h. For a 5-year return period, the maximum flooding depth increases from 2 to 5 h and decreases after that, with the maximum value of 0.59 m at 5 h; the maximum flooding area increases before 7 h and decreases from 7 to 14 h, with the maximum value of 2.32 km^2; the flooding area remains almost unchanged after 14 h. When meeting the 10-year return period, the maximum flooding depth increases from 2 to 5 h, and decreases during 5–10 h, reaching the maximum value of 0.73 m; the maximum flooding area increases from 2 to 8 h, decreasing from 8 to 15 h, with the maximum value of 3.14 km^2 at 8 h. As for the 20-year return period, the maximum flooding depth increases over time before 8 h, but gradually decreases from after that, with the maximum value of 0.91 m and the flooding depth is relatively stable after 14 h; the maximum flooding area also increases over time before 7 h, decreasing from 7 to 16 h, reaching the maximum value of 4.03 km^2 and remaining unchanged after 16 h (Figure 5).

The spatial flooding situations for different return periods are shown in Figure 6. In combination of the results from Figure 5, it can be found that the current flood bear capacity of SheGong River does not reach the standard with a 20-year return period; the flooding area increases from 0.93 km^2 to 4.03 km^2 with the return periods of 2, 5, 10, and 20 years, and the maximum flooding depths increases from 0.32 to 0.91 m. The inundated area is gradually increasing from midstream to downstream of SheGong River with the return periods of 2, 5, 10, and 20 years, respectively (Figure 6). When under the 10-year and 20-year return period scenarios, the flooding depth of the downstream of SheGong River is higher than 0.4 m. However, the TieShan River can bear flooding with a 20-year return period safely, and the downstream flood discharge capacity is still at a surplus. It can also be seen from Figure 6 that because the section width of culvert is too narrow, flood overflowed from river channel at this location, while the TieShan River can bear the flooding with a 20-year return period safely, and the flood discharge capacity of downstream is still at a surplus. Therefore, under this condition, how to solve the problem of insufficient flood control capacity in the section of the culvert should be the focus of regulation schemes.

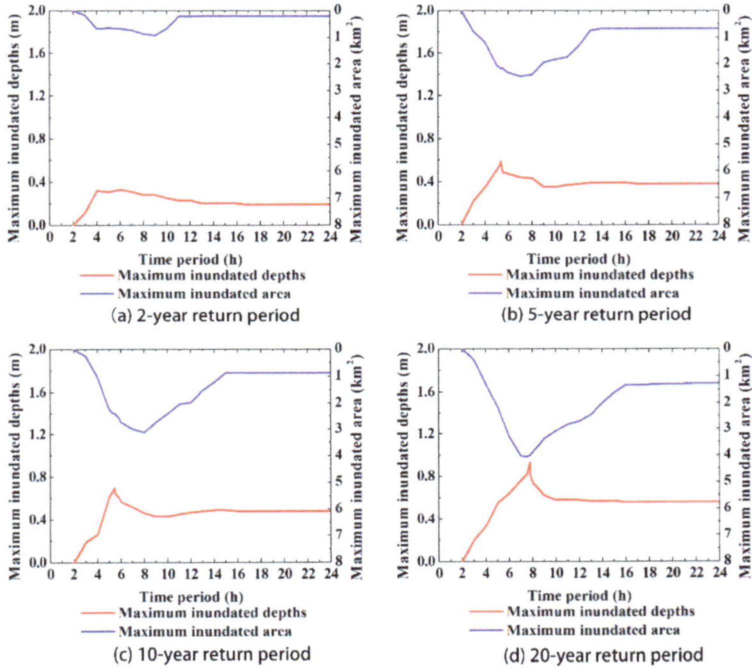

Figure 5. Maximum flooding depths and areas.

4.2. River Regulation Schemes

4.2.1. Formulation of River Regulation Schemes

According to the results of flooding analysis and survey results in the TieShan River basin, the main problems are as follows:

A. The banks at the south of the SG5 + 600 pile are seriously invaded, resulting in the flood control capacity below the standard with a 5-year return period; developed urban areas in the vicinity are affected by flood seriously;

B. At the north of the SG3 + 750 pile in SheGong River, there is a culvert with narrow section and shallow depth, which results in not enough flood discharge capacity of the river channel for flooding with a 2-year return period. Moreover, both sides of this culvert are construction lands; it is difficult to widen the culvert by land acquisition and demolition.

In order to solve these problems, the following three regulation schemes are proposed:

(1) Scheme 1: Widening the downstream width and heightening dikes on both sides of the TieShan River to improve flood control capacity to reach the standard of a 20-year return period.

(2)   Scheme 2: Set a sluice at the SG5 + 600 section in the downstream of SheGong River, and open up the tributary-1 in front of the sluice, transferring flood water from the upstream of SheGong River to the downstream of TieShan River; meanwhile, widen the culvert at the north of the SG3 + 750 pile in the SheGong River.

(3)   Scheme 3: As in Scheme 2, i.e., set up sluice and open up tributary-1 in SheGong River; for the culvert, widen it slightly, and open up tributary-2 in front of the culvert, which leads part of the water flow into TieShan River before entering the culvert.

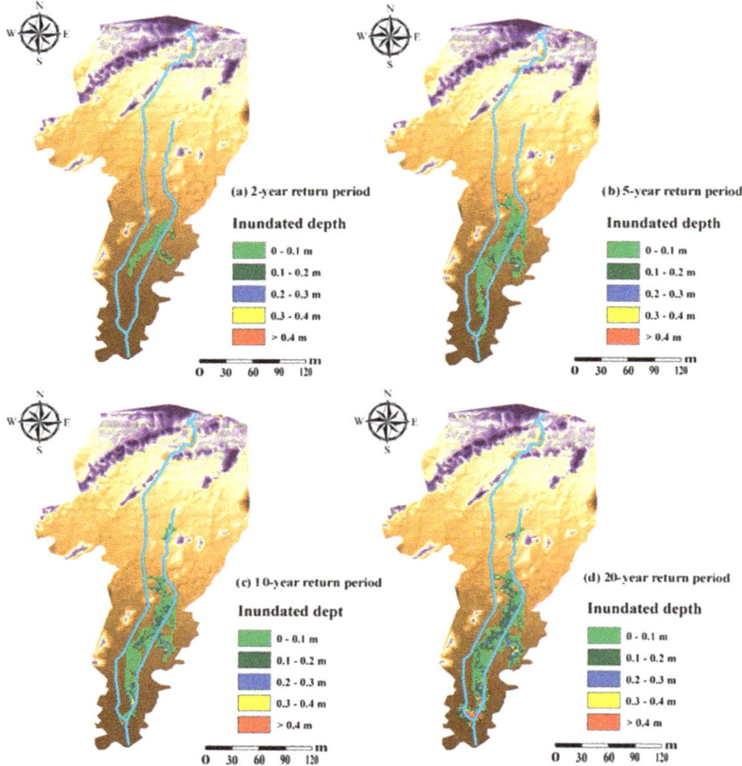

Figure 6. Flooding situation in the study area.

4.2.2. Comparison and Selection of Three Schemes

For Scheme 1, considering the difficulty and high cost in land acquisition and demolition, the general practice is to build dikes on both sides of the SheGong River, maintaining the existing river width. However, according to the results of the water surface calculated by the simulation model (Figure 7), if we only build dikes to improve the flood control capacity to reach the standard of a 20-year return period, the elevation at the top of dikes must be 1.5 m above the ground at least (the safe height above the water surface is taken as 0.6 m), which is probably not allowed for the surrounding environment. Therefore, Scheme 1 that only adopts general engineering measures is not recommended here.

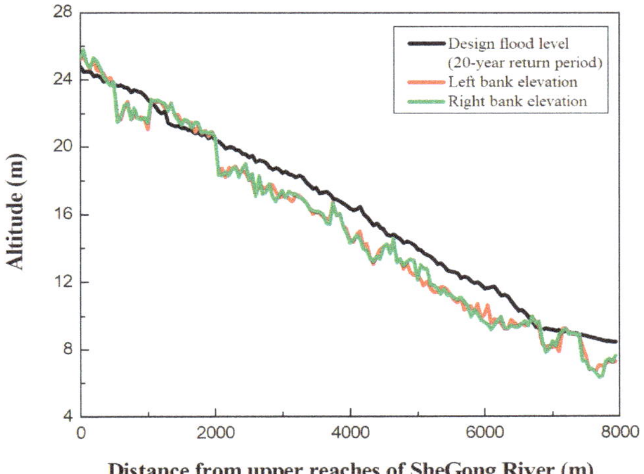

Figure 7. Comparison of current bank elevation and design flood level (20-year return period).

If analyzing from the whole region perspective, it can be found that the surplus of flood discharge capacity in TieShan River is the key of the solution. In Schemes 2 and 3, the construction of tributary-1 is to take advantage of this convenience, and it can play a role in alleviating the flood pressure of SheGong River and avoid the construction of dikes. At the same time, taking the construction cost of engineering projects into account, it is still more reasonable to open up the tributary than to build dikes. Therefore, the next content is mainly to compare Scheme 2 and Scheme 3 (Figure 8).

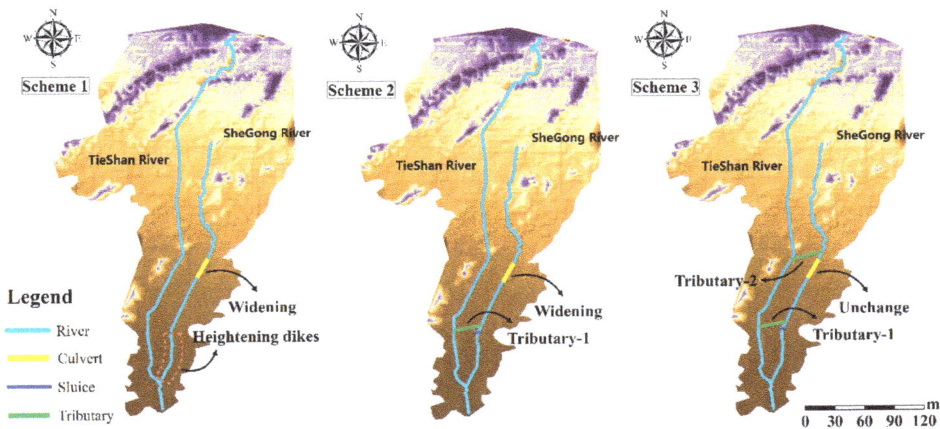

Figure 8. Schematic diagram of Scheme 2 and Scheme 3.

Through simulation, it can be determined that after the adoption of Scheme 2 or Scheme 3, there is no flooding with the 20-year return period. The difference between Scheme 2 and Scheme 3 is the treatment of the culvert. In Scheme 2, the culvert is to be widened; in Scheme 3, a part of water flows that should have passed through the culvert is transferred to the TieShan River through tributary-3. Considering the land acquisition and demolition, the costs of Schemes 2 and 3 are shown in Table 3.

Table 3. Cost of land acquisition and house demolition for Scheme 2 and Scheme 3.

Scheme	Area of Land Acquisition and Demolition (m^2)		Unit Price ($/m^2)		Total Cost ($)
	Construction Land	Farmland	Construction Land	Farmland	
#2	11,861	64,989	24.31	4.29	558,702
#3	3385	83,915			439,869

The simulation results of water surface lines with Scheme 2 and Scheme 3 are shown in Figure 9. It can be seen from Figure 9 that when flooding with a 20-year return period occurs, the water level of the culvert is below safe altitude; the water level of the culvert in Scheme 2 is 0.4 m lower than the water level in Scheme 3 on average, but in the downstream of SheGong River, the water level is significantly higher than that in Scheme 3.

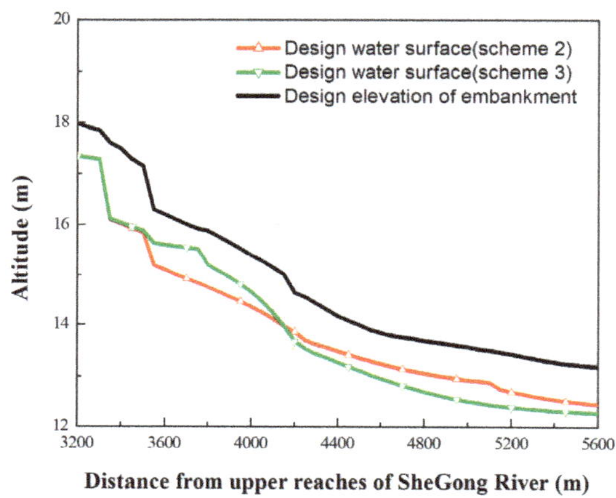

Distance from upper reaches of SheGong River (m)

Figure 9. Simulation results of design water surface line.

In terms of the cost of land acquisition and house demolition, Scheme 3 is lower than Scheme 2. When looking into the water level reduction effect, Scheme 2 has a better reduction effect at the culvert, but Scheme 3 has a better reduction effect in the downstream area. Considering that the downstream area has a higher grade in the development planning, the final regulation scheme is to adopt Scheme 3, which is easier to implement in terms of economic cost and is more effective in reducing flood hazards in the downstream area.

Through the comparison of Schemes 1 and 2, it can be found that a sluice in the downstream of SheGong River can be set and a tributary in front of the sluice transferring floodwater from the upstream of SheGong River to the downstream of TieShan River can be opened, avoiding the widening of the downstream width and heightening dikes on both sides of the SheGong River. In Scheme 2, the downstream of SheGong River could remain the original width but the culvert would need to be widened, while in Scheme 3, the culvert needs to be slightly widened, or can even remain unchanged after opening up a tributary in front of the culvert. Considering the land acquisition and demolition, along with the simulation results of water surface lines, Scheme 3 is the optimal one.

5. Discussion

Due to high urbanization development, there is nearly no extra land around the downstream of the SheGong River that is surrounded by various kinds of buildings. The mid- and the downstream of the river are also the reaches with extremely serious flooding. Therefore, there is not enough space for flood control construction in the SheGong River. Large land occupation, high cost of land acquisition and demolition, difficult demolition work, local resident disputes have increased the difficulty of traditional river flood control measures. If one wants to reduce land acquisition and demolition and keep the original river width, then the embankment height should be at least 1.5 m above the ground, which is disharmonious to the surrounding environment. However, if one makes full use of the flood bearing capacity of the downstream of the Tieshan River and develops new tributaries, it would be an efficient and economical solution. The scope of land acquisition, which is mainly farmland, will be minimized, and the cost of compensation for land acquisition and demolition will be at its lowest. The reduction of land acquisition, demolition and compensation disputes, and the avoidance of raising river embankment make the flood control measures of the river simpler and reduce flooding risk. According to the flood simulation analysis, the flood bearing capacity of the TieShan River is still at a surplus. Flood water in the SheGong River can be transferred into the TieShan River so that the buildings around the SheGong River can remain and the work is greatly simplified, whilst the flood pressure in the mid- and downstream of the SheGong River is effectively reduced. This is from a whole region perspective, given that flood control of the SheGong River is not conducted independently in the SheGong River.

In this study, three possible flood control schemes are proposed. Scheme 1 is to simulate the design flood condition of the SheGong River without changing the existing river width. However, the bank elevation of the whole river in the scheme is lower than the design flood level; consequently the river channel will seriously flood. Due to the lack of flood bearing capacity in the middle and lower reaches of the SheGong River, the downstream of the SheGong River will suffer from flooding. A tributary can be opened in the downstream. There is a culvert in the midstream, which can be uncovered and widened or a tributary can be opened in front of the culvert. In Scheme 2, a flood gate will be built at the downstream of the SheGong River, and a tributary will be built in front of the gate to transfer floodwater to the downstream of the TieShan River. Meanwhile, the culvert at the midstream will be uncovered and widened. Scheme 3 is to open the tributary at the downstream of the SheGong River with the culvert at the midstream unchanged and build the second tributary in front of the culvert. Compared with Scheme 1, Scheme 2 effectively reduces flooding in the downstream of the SheGong River. Comparing Scheme 2 with Scheme 3, the design flood level of the original culvert in Scheme 2 is 0.4 m lower than that in Scheme 3 on average, and the design flood level of the downstream section is higher due to large flows, while tributary 2 in Scheme 3 can effectively reduce flooding pressure of the downstream of the SheGong River. Scheme 3 needs to build a new tributary because water flows in the bottom of the river in Scheme 2 are relatively large and the river needs to be widened. In terms of the cost of land acquisition and house demolition, Scheme 3 is lower than Scheme 2; for a water level reduction effect, Scheme 2 has a better reduction effect at the culvert, but Scheme 3 has a better reduction effect in the downstream. Considering that the downstream area has a higher grade in the development planning and that the cost of land acquisition and demolition cannot be neglected, Scheme 3 is preferable. This scheme makes full use of the surplus flood discharge capacity of the TieShan River from a whole basin perspective. Transferring river floodwater not only avoids a lot of land acquisition and demolition but also solves flooding in the SheGong River economically and effectively. With the rapid development of high urbanization in developing countries around the world, the problem of flood control and drainage in small and medium-scale river basins has become increasingly prominent. Therefore, this study can provide a feasible flood control measure for small and medium-scale rivers in other developing countries.

While this study provides a new idea of flood control, there are some caveats when the idea is applied in other regions. For one, MIKE 21 has high requirements for data; for tributaries and rivers, when simulating small-scale areas using the model, it needs a precise spatial resolution, whilst it is unable to predict elevation differences in floodplain modeling, which has its own significance in accurate results [54,55]. Second, there are some differences between the data obtained in the research area and the requirements of the model, and considering errors such as the generalization of the river network and grids as well as the changes in working conditions in the model, the verification of the one-dimensional model and the two-dimensional model is seemingly simple. Third, we only tested three schemes in the study which are seemingly inadequate given that flood conditions across different regions are rather complicated, and therefore a variety of schemes should be tested, if allowed, for other regions. Last but not least, the development of tributaries will somewhat impact the ecological environment, which needs further study.

6. Conclusions

To explore the most effective flood control strategy for small and medium-scale rivers in highly urbanized areas, this study took the SheGong River (the tributary of the TieShan River) basin as the case to demonstrate how flooding across small and medium-scale river basins can be controlled from a whole region perspective. The MIKE FLOOD model was coupled with MIKE 11 and MIKE 21 to simulate flooding in the basin for different return periods, and the results showed that the mid- and downstream areas of the basin suffer from serious flooding with a 20-year return period, yet the downstream of the mainstream, i.e., TieShan River, can still sufficiently bear more flooding. Therefore, it is possible to deal with flooding over the SheGong River basin by consideration of the flood bearing capacity of the mainstream TieShan River from a whole region perspective. Further, three possible flood control schemes were compared to determine the most cost-effective one. It showed that the construction of two new tributaries transferring floodwater in the mid- and downstream of the SheGong River into the downstream of the TieShan River is the best way to reduce flooding pressure in the SheGong River basin. The scheme of opening up new tributaries to replace the reconstruction of the original river channel can reduce a large number of land acquisition and demolition projects and in turn, reduces the compensation cost of land acquisition and demolition. In highly urbanized areas, the rivers with abundant flood bearing capacity can be fully utilized to control floods. When the flood bearing capacity of small and medium-scale rivers is inadequate, floods in the middle and lower reaches of the river can be transferred to the river with abundant flood-carrying capacity. The new idea of dealing with flooding from a whole region perspective can consider and make full use of the flood bearing capacities of both the mainstream and tributaries. This can not only avoid a large number of land acquisitions, demolitions and compensation fees in highly urbanized areas but also reduce the risk of river flood control, which is particularly useful for highly urbanized areas in developing countries around the world.

Author Contributions: Conceptualization, Z.L. and Y.C.; methodology, Z.L.; software, S.W. and F.L.; validation, Y.C.; formal analysis, Z.L. and Y.C.; investigation, S.W.; resources: X.W.; data curation, F.L.; writing-original draft preparation, Z.L. and Y.C.; writing-review and editing, X.W.; visualization, S.W.; supervision, X.W.; project administration, Z.L. and X.W.; funding acquisition, Z.L. All authors have read and agreed to the published version of the manuscript.

Funding: The research was partially funded by the national key research and development project, Research and Demonstration of Key Technologies for Real-time Flood Control and Drainage Operation in Highly Urbanized Areas", and partially by Guangdong Water Conservancy Science and Technology Project and Guangzhou Water Science and Technology Project.

Acknowledgments: The authors wish to express their gratitude to all authors of the numerous technical reports used for this paper.

Conflicts of Interest: The authors declare no conflicts of interest.

References

1. Mileti, D.; Gailus, J.L. Sustainable development and hazards mitigation in the United States: Disasters by design revisited. *Mitig. Adapt. Strateg. Glob. Chang.* **2005**, *10*, 491–504. [CrossRef]
2. Wu, X.; Wang, Z.; Guo, S.; Liao, W.; Zeng, Z.; Chen, X. Scenario-based projections of future urban inundation within a coupled hydrodynamic model framework: A case study in Dongguan City, China. *J. Hydrol.* **2017**, *547*, 428–442. [CrossRef]

3. Chen, M.; Pang, J.; Wu, P. Flood Routing Model with Particle Filter-Based Data Assimilation for Flash Flood Forecasting in the Micro-Model of Lower Yellow River, China. *Water* **2018**, *10*, 1612. [CrossRef]

4. Wu, X.; Guo, S.; Liu, D.; Hong, X.; Liu, Z.; Liu, P.; Chen, H. Characterization of rainstorm modes along the upper mainstream of Yangtze River during 2003–2016. *Int. J. Climatol.* **2018**, *38*, 1976–1988. [CrossRef]

5. Jung, I.W.; Chang, H.; Moradkhani, H. Quantifying uncertainty in urban flooding analysis considering hydro-climatic projection and urban development effects. *Hydrol. Earth Syst. Sci.* **2011**, *15*, 617–633. [CrossRef]

6. Wang, Z.; Lai, C.; Chen, X.; Yang, B.; Zhao, S.; Bai, X. Flood hazard risk assessment model based on random forest. *J. Hydrol.* **2015**, *527*, 1130–1141. [CrossRef]

7. Chang, L.C.; Amin, M.; Yang, S.N.; Chang, F.J. Building ANN-Based Regional Multi-Step-Ahead Flood Inundation Forecast Models. *Water* **2018**, *10*, 1283. [CrossRef]

8. Xi, F.; He, H.S.; Clarke, K.C.; Hu, Y.; Wu, X.; Liu, M.; Gao, C. The potential impacts of sprawl on farmland in Northeast China-Evaluating a new strategy for rural development. *Landsc. Urban Plan.* **2012**, *104*, 34–46. [CrossRef]

9. Chang, L.C.; Chang, F.J.; Yang, S.N.; Kao, I.; Ku, Y.Y.; Kuo, C.L.; Amin, I. Building an Intelligent Hydroinformatics Integration Platform for Regional Flood Inundation Warning Systems. *Water* **2019**, *11*, 9. [CrossRef]

10. Yin, J.; Yin, Z.; Wang, J.; Xu, S. National assessment of coastal vulnerability to sea-level rise for the Chinese coast. *J. Coast Conserv.* **2012**, *16*, 123–133. [CrossRef]

11. Jabbari, A.; Bae, D.-H. Application of Artificial Neural Networks for Accuracy Enhancements of Real-Time Flood Forecasting in the Imjin Basin. *Water* **2018**, *10*, 1626. [CrossRef]

12. Lai, C.; Shao, Q.; Chen, X.; Wang, Z.; Zhou, X.; Yang, B.; Zhang, L. Flood risk zoning using a rule mining based on ant colony algorithm. *J. Hydrol.* **2016**, *542*, 268–280. [CrossRef]

13. Huong, H.T.L.; Pathirana, A. Urbanization and climate change impacts on future urban flooding in Can Tho city, Vietnam. *Hydrol. Earth Syst Sci.* **2013**, *17*, 379–394. [CrossRef]

14. Lai, C.; Chen, X.; Chen, X.; Wang, Z.; Wu, X.; Zhao, S. A fuzzy comprehensive evaluation model for flood risk based on the combination weight of game theory. *Nat. Hazards* **2015**, *77*, 1243–1259. [CrossRef]

15. Quan, R. Rainstorm waterlogging risk assessment in central urban area of Shanghai based on multiple scenario simulation. *Nat. Hazard* **2014**, *73*, 1569–1585. [CrossRef]

16. Mishra, B.K.; Herath, S. Assessment of future floods in the Bagmati River basin of Nepal using bia-corrected daily GCM precipitation data. *J. Hydrol. Eng.* **2015**, *20*, 8. [CrossRef]

17. Saraswat, C.; Kumar, P.; Mishra, B.K. Assessment of stormwater runoff management practices and governance under climate change and urbanization: An analysis of Bangkok, Hanoi and Tokyo. *Environ. Sci. Policy* **2016**, *64*, 101–117. [CrossRef]

18. Wu, X.; Guo, S.; Yin, J.; Yang, G.; Zhong, Y.; Liu, D. On the event-based extreme precipitation across China: Time distribution patterns, trends, and return levels. *J. Hydrol.* **2018**, *562*, 305–317. [CrossRef]

19. Di Paola, F.; Ricciardelli, E.; Cimini, D.; Romano, F.; Viggiano, M.; Cuomo, V. Analysis of Catania flash flood case study by using combined microwave and infrared technique. *J. Hydrometeorol.* **2014**, *15*, 1989–1998. [CrossRef]

20. Fu, G.; Butler, D.; Khu, S.T.; Sun, S.A. Imprecise probabilistic evaluation of sewer flooding in urban drainage systems using random set theory. *Water Resour. Res.* **2011**, *47*, W02534. [CrossRef]

21. Hammond, M.J.; Chen, A.S.; Djordjević, S.; Butler, D.; Mark, O. Urban flood impact assessment: A state-of-the-art review. *Urban Water J.* **2015**, *12*, 14–29. [CrossRef]

22. Vacondio, R.; Aureli, F.; Ferrari, A.; Mignosa, P.; Dal Palu, A. Simulation of the January 2014 flood on the Secchia River using a fast and high-resolution 2D parallel shallow-water numerial scheme. *Nat. Hazards* **2016**, *80*, 103–125. [CrossRef]

23. Yang, L.; Smith, J.; Baeck, M.L.; Smith, B.; Tian, F.; Niyogi, D. Structure and evolution of flash flood producing storms in a small urban watershed. *J. Geophys. Res. Atmos.* **2016**, *121*, 3139–3152. [CrossRef]

24. Bai, T.; Wei, J.; Yang, W.; Huang, Q. Multi-Objective Parameter Estimation of Improved Muskingum Model by Wolf Pack Algorithm and Its Application in Upper Hanjiang River, China. *Water* **2018**, *10*, 1415. [CrossRef]

25. van den Honert, R.C.; McAneney, J. The 2011 Brisbane Flood: Causes, Impacts and Implications. *Water* **2011**, *3*, 1149–1173. [CrossRef]

26. Rakwatin, P.; Sansena, T.; Marjang, N.; Rungsipanich, A. Using multi-temporal remote-sensing data to estimate 2011 flood area and volume over Chao Phraya River basin, Thailand. *Remote Sens. Lett.* **2012**, *4*, 243–250. [CrossRef]

27. Zhang, H.; Ma, W.C.; Wang, X.R. Rapid urbanization and implications for flood risk management in hinterland of the Pearl River Delta, China: The Foshan study. *Sensors* **2008**, *8*, 2223–2239. [CrossRef]

28. Wang, K.; Wang, L.; Wei, Y.-M.; Ye, M. Beijing storm of July 21, 2012: Observations and reflections. *Nat. Hazards* **2013**, *67*, 969–974. [CrossRef]

29. Huang, Y.; Chen, S.; Cao, Q.; Hong, Y.; Wu, B.; Huang, M.; Qiao, L.; Zhang, Z.; Yang, X. Evaluation of Version-7 TRMM Multi-Satellite Precipitation Analysis Product during the Beijing Extreme Heavy Rainfall Event of 21 July 2012. *Water* **2013**, *6*, 32–44. [CrossRef]

30. Jha, A.K.; Bloch, R.; Lamond, J. Cities and Flooding: A Guide to Integrated Urban Flood Risk Management for the 21st Cemtury. *J. Reg. Sci.* **2012**, *52*, 885–887. [CrossRef]

31. Batisani, N.; Yarnal, B. Urban expansion in Centre County, Pennsylvania: Spatial dynamics and landscape transformations. *Appl. Geogr.* **2009**, *29*, 235–249. [CrossRef]

32. Wu, X.; Wang, Z.; Zhou, X.; Lai, C.; Lin, W.; Chen, X. Observed changes in precipitation extremes across 11 basins in China during 1961–2013. *Int. J. Climatol.* **2016**, *36*, 2866–2885. [CrossRef]

33. Hassan, M.M.; Nazem, M.N.I. Examination of land use/land cover changes, urban growth dynamics, and environmental sustainability in Chittagong city, Bangladesh. *J. Environ. Dev. Sustain.* **2016**, *18*, 697–716. [CrossRef]

34. Miguez, M.G.; Veról, A.P.; De Sousa, M.M.; Rezende, O.M. Urban floods in lowlands—Levee systems, unplanned urban growth and river restoration alternative: A case study in Brazil. *Sustainability* **2015**, *7*, 11068–11097. [CrossRef]

35. Bruwier, M.; Mustafa, A.; Aliaga, D.G.; Archambeau, P.; Erpicum, S.; Nishida, G.; Zhang, X.; Pirotton, M.; Teller, J.; Dewals, B. Influence of urban pattern on inundation flow in floodplains of lowland rivers. *Sci. Total Environ.* **2018**, *622*, 446–458. [CrossRef]

36. Cho, S.Y.; Chang, H. Recent research approaches to urban flood vulnerability, 2006–2016. *Nat. Hazards* **2017**, *88*, 633–649. [CrossRef]

37. Luino, F.; Turconi, L.; Petrea, C.; Nigrelli, G. Uncorrected land-use planning highlighted by flooding: The Alba case study (Piedmont, Italy). *Nat. Hazards Earth Syst. Sci.* **2012**, *12*, 2329–2346. [CrossRef]

38. Li, C.; Cheng, X.; Li, N.; Du, X.; Yu, Q.; Kan, G. A Framework for Flood Risk Analysis and Benefit Assessment of Flood Control Measures in Urban Areas. *Int. J. Environ. Res. Public Health* **2016**, *13*, 787. [CrossRef]

39. Wu, X.; Wang, Z.; Guo, S.; Lai, C.; Chen, X. A simplified approach for flood modeling in urban environments. *Hydrol. Res.* **2018**, *49*, 1804–1816. [CrossRef]

40. Dutta, D.; Herath, S.; Musiake, K. Flood inundation simulation in a river basin using a physically based distributed hydrologic model. *Hydrol. Process.* **2000**, *14*, 497–519. [CrossRef]

41. Yazdi, J.; Neyshabouri, S.S. Optimal design of flood-control multi-reservoir system on a watershed scale. *Nat. Hazards* **2012**, *63*, 629–646. [CrossRef]

42. Miller, J.D.; Kim, H.; Kjeldsen, T.R.; Packman, J.; Grebby, S.; Dearden, R. Assessing the impact of urbanization on storm runoff in a peri-urban catchment using historical change in impervious cover. *J. Hydrol.* **2014**, *515*, 59–70. [CrossRef]

43. Kong, F.; Ban, Y.; Yin, H. Modeling stormwater management at the city district level in response to changes in land use and low impact development. *Environ. Model. Softw.* **2017**, *95*, 132–142. [CrossRef]

44. Garg, V.; Aggarwal, S.P.; Gupta, P.K. Assessment of land use land cover change impact on hydrological regime of a basin. *Environ. Earth Sci.* **2017**, *76*, 635. [CrossRef]

45. Li, J.; Wang, Z.; Lai, C. Severe drought events inducing large decrease of net primary productivity in mainland China during 1982–2015. *Sci. Total Environ.* **2019**, in press. [CrossRef]

46. Shokoohi, A.R. Assessment of Urban Basin Flood Control Measures Using Hydrogis Tools. *J. Appl. Sci.* **2007**, *7*, 1726–1733. [CrossRef]

47. Miguez, M.G.; Mascarenhas, F.C.B.; Canedo de Magalhães, L.P.; D'Alterio, C.F.V. Planning and Design of Urban Flood Control Measures: Assessing Effects Combination. *J. Urban Plan. Dev.* **2009**, *135*, 100–109. [CrossRef]

48. Gül, G.O.; Harmancıoğlu, N.; Gül, A. A combined hydrologic and hydraulic modeling approach for testing efficiency of structural flood control measures. *Nat. Hazards* **2009**, *54*, 245–260. [CrossRef]

49. Pilarczyk, K.W.; Nuoi, N.S. Experience and Practices on Flood Control in Vietnam. *Water Int.* **2005**, *30*, 114–122. [CrossRef]

50. Konrad, C.P. *Effects of Urban Development on Floods*; 076–03; U.S. Geol. Surv.: Washington, DC, USA, 2003.

51. Liu, Z.; Zhang, H.; Liang, Q. A coupled hydrological and hydrodynamic model for flood simulation. *Hydrol. Res.* **2018**, *50*, 589–606. [CrossRef]

52. Hu, C.Y.; Wang, J.X. *Watershed Runoff Model and Hydrological Model*; The Yellow River Water Conservancy Press: Zhengzhou, China, 2010.

53. Raghunath, H.M. *Hydrology Principles·Analysis·Design*; New Age International Publishers: Manipal, Karnataka, India, 2006.

54. Merz, B. Floods and climate: Emerging perspectives for flood risk assessment and management. *Nat. Hazards Earth Syst. Sci.* **2014**, *2*, 1559–1612. [CrossRef]

55. Fan, Y.; Ao, T.; Yu, H.; Huang, G.; Li, X. A Coupled 1D-2D Hydrodynamic Model for Urban Flood Inundated. *Adv. Meteorol.* **2017**, *2017*, 2819308. [CrossRef]

Case Report

Emergency Disposal Solution for Control of a Giant Landslide and Dammed Lake in Yangtze River, China

Guiya Chen [1,2], Xiaofeng Zhao [3], Yanlai Zhou [2,4,*], Shenglian Guo [2], Chong-Yu Xu [4] and Fi-John Chang [5]

1   Changjiang Water Resources Commission, Wuhan 430010, China; chengy@cjh.com.cn
2   State Key Laboratory of Water Resources and Hydropower Engineering Science, Wuhan University, Wuhan 430072, China; slguo@whu.edu.cn
3   Hubei Provincial Water Resources and Hydropower Planning Survey and Design Institute, Wuhan 430064, China; zyl23bulls@163.com
4   Department of Geosciences, University of Oslo, P.O. Box 1047 Blindern, N-0316 Oslo, Norway; c.y.xu@geo.uio.no
5   Department of Bioenvironmental Systems Engineering, National Taiwan University, Taipei 10617, Taiwan; changfj@ntu.edu.tw
*   Correspondence: yanlai.zhou@whu.edu.cn

Received: 11 August 2019; Accepted: 16 September 2019; Published: 18 September 2019

Abstract: Although landslide early warning and post-assessment is of great interest for mitigating hazards, emergency disposal solutions for properly handling the landslide and dammed lake within a few hours or days to mitigate flood risk are fundamentally challenging. In this study, we report a general strategy to effectively tackle the dangerous situation created by a giant dammed lake with 770 million cubic meters of water volume and formulate an emergency disposal solution for the 25 million cubic meters of debris, composed of engineering measures of floodgate excavation and non-engineering measures of reservoirs/hydropower stations operation. Such a disposal solution can not only reduce a large-scale flood (10,000-year return period, 0.01%) into a small-scale flood (10-year return period, 10%) but minimize the flood risk as well, guaranteeing no death raised by the giant landslide.

Keywords: water resources management; landslide; dammed lake; flood risk

---

1. Introduction

Landslides can be attributed to rainstorms [1–3], earthquakes [4,5], avalanches, human construction activities, land-use change [6–8] as well as natural processes of erosion that ruin land slopes [9,10]. Furthermore, large landslides often block river vales, giving rise to the formation of giant dams, so that catastrophic debris flows and floods would easily take place in case of dam break [11].

Early warning, in close association with theoretical considerations, predicts that the intensity of giant landslides will increase in a climate and anthropogenic changes setting according to physical experiments and post-assessment [12,13]. The potential for a giant landslide intensification with earthquake, rainfall-induced storm runoff, and avalanche is of significant societal concern, with a huge dammed lake inducing dam break, flash floods and debris being one of the costliest and dangerous natural hazards worldwide [14]. Landslides are one of the most difficult natural disasters to predict since the factors that affect slope stability vary dramatically in both space and time. An emergency disposal solution for adequately handling the landslide and dammed lake within a few hours or days to reduce flood risk is fundamentally challenging, despite landslide early warning and post-assessment being of great interest for mitigating hazards. Nevertheless, the expected general strategy in response to landslides and dammed lake extreme intensification is still unclear. Hence, we propose that more attention should be paid to increase infrastructure resilience to our changing environment, as observations and field surveys of storm runoff extremes, earthquakes and avalanches show that they would cause major challenges for existing infrastructure systems.

In the stream nearby the Baige Village of the Tibet Autonomous Region, which suffers from the obstacles caused by massive landslides, bulldozers are excavating a diversion channel to reduce the risk of flooding due to the fast-rising water levels of the blocked Jinsha River, which is situated in the upper reach of the Yangtze River (Figure 1). On 3 November, 2018, the landslide following the first landslide, which occurred on 11 October, 2018, caused about 25 million cubic meters of debris to hurtle down a mountainside, which created a natural dam that was 58.24 m high, 195 m wide and 273 m long, situated above the elevation of 2966 m along the Jinsha River. "I was surprised that the water level of the lake increased so fast at a speed of 0.7 m per hour" said one of the hydrologists from the Changjiang Water Resources Commission (CWRC) who first glimpsed the dammed lake. In the next few days, the water volume of this dammed lake increased from 300 million cubic meters and reached 770 million cubic meters.

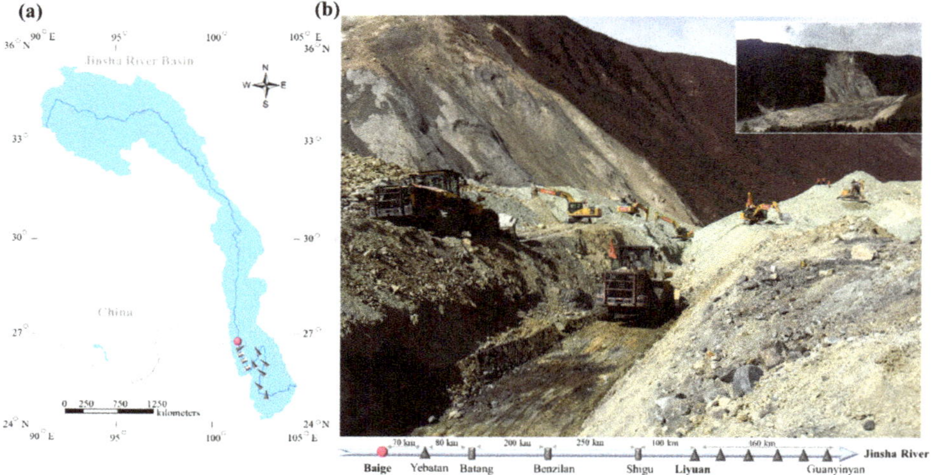

Figure 1. Location of landslide and excavation of floodgate. (**a**) Location of landslide; (**b**) workers carved out a diversion channel through the landslide blocking the Jinsha River. Baige is the place where the landslide occurred, and Yebatan is a hydropower station. Batang, Benzilan, and Shigu are hydrological stations. Liyuan and Guanyinyan are the first and the last reservoirs of the six cascade reservoirs in the midstream of the Jinsha River, respectively.

In this study, we propose an emergency disposal solution combining structural and non-structural measures to reduce the flood risk encountered in a giant landslide and dammed lake. The rest of this study is arranged as below. Section 2 introduces the used methods, including the structural measures and the non-structural measures, respectively. Section 3 shows the results on the methods employed to control the landslide and dammed lake while discussing the future challenges of landslide management in the Yangtze River.

2. Methods

The goal of this study is to propose an emergency disposal solution for landslide control, including structural and non-structural measures and reducing flood risk to a safe range. Figure 2 illustrates the emergency disposal solution architecture that integrates the structural measures (Figure 2a) with the non-structural measures (Figure 2b). The used methods are briefly described below.

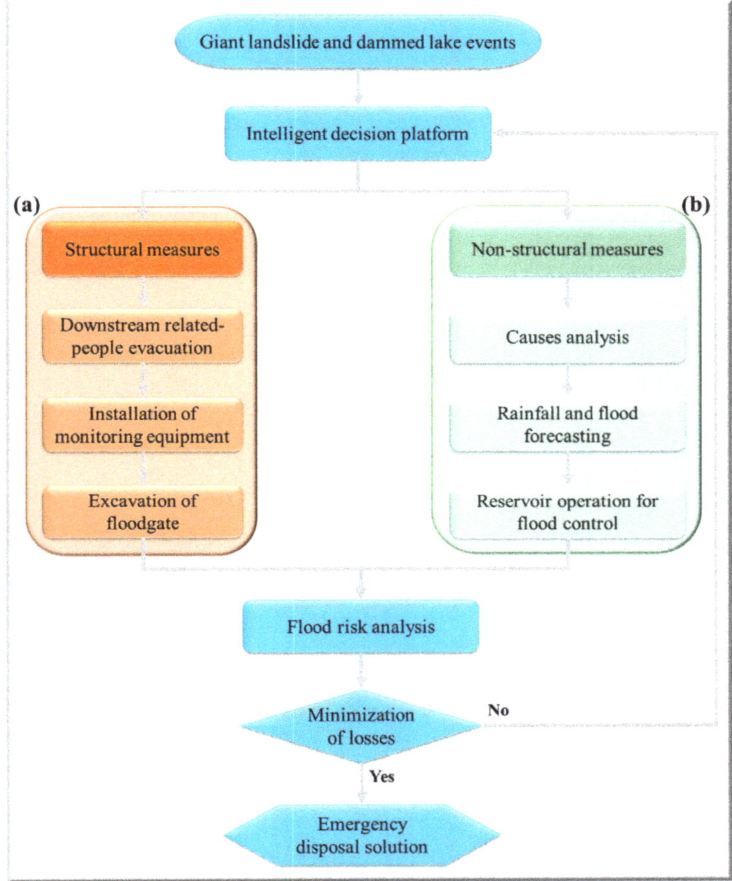

Figure 2. Emergency disposal solution for landslide control. (**a**) Structural measures; (**b**) non-structural measures.

2.1. Structural Measures

The most commonly used structural measures are classified by the sort of slope stabilization methods used, including the geometric method [15,16], chemical and mechanical method [17], and hydrogeological method [18]. The former two methods are preferentially adopted to mitigate potential landslides, while the third method is preferentially adopted to handle breakout landslides [19]. Therefore, in this study, a package of hydrogeological methods is employed to mitigate landslide and dammed lake. The hydrogeological methods consist of downstream related-people evacuation, installation of monitoring equipment, as well as excavation of floodgate. The downstream related-people evacuation can effectively lower life and property losses ahead of natural disasters extension. Mobile hydrological monitoring equipment is installed to collect real-time rainfall and flood information. The excavation of floodgate can be used to rapidly decrease the water volume of the dammed lake after implementing landslide surface protection and control of surface erosion mitigation measures.

2.2. Non-Structural Measures

The non-engineering measures consist of analysis of the causes, rainfall and flood forecasting and reservoir operation for flood prevention. Firstly, a clear understanding of the processes that caused the landslide (i.e., cause analysis) plays a vital role in the selection and design of appropriate, cost-effective remedial measures. While the destabilizing processes are grouped into slow changing (e.g., weathering, erosion) and fast-changing processes (e.g., earthquake, drawdown) [20]. Secondly, the early warning system for rainfall and flood forecasting can make it possible to implement a regional real-time assessment of landslide hazard threats [21]. Last but not least, flood prevention and pre-discharge measures for reservoirs (hydropower stations) can be used to decrease the flood risk of downstream area—especially when the breakout landslide has further caused a dammed lake in the river.

After several structural and non-structural measures are formulated and designed to cope with the breakout landslide, it is imperative to undertake a flood risk analysis [22–25] for selecting an appropriate and cost-effective emergency disposal solution from the standpoint of minimizing losses.

3. Results and Discussion

3.1. Causes Analysis for Giant Landslide

First—consdering the earthquake factor—Tibet in China is an earthquake-prone area, and a great number of earthquakes (642, respectively) larger than magnitude 5.0 occurred in Tibet during 1900 and 2017. Out of these earthquakes, there were 503 measuring a magnitude of 5.0–5.9, 130 measuring 6.0–6.9, 7 measuring 7.0–7.9, and two measuring 8.0–8.9 (Figure 3).

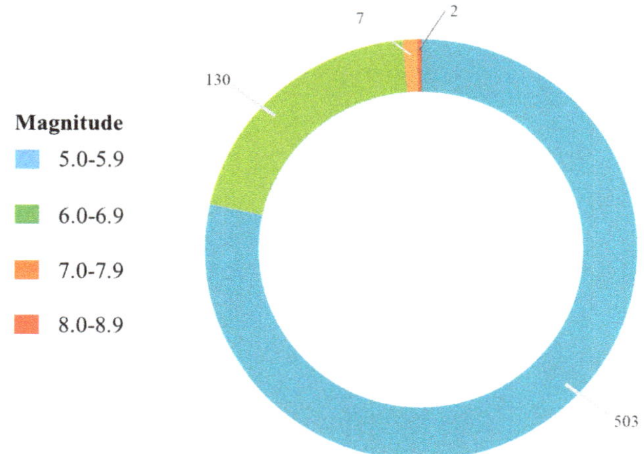

Figure 3. Data on earthquakes in Tibet. Data on earthquakes larger than magnitude 5.0 in Tibet during 1900 and 2017, extracted from the United States Geological Survey (USGS) [26] and the China Earthquake Networks Center (CENC) [27].

According to the statistical data released by the USGS, the Chayu County of Tibet underwent a magnitude 8.6 earthquake on 15 August, 1950—which was the most massive earthquake in China to date. Besides, the Naqu City of Tibet was hit by two earthquakes of magnitudes 4.1 and 4.2 on the 6 and 7 October, 2018, respectively. The epicenters of the two earthquakes were 1000 km away from the Baige Village. Second—considering the weather factor—between 11 October and 3 November 2018, some heavy rainfalls (100–250 mm/day) occurred in succession. Therefore, in consequence of the two factors associated with earthquake and storm precipitation, a landslide struck the Baige Village and initiated the formation of a giant dammed lake.

3.2. Emergency Disposal Solution

The director of the Changjiang Water Resources Commission (CWRC) and experts from the Changjiang River Defense General formulated an emergency disposal plan for the dammed lake to properly handle disasters for minimizing losses, as addressed below: (1) organize an evacuation of the affected areas in Tibet, Sichuan, and Yunnan Provinces. (2) Undertake hydrological emergency monitoring, forecasting, and early warning in the upper and lower reaches of the dammed lake. (3) Quickly tackle the dangerous situation raised by the dammed lake and formulate an emergency disposal plan for the stagnation. (4) Implement flood prevention and pre-discharge measures for reservoirs (hydropower stations) that are operational and under construction in the lower reaches of the dammed lake.

On 4 and 5 November, 2018, combining the aerial videos, images and in-situ measurement data collected by drones with the development of the water storage in the dammed lake, it was concluded that without human intervention, the possibility for the dammed lake to collapse would be very high. Besides, the anticipated magnitude of the flood-induced by such a collapse would be enormous, and would bring significant economic losses to the downstream. During 4 and 5 November, 2018, after a comprehensive site-specific survey of the landslide (Figure 4), the director of the CWRC said, "in such emergency case, we only have a few days to assess landslide hazards and figure out what to do".

(a)   **(b)**

Figure 4. Field survey of the Baige landslide. Engineering geologists and hydrologists carried out a field survey of the Baige landslide during (**a**) 4 November, 2018 and (**b**) 5 November, 2018.

The CWRC dispatched 245 people to participate in the disposal of the dammed lake. On 9 November, 2018, three kinds of diversion channels and flood risks were analyzed and formulated, based on three scenarios designed to reduce the lake water level by 5, 10, and 15 m, respectively. Flow peaks of the main control sites along the downstream of the dammed lake were predicted in case of a dam break as well. On 10 November, 2018, about 80 sets of equipment—including automatic gauges designed for high water level and water gauges—were installed and the forecasts of water level, flow, and possible inundated areas were made (Figure 5). On 11 November, 2018, the engineering team succeeded in digging out a diversion channel with a length of 220 m, the maximum depth of 15 m, a width of 42 m at the top, and a width of 4 m at the base so that floodwater could be released for decreasing the lake water level as well as reducing the flood risks at downstream areas (Supplementary Video S1). Meanwhile, the government quickly evacuated more than 50,000 people downstream of the Baige area and at least 18,657 people in Lijiang City of Yunnan Province.

(a)   **(b)**

Figure 5. Hydrological monitoring in the downstream of Baige landslide. (**a**) Installation of water level ruler; (**b**) testing hydrological monitoring equipment.

Before the dam broke on 13 November, 2018 (Supplementary Video S2), four operational commands on cascade reservoirs were issued by the CWRC to cope with the dam break based on in-situ survey and hydrological forecasting, as described below. To deal with the floods caused by dam break, a series of reservoir operation was implemented started from 5 November, 2018. Three cascade reservoirs (Liyuan, Ahai, and Jinanqiao) vacated a total storage capacity of 0.62 billion cubic meters on 5 November, 2018. It followed six cascade reservoirs (Liyuan, Ahai, Jinanqiao, Longkaikou, Ludila, and Guanyinyan) that vacated a total storage capacity of 1.03 billion cubic meters on 10 November, followed by 1.189 billion cubic meters on 14 November, and finally reached 1.3 billion cubic meters on 15 November.

A few hours after the dam break, the peak flow discharge reached 28,300 cubic meters per second (cms) at the Yebatan hydropower station (19:00, 13 November), 20,900 cms at the Batang station (02:00, 14 November), 15,700 cms at the Benzilan station (13:00, 14 November), 7170 cms at the Shigu station (08:00, 15 November), and 7410 cms in the Liyuan Reservoir (10:00, 16 November), respectively. We noticed that reservoir inflow increased by 30% to 60% at the above-mentioned upstream reservoirs, whereas it only increased 5% to 20% at downstream stations, as compared with historical maximal peak flows. Taking the Liyuan Reservoir as an example, the decreasing rate of the flood peak reached 40%, based on the maximal inflow (7410 cms) and outflow (4490 cms) (Figure 6 and Supplementary Video S3). In other words, the joint operation of the six cascade reservoirs released a total of 1.3 billion cubic meters in storage capacity for flood control would pave the way not only to reduce a large-scale (10,000-year return period) flood into a small-scale (10-year return period) flood but to minimize the flood risk raised by the Baige landslide as well.

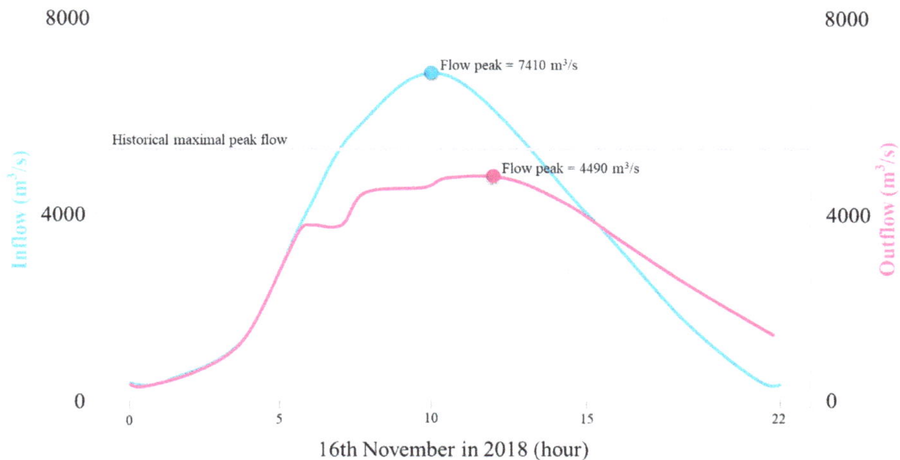

Figure 6. Sketch map of reservoir operation for releasing water in the Liyuan Reservoir.

The results indicate that the flood risk and life-property loss are significantly reduced if the structural and non-structural measures are contained and a wide scope of stakeholders are involved in response to the destruction caused by water-related natural catastrophes. The effectiveness of the structural and non-structural measures in such a case is due to a variety of reasons: the downstream related-people evacuation can play a vital role in helping avert losses from flooding and debris; the fast installation of mobile hydrological monitoring equipment can provide first-hand information of real-time hydro-meteorological data for flood forecasting; the excavation of floodgate can effectively address floods resulting from the failure of the dammed lake; the physical cause analysis results can help managers and decision-makers to figure out remedial measures according to the causes and effects of landslides; the early warning system based on hydrological forecasts can make sufficient lead time for diverse sectors and stakeholders to respond if flooding is severe, widespread and sudden; and reservoir operation (e.g., pre-discharge measures) can provide flood control capacity for coping with the floods, in case the dam resulting from the landslide breaks.

In the coming months, the director of the CWRC and his team will assess landslide hazards, reconstruct the damaged monitoring facilities, and revise the safety standards against landslides and dammed lakes. In addition, follow-up studies will pay special attention to predicting rainfall and flood used machine learning techniques [28–30], simulating and/or optimizing reservoirs operation based on advanced intelligent algorithms [31,32], and assessing the impacts of dammed lakes and dam collapse on the hydrological conditions [33,34], sediment deposition and erosion [35,36] as well as biodiversity [37] of the Yangtze River. After these studies were accomplished, the director of the CWRC said, "new achievements will contribute to landslide management in other countries or regions of the world". An integration of landslide-risk-management measures demands the participation of diverse sectors (water management, spatial planning, and emergency management) and diverse governmental levels to offer appropriate policy instruments. Improving cooperation and coordination between policy sectors and administrative levels will benefit to cope with natural hazards. Besides, from the standpoint of methodological transferability, the package of the structural and non-structural measures—apart from the reservoir operation—can be directly applied for the management of other landslide cases or studies, whereas the reservoir operation can only be considered as a special measure against landslide in response to the landslide and a dammed lake raised in the river.

4. Conclusions

Sudden landslides with large rock avalanches often block river valleys and result in the formation of large dammed lakes, which would cause devastating debris flows and floods once the dams collapse. The loss of life and destruction of property caused by catastrophic debris flows around the world will most likely continue intensifying as the world population increases, urban development boosts, deforestation expands and land-use alternates. In this case report, we proposed an emergency disposal solution for adequately processing the landslide and dammed lake. The Baige landslide and dammed lake upstream of the Yangtze River of China was selected as a study case. The contributions of the proposed solution consisted of: (1) the emergency disposal solution's ability to integrate the structural and non-structural measures; and (2) that the emergency disposal solution can significantly reduce the flood risk and property losses caused by the landslide and dammed lake.

Despite the Baige landslide created a giant dammed lake with 770 million cubic meters of water volume, the formulated emergency disposal solution can adequately remove the 25 million cubic meters of debris and effectively release the water volume. Furthermore, the hybrid of engineering measures of floodgate excavation and non-engineering measures of reservoirs/hydropower stations operation can decrease a large-scale flood (10,000-year return period, 0.01%) into a small-scale flood (10-year return period, 10%) to minimize the flood risk.

Within only two weeks, China succeeded in coping with the life-threatening event of the giant landslide-induced dammed lake that took place in the Baige Village along the Yangtze River, guaranteeing that no death was caused by the landslide. Besides, this provided valuable experiences and effective strategies to minimize disaster damages, and the losses of life and property resulted from giant landslides stimulated by climate and anthropogenic changes.

Supplementary Materials: The following are available online at http://www.mdpi.com/2073-4441/11/9/1939/s1, Video S1: Succeeded in excavating the diversion channel. After working round the clock for one week, workers succeeded in excavating the diversion channel through the Baige Dam on 11 November 2018. Video S2: Break of the Baige dam. After the human intervention, the Baige dam started to break on 13 November 2018. Video S3: A swollen Jinsha River gush through the Liyuan reservoir. In response to the Baige dam break, the Liyuan Reservoir started to release floodwaters in advance on 13 November 2018. Data and materials availability: The data of global earthquakes can be downloaded from the United States Geological Survey (USGS, https://earthquake.usgs.gov) and the China Earthquake Networks Center (CENC, http://www.cenc.ac.cn, Chinese). All data needed to evaluate the results and conclusions in the manuscript are provided in the manuscript or the Supplementary Materials.

Author Contributions: G.C., X.Z. and Y.Z. carried out the analysis and wrote the article, S.G., C.X. and F.C. provided technical assistance and contributed in writing the article, G.C. carried out the field survey.

Funding: This research was funded by the National Key Research and Development Program of China (2018YFC0407904), the Research Council of Norway (FRINATEK Project 274310) and the Excellent Youth Science Foundation of NSFC (Number 51822908).

Acknowledgments: We thank the Changjiang Water Resources Commission (CWRC) of China for providing the monitoring data and video materials of field survey.

Conflicts of Interest: The authors declare no conflict of interest.

References

1. Dikshit, A.; Sarkar, R.; Pradhan, B.; Acharya, S.; Dorji, K. Estimating rainfall thresholds for landslide occurrence in the Bhutan Himalayas. *Water* **2019**, *11*, 1616. [CrossRef]
2. Zhang, K.; Wang, S.; Bao, H.; Zhao, X. Characteristics and influencing factors of rainfall-induced landslide and debris flow hazards in Shaanxi Province, China. *Nat. Hazards Earth Syst. Sci.* **2019**, *19*, 93–105. [CrossRef]
3. Roccati, A.; Faccini, F.; Luino, F.; Ciampalini, A.; Turconi, L. Heavy rainfall triggering shallow landslides: A susceptibility assessment by a GIS-approach in a Ligurian Apennine Catchment (Italy). *Water* **2019**, *11*, 605. [CrossRef]
4. McPhillips, D.; Bierman, P.R.; Rood, D.H. Millennial-scale record of landslides in the Andes consistent with earthquake trigger. *Nat. Geosci.* **2014**, *7*, 925.

5. Cao, B.; Yang, S.; Ye, S. Integrated application of remote sensing, GIS and hydrological modeling to estimate the potential impact area of earthquake-induced dammed lakes. *Water* **2017**, *9*, 777. [CrossRef]

6. Klimeš, J.; Novotný, J.; Novotná, I.; de Urries, B.J.; Vilímek, V.; Emmer, A.; Strozzi, T.; Kusák, M.; Rapre, A.C.; Hartvich, F.; et al. Landslides in moraines as triggers of glacial lake outburst floods: Example from Palcacocha Lake (Cordillera Blanca, Peru). *Landslides* **2016**, *13*, 1461–1477. [CrossRef]

7. Croissant, T.; Lague, D.; Steer, P.; Davy, P. Rapid post-seismic landslide evacuation boosted by dynamic river width. *Nat. Geosci.* **2017**, *10*, 680. [CrossRef]

8. Yin, J.; Gentine, P.; Zhou, S.; Sullivan, S.C.; Wang, R.; Zhang, Y.; Guo, S. Large increase in global storm runoff extremes driven by climate and anthropogenic changes. *Nat. Commun.* **2018**, *9*, 4389. [CrossRef]

9. Borrelli, P.; Robinson, D.A.; Fleischer, L.R.; Lugato, E.; Ballabio, C.; Alewell, C.; Bagarello, V. An assessment of the global impact of 21st century land use change on soil erosion. *Nat. Commun.* **2017**, *8*, 2013. [CrossRef]

10. Yan, R.X.; Peng, J.B.; Huang, Q.B.; Chen, L.J.; Kang, C.Y.; Shen, Y.J. Triggering influence of seasonal agricultural irrigation on shallow loess landslides on the south Jingyang Plateau, China. *Water* **2019**, *11*, 1474. [CrossRef]

11. Latrubesse, E.M.; Arima, E.Y.; Dunne, T.; Park, E.; Baker, V.R.; d'Horta, F.M.; Wight, C.; Wittmann, F.; Zuanon, J.; Baker, P.A.; et al. Damming the rivers of the Amazon basin. *Nature* **2017**, *546*, 363. [CrossRef] [PubMed]

12. Kuo, H.-L.; Lin, G.-W.; Chen, C.-W.; Saito, H.; Lin, C.-W.; Chen, H.; Chao, W.-A. Evaluating critical rainfall conditions for large-scale landslides by detecting event times from seismic records. *Nat. Hazards Earth Syst. Sci.* **2018**, *18*, 2877–2891. [CrossRef]

13. Chiu, Y.Y.; Chen, H.E.; Yeh, K.C. Investigation of the influence of rainfall runoff on shallow landslides in unsaturated soil using a mathematical model. *Water* **2019**, *11*, 1178. [CrossRef]

14. Segoni, S.; Piciullo, L.; Gariano, S.L. A review of the recent literature on rainfall thresholds for landslide occurrence. *Landslides* **2018**, *15*, 1483–1501. [CrossRef]

15. Gullà, G.; Peduto, D.; Borrelli, L.; Antronico, L.; Fornaro, G. Geometric and kinematic characterization of landslides affecting urban areas: The Lungro case study (Calabria, Southern Italy). *Landslides* **2017**, *14*, 171–188. [CrossRef]

16. Aryal, A.; Brooks, B.A.; Reid, M.E. Landslide subsurface slip geometry inferred from 3-D surface displacement fields. *Geophys. Res. Lett.* **2015**, *42*, 1411–1417. [CrossRef]

17. Fan, X.; Xu, Q.; Scaringi, G.; Li, S.; Peng, D. A chemo-mechanical insight into the failure mechanism of frequently occurred landslides in the Loess Plateau, Gansu Province, China. *Eng. Geol.* **2017**, *228*, 337–345. [CrossRef]

18. Bogaard, T.A.; Greco, R. Landslide hydrology: From hydrology to pore pressure. *Wiley Interdiscip. Rev. Water* **2016**, *3*, 439–459. [CrossRef]

19. Maes, J.; Kervyn, M.; de Hontheim, A.; Dewitte, O.; Jacobs, L.; Mertens, K.; Poesen, J. Landslide risk reduction measures: A review of practices and challenges for the tropics. *Prog. Phys. Geogr.* **2017**, *41*, 191–221. [CrossRef]

20. Froude, M.J.; Petley, D.N. Global fatal landslide occurrence from 2004 to 2016. *Nat. Hazards Earth Syst. Sci.* **2018**, *18*, 2161–2181. [CrossRef]

21. Segoni, S.; Rosi, A.; Fanti, R.; Gallucci, A.; Monni, A.; Casagli, N. A regional-scale landslide warning system based on 20 years of operational experience. *Water* **2018**, *10*, 1297. [CrossRef]

22. Cruden, D. *Landslide Risk Assessment*; Routledge: Abingdon, UK, 2017.

23. Xu, F.G.; Yang, X.G.; Zhou, J.W. Dam-break flood risk assessment and mitigation measures for the Hongshiyan landslide-dammed lake triggered by the 2014 Ludian earthquake. *Geomat. Nat. Hazards Risk* **2017**, *8*, 803–821. [CrossRef]

24. Zhou, Y.; Guo, S. Risk analysis for flood control operation of seasonal flood-limited water level incorporating inflow forecasting error. *Hydrol. Sci. J.* **2014**, *59*, 1006–1019. [CrossRef]

25. Zhou, Y.; Guo, S.; Xu, J.; Zhao, X.; Zhai, L. Risk analysis for seasonal flood-limited water level under uncertainties. *J. Hydro Environ. Res.* **2015**, *9*, 569–581. [CrossRef]

26. United States Geological Survey (USGS). Available online: https://earthquake.usgs.gov (accessed on 18 September 2019).

27. China Earthquake Networks Center (CENC). Available online: https://earthquake.usgs.gov (accessed on 18 September 2019).

28. Chang, L.C.; Chang, F.J.; Yang, S.N.; Kao, I.; Ku, Y.Y.; Kuo, C.L.; Amin, I. Building an intelligent hydroinformatics integration platform for regional flood inundation warning systems. *Water* **2019**, *11*, 9. [CrossRef]

29. Chang, F.J.; Tsai, M.J. A nonlinear spatio-temporal lumping of radar rainfall for modeling multi-step-ahead inflow forecasts by data-driven techniques. *J. Hydrol.* **2016**, *535*, 256–269. [CrossRef]

30. Zhou, Y.; Guo, S.; Chang, F.J. Explore an evolutionary recurrent ANFIS for modelling multi-step-ahead flood forecasts. *J. Hydrol.* **2019**, *570*, 343–355. [CrossRef]

31. Zhou, Y.; Guo, S.; Chang, F.J.; Liu, P.; Chen, A.B. Methodology that improves water utilization and hydropower generation without increasing flood risk in mega cascade reservoirs. *Energy* **2018**, *143*, 785–796. [CrossRef]

32. Zhou, Y.; Guo, S.; Chang, F.J.; Xu, C.Y. Boosting hydropower output of mega cascade reservoirs using an evolutionary algorithm with successive approximation. *Appl. Energy* **2018**, *228*, 1726–1739. [CrossRef]

33. Jane, Q. Landslides pose threat to Himalayan hydropower dream. *Nature* **2018**, *547*, 241.

34. Larsen, I.J.; Montgomery, D.R. Landslide erosion coupled to tectonics and river incision. *Nat. Geosci.* **2012**, *5*, 468. [CrossRef]

35. Zhu, X.; Peng, J.; Jiang, C.; Guo, W. A preliminary study of the failure modes and process of landslide dams due to upstream flow. *Water* **2019**, *11*, 1115. [CrossRef]

36. Tibaldi, A.; Oppizzi, P.; Gierke, J.S.; Oommen, T.; Tsereteli, N.; Gogoladze, Z. Landslides near Enguri dam (Caucasus, Georgia) and possible seismotectonic effects. *Nat. Hazards Earth Syst. Sci.* **2019**, *19*, 71. [CrossRef]

37. Cheng, S.T.; Tsai, W.P.; Yu, T.C.; Herricks, E.E.; Chang, F.J. Signals of stream fish homogenization revealed by AI-based clusters. *Sci. Rep.* **2018**, *8*, 15960. [CrossRef] [PubMed]

MDPI
St. Alban-Anlage 66
4052 Basel
Switzerland
Tel. +41 61 683 77 34
Fax +41 61 302 89 18
www.mdpi.com

Water Editorial Office
E-mail: water@mdpi.com
www.mdpi.com/journal/water